# Collective Diffusion on Surfaces: Correlation Effects and Adatom Interactions

# NATO Science Series

*A Series presenting the results of scientific meetings supported under the NATO Science Programme.*

The Series is published by IOS Press, Amsterdam, and Kluwer Academic Publishers in conjunction with the NATO Scientific Affairs Division

*Sub-Series*

| | |
|---|---|
| **I. Life and Behavioural Sciences** | IOS Press |
| **II. Mathematics, Physics and Chemistry** | Kluwer Academic Publishers |
| **III. Computer and Systems Science** | IOS Press |
| **IV. Earth and Environmental Sciences** | Kluwer Academic Publishers |

The NATO Science Series continues the series of books published formerly as the NATO ASI Series.

The NATO Science Programme offers support for collaboration in civil science between scientists of countries of the Euro-Atlantic Partnership Council. The types of scientific meeting generally supported are "Advanced Study Institutes" and "Advanced Research Workshops", and the NATO Science Series collects together the results of these meetings. The meetings are co-organized bij scientists from NATO countries and scientists from NATO's Partner countries – countries of the CIS and Central and Eastern Europe.

**Advanced Study Institutes** are high-level tutorial courses offering in-depth study of latest advances in a field.
**Advanced Research Workshops** are expert meetings aimed at critical assessment of a field, and identification of directions for future action.

As a consequence of the restructuring of the NATO Science Programme in 1999, the NATO Science Series was re-organized to the four sub-series noted above. Please consult the following web sites for information on previous volumes published in the Series.

http://www.nato.int/science
http://www.wkap.nl
http://www.iospress.nl
http://www.wtv-books.de/nato-pco.htm

**Series II: Mathematics, Physics and Chemistry – Vol. 29**

# Collective Diffusion on Surfaces: Correlation Effects and Adatom Interactions

edited by

M.C. Tringides

Department of Physics and Astronomy,
Iowa State University,
Ames, Iowa, U.S.A.

and

Z. Chvoj

Institute of Physics,
Academy of Sciences of the Czech Republic,
Prague, Czech Republic

**Kluwer Academic Publishers**

Dordrecht / Boston / London

Published in cooperation with NATO Scientific Affairs Division

Proceedings of the NATO Advanced Research Workshop on
Collective Diffusion on Surfaces: Correlation Effects and Adatom Interactions
Prague, Czech Republic
2–6 October 2000

A C.I.P. Catalogue record for this book is available from the Library of Congress.

ISBN 0-7923-7115-1

---

Published by Kluwer Academic Publishers,
P.O. Box 17, 3300 AA Dordrecht, The Netherlands.

Sold and distributed in North, Central and South America
by Kluwer Academic Publishers,
101 Philip Drive, Norwell, MA 02061, U.S.A.

In all other countries, sold and distributed
by Kluwer Academic Publishers,
P.O. Box 322, 3300 AH Dordrecht, The Netherlands.

*Printed on acid-free paper*

---



Printed in the Netherlands.

# TABLE OF CONTENTS

Preface ........ ix

## I. Surface Diffusion at Equilibrium

Quasielastic Helium Scattering Studies of Surface Diffusion: The Diffusion Barrier ........ 1
A. P. Graham and J. P. Toennies

Experimental Aspects of Metal Particle Diffusion on a Silicon Surface ........ 11
Jiri Slezak, Pavel Schukrinov, Alexei Savchenkov, Pingo Mutombo, and Vladimir Chab

Surface Dynamics of Stepped Si(001) Studied by Temporal LEED Spectroscopy ........ 23
K. Kammler, M. Horn Von Högen, N. Voss, M. Tringides, A. Menzel, and E. H. Conrad

Electric Current Induced Asymmetry of Surface Diffusion—Instability of Vicinal Crystal Surfaces ........ 33
S. S. Stoyanov, J. J. Métois, and V. Tonchev

Memory Effects and Memory Functions in Surface Diffusion ........ 47
T. Hjelt, E. Kuusela, J. M. Lahtinen, T. Ala-Nissila, I. Vattulainen, and S. C. Ying

Driven Diffusion in a Model of the O/W(110) System ........ 59
M. A. Zaluska-Kotur, Stanislaw Krukowski, and Lukasz A. Turski

Surface Diffusion and Real-Space Renormalization Group ........ 71
A. A. Tarasenko, L. Jastrabik, F. Nieto, and C. Uebing

Surface Diffusion Near Phase Transitions ........ 83
F. Nieto, A. Tarasenko, and C. Uebing

## II. Surface Diffusion from Profile Evolution Methods

Surface Heterodiffusion in Adsorbed and Coadsorbed Overlayers of Li, Sr, and Cu on the W and Mo (112) Surfaces ........ 97
A. T. Loburets, N. B. Senenko, Yu. S. Vedula, and A. G. Naumovets

Metal Heterodiffusion on Metallic Surfaces: Case of Pb on Cu ........ 107
C. Cohen, G. Prévot, and D. Schmaus

Step Effects on Surface Diffusion: A Comprehensive Study of CO on Pt(111) ........ 117
Xudong Xiao

Adsorption, Desorption, and Diffusion of Potassium on Metal and Oxide Surfaces ........ 129
G. Kerner, I. M. Danziger, W. Zhao, and M. Asscher

## III. Surface Diffusion Under Non-equilibrium Conditions

Dynamics of Interacting Adparticles Under Non-Equilibrium Conditions ........ 143
Z. Chvoj

Statistical-Mechanical Description of Nonequilibrium Processes in Interacting Lattice Gases ........ 157
V. S. Vikhrenko, G. S. Bokun, and Ya. G. Groda

Model Studies of Collective Diffusion ........ 169
I. Vattulainen

Study of Non-Equilibrium Surface Diffusion From an Initial Step-Concentration Profile With a New Probability-Based Continuum Method ........ 179
M. R. Dudek and M. C. Tringides

**IV. Diffusion Mechanisms in Growth Processes**

Diffusion on and IN Surfaces: The Atomic Slide Puzzle ........ 191
J. W. M. Frenken, R. Van Gastel, S. B. Van Albada, E. Somfai, and W. Van Saarloos

Decay of Two-Dimensional Islands on Ag(110) ........ 201
K. Morgenstern, E. Lægsgaard, and F. Besenbacher

Surface Morphology and dynamics: Using *Ab-initio* Total Energies to make the Most of the STM Data ........ 213
Peter J. Feibelman

Substrate-Mediated Interaction on Ag(111) Surfaces From First Principles........ 225
Kristen A. Fichthorn and Matthias Scheffler

Diffusion of Adatoms and Small Clusters on Missing-Row-Reconstructed Surfaces........ 237
F. Montalenti and R. Ferrando

Effects of Adsorbates on Submonolayer Growth ........ 247
Miroslav Kotrla, Joachim Krug, and Pavel Šmilauer

Surfactants in Semiconductor Heteroepitaxy: Thermodynamics and/or kinetics? ........ 259
Ivan Markov

**V. Substrate Effects and Surface Diffusion**

Nonlinear Diffusion and Sliding Friction ........ 273
S. C. Ying, E. Granato, and T. Ala-Nissila

Selective Activation Dynamics: A Minimal Path Approach ........ 285
L. Y. Chen, P. L. Nash, and S. C. Ying

Long Jumps in Surface Diffusion Investigated by Stochastic Equations ........ 295
R. Ferrando, R. Spadacini, and G. E. Tommei

Analytical Kinetic Theory of Single-Particle and Collective Surface Diffusion ........ 305
S. Yu. Krylov

Adatom Island Diffusion on Metal FCC(100) Surfaces ........ 317
O. S. Trushin, J. Hirvonen, J. Heinonen, P. Salo, M. Alatalo, T. Ala-Nissila, I. Koponen, and J. Merikoski

Paths, Barriers, and Prefactors for Adatom Descent from Ag Clusters on Ag(111) ........ 327
T. S. Rahman, A. Kara, A. Karim, A. Al-Rawi

Correlated Diffusion of Adatoms and Islands on the Low Indexed Surfaces of Noble Metals and Alloys ........ 339
A. E. Evangelakis

Author Index ........ 351

Subject Index ........ 353

## Preface

As materials research focuses into finding ways to control the growth of atomic scale structures, there is correspondingly increasing emphasis on to the problem of surface diffusion. Clearly surface diffusion is the key process, which determines how atoms move on the surface. Controlling this motion can lead to the easy fabrication of well-controlled nanostructures broadening the present possibilities in nanotechnology.

The paradigm of surface diffusion has outgrown its standard textbook description as a random walk on a rigid substrate. In real systems for more complex situations are encountered: interacting atoms are commonly present on the surface with their motions highly correlated, different phases form on the surface with different dynamics, large concentration gradients drive the system far away from the linear response regime, rich metastable structures form as a result of balanced interplay between different kinetic processes, substrate relaxation can change the energy landscape and the diffusion barriers, etc.

The motivation behind this ARW was to bring together the international community working on these problems. We felt that the large number of researchers, new results, and well-formulated open questions in this area require some form of integration in a single forum. The ARW and the upcoming proceedings book with papers by the majority of the participants has provided this forum.

The meeting was not planned as a continuation of the earlier NATO ASI in Rhodes in 1996, although several people have participated in both meetings. The goal of the Rhodes meeting was to bring together workers on the atomistic aspects of surface diffusion with workers on many-body, collective effects to initiate a dialogue between the two groups. The Prague meeting has concentrated more on the latter.

The meeting was held in the beautiful city of Prague, which offers an unparallel combination of physical beauty, historical treasures, cultural life, and cosmopolitan spirit. After the rigor of the scientific lectures, it was relaxing to stroll along the banks of the river Vltava or to climb up the unending steps to the top of the castle to be rewarded with the panoramic view of the city. Maybe the only uncooperative element was the weather, to ensure that the experience was real.

It is impossible in the short preface to give justice to the excitement of the presentations and discussions of the meeting. The excitement can be found by reading the chapters that follow. We are limited in simply making a general observation that we believe is shared by most practitioners in the field: how the gap between theory and experiment is narrowing. This narrowing between theory and experiment can be seen from several advances. Realistic simulations of complicated physical phenomena can include many competitive microscopic processes over longer time scales; analytic methods have been developed to describe surface diffusion in far-from-equilibrium experiments; the behavior of the diffusion coefficient close to a phase transition has been predicted and measured in similar systems; new experimental methods can probe the system close to equilibrium to test fundamental predictions of statistical mechanics.

The success of the meeting is a result of the hard work of several people. We would like to first thank the members of the Organizing Committee (Professors V. Cháb and R. Ferrando) who have been full of ideas about the scientific and social program. The meeting would not have run so smoothly without the dedicated commitment of the student support (M. Ma_ín, H. Chvojová, P.Jelínek). The professionalism and expertise of the veteran of several NATO meetings, Rebecca Shivvers, is greatly appreciated for the high quality of the proceedings volume. The great hospitality displayed by the Czech people involved with the meeting (both scientists and non-scientists) has flooded the meeting with unbounded warmth and care.

Of course, no meeting is possible without generous financial support. We greatly thank the NATO Scientific Division for providing us the majority of the funds for the meeting, in their tradition of supporting so many high quality scientific meetings year-after-year. We thank the support of the Czech Academy of Science and the Helsinki Institute of Technology, which have offered additional financial support to cover the cost of the facilities and support student participation.

We believe the meeting has catalyzed the formation of a cohesive cluster of scientists working on the collective aspects of surface diffusion. The strong interactions between the participants displayed at the meeting proves that the cluster is highly stable beyond critical size. Future meetings to follow-up on the goals and vision of the Prague meeting should be anticipated.

Ames and Prague, March 2001.

# QUASIELASTIC HELIUM SCATTERING STUDIES OF SURFACE DIFFUSION: THE DIFFUSION BARRIER

A. P. GRAHAM[1] AND J. P. TOENNIES[2]
*Max-Planck Insitut für Strömungsforschung,*
*Bunsenstraße 10,*
*D-37073 Göttingen,*
*Germany*

**Abstract.** Some results of recent quasielastic helium scattering (QHAS) on the surface diffusion of adatoms and adsorbates are reviewed. Particular emphasis is placed on the evaluation of the effective diffusion barrier energies measured and their dependence on the system and temperature range.

## 1. Introduction

During the last 12 years, improvements in the helium atom scattering (HAS) method, particularly in resolution, intensity, and stability, have permitted the quasielastic broadening of the elastic peak due to adsorbate diffusion to be determined with high accuracy and reliability [1]. The dependence of the quasielastic width on momentum transfer (or wave vector) and temperature have, typically, been analyzed using analogies to quasielastic neutron scattering (QNS) [2], enabling the diffusion of a large number of adsorbate/adatom-substrate systems to be measured [1, 3]. With only two exceptions, namely Pb/Pb(110) [4] and Ni/Ni(110) [1], the effective diffusion barriers obtained with quasielastic helium atom scattering (QHAS) have been significantly lower than anticipated, prompting speculation that the QHAS results have been incorrectly interpreted [5]. Consequently, it is appropriate to discuss the diffusion parameters obtained with QHAS, particularly the barrier to diffusion, and how they relate to the results of other

[1]Present address: Infineon Technologies AG, Corporate Research, Otto-Hahn-Ring 6, D-81739 Munich, Germany.
[2]Corresponding author: jtoenni@gwdg.de



*M.C. Tringides and Z. Chvoj (eds.),*
*Collective Diffusion on Surfaces: Correlation Effects and Adatom Interactions,* 1–10.

methods and the static adiabatic (zero temperature) potential surfaces estimated from *ab initio* calculations.

The theoretical description of the relationship between the adatom diffusional motion and the QHAS peak width dependence on momentum transfer and temperature have been covered extensively elsewhere [6] and will not be presented here. Rather, several quite different example systems will be discussed, namely Ni/Ni(110) [1], Na/Cu(001) [7, 8] and Na/Pt(111) [9]. Before beginning it is important to present the terms used to describe the activation barriers for diffusion, since they are one of the major sources of confusion when considering the QHAS results.

- *Adiabatic or static* activation energy $E_{\mathrm{ad}}$: This is the energy difference between the global minima of the lateral potential energy surface (PES) and the energy at the saddle, or crossing, point between adsorption sites determined from molecular dynamics (MD) simulations of experimental diffusion results or *ab initio* calculations. The adiabatic barrier energy difference takes no account of the zero-point motions of the adatoms in the PES or the thermal motions in the potential minima.
- *Dynamic* activation energy $E_{\mathrm{dyn}}$: The energy required for a transition from one adsorption site to another including the zero-point and thermal motions of the adatoms. The dynamic barrier is usually invoked to describe uncorrelated successful attempts to jump over the barrier, i.e., for rare diffusion events where the thermal energy is much lower than the adiabatic activation energy.
- Effective activation energy $E_{\mathrm{eff}}$: The energy obtained from an Arrhenius fit to the experimental results *over the temperature range of the measurements.* For many of the QHAS results which have been presented to date, the thermal energy corresponding to the temperature range of the measurements is a considerable fraction of the adiabatic activation energy and, consequently, a series of successful jump attempts may be correlated.

## 2. Experimental

To date, all of the reported QHAS measurements have been undertaken on four slightly different HAS apparatus in Göttingen which follow the same basic design [10]. The neutral helium atom beam is produced by supersonic expansion from high pressure ($\approx$ 100 bar) through a 10 $\mu$m nozzle and collimated first by a conical skimmer and later by a series of apertures connected to several differential pumping stages. The background helium pressure is thus reduced from $\sim 10^{-3}$ mbar in the nozzle chamber to $< 10^{-10}$ mbar in the target chamber. The helium atoms scattered from the target crystal, which is mounted in ultra-high vacuum ($< 10^{-10}$ mbar) on a six-axis manipulator, through a fixed angle $\theta_{\mathrm{SD}}$ ($\approx 90°$), are detected by a homemade magnetic-sector mass-spectrometer following several further differential pumping stages, which reduce the background He partial pressure at the detector to $< 10^{-15}$ mbar. The energy transfer during the interaction of the helium atoms with the sample is determined using a time-of-flight (TOF) technique in which the flight times of pulses of helium atoms, produced by a mechanical chopper typically located before the target, to the mass-spectrometer detector are measured [10].

The key components of the HAS systems essential for QHAS measurements are high resolution and stability in addition to high intensity. The high intensities are required so that the time required for a series of measurements is short and the target surface does not become contaminated. This is particularly important for measurements at high temperatures, in which contamination from the bulk sample diffuses to the surface, or for reactive systems such as adsorbed alkali metals. High energy transfer resolution and helium source stability are necessary in order that the quasielastic peak broadening can be extracted accurately. Typical conditions for QHAS measurements on the HUGO II HAS apparatus are an incident helium energy of $E_i = 11$ meV (corresponding to a source temperature of 50 K), a total energy transfer resolution of 250-300 $\mu$eV (2.2-2.7%) and measurement times of the order 5-30 min. The source temperature is stabilized to better than $< 0.05$ K over the course of the experiments, so that it has a negligible effect on the total energy resolution, i.e., $< 0.1\%$ or $\approx 10$ $\mu$eV. This value is about the same as the error involved in extracting the peak broadening by deconvolution, which is discussed in detail in Ref. [1], and determines a lower limit on the measurable diffusion rate of $D \approx 5 \times 10^{-7}$ cm$^2$/s, assuming diffusion by single jumps to adjacent adsorption sites.

## 3. Ni/Ni(110)

The self-diffusion of nickel on a Ni(110) surface has been investigated with several different methods. In 1967, Maiya and Blakely used a photo-etching technique to create micrometer scale periodic height variations on a Ni(110) and followed the annealing out of the profiles with optical interference methods for a range of temperatures to obtain an activation barrier for diffusion of 1.85 eV along the $[1\bar{1}0]$ azimuth [11]. Later, in 1978, Bonzel and Latta also used the decay of a sinusoidal profile at different temperatures to obtain an activation energy of 0.76 eV along the $[1\bar{1}0]$ direction for temperatures less than 1100 K, and 1.95 eV in the range $T_s = 1200 - 1600$ K [12]. Shortly thereafter, in 1980, Tung and Graham obtained a barrier of 0.32 eV using atomic resolution field ion microscopy (FIM) similar to the other surfaces of nickel which they studied [13]. The surface melting and pre-melting processes on Ni(110) have also been investigated by low-energy

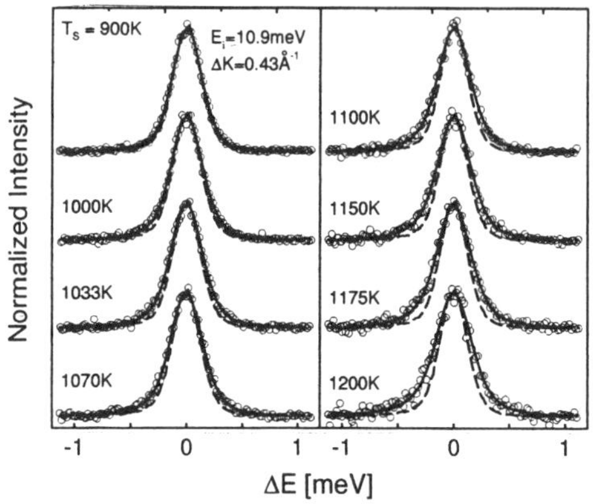


*Figure 1.* Typical time-of-flight spectra of the quasielastic peak for scattering from Ni atoms diffusing on the Ni(110) surface along the [1$\bar{1}$0] azimuth at an incident angle of $\theta_i = 43.9°$ (total scattering angle $\theta_{SD} = 95.8°$) and an energy of $E_i = 11.0$ meV for 8 different surface temperatures $T_s$. The best-fit of a convolution of the instrument response function (dashed line) with a Lorentzian peak profile (solid line) is compared with the experimental data points (open circles) after subtraction of a linear background.

electron diffraction (LEED) [14] and MD [15] methods. A detailed review of these effects has been presented by Kern [16].

The onset of significant thermal adatom creation on Ni(110) occurs at temperatures above 900K, as determined by HAS [17]. A transition at a similar temperature was found by Cao and Conrad using LEED, although it was assigned to substantial anharmonic effects due to the lack of an observable increase in the diffuse background [14]. The most energetically favorable adatom formation process in this temperature range is step edge detachment rather than terrace adatom-vacancy production, due to the lower number of nearest neighbors at the step edge [17]. Thus, significant self-diffusion on the Ni(110) surface is anticipated for $T_s > 900$ K with an activation barrier between 0.32 eV [13] and 0.76 eV [12].

Figure 1 presents a series of HAS TOF spectra from the Ni(110) surface for a momentum transfer $\Delta K = 0.43$ Å$^{-1}$ along the [1$\bar{1}$0] azimuth in the temperature range 900-1200 K. At 900 K the experimental peak (open circles) is only slightly broader than the measured instrument response function IRF (dashed curve). Consequently, the best-fit deconvolution of the IRF with a Lorentzian [1] (solid curve) yields a small quasielastic broadening of $\Gamma = 15$ $\mu$eV. At higher temperatures the broadening is more apparent, reaching almost 150 $\mu$eV at 1200 K. The best-fit deconvolution results for QHAS measurements at a number of temperatures in this range are shown in Fig. 2 for both the [1$\bar{1}$0] and [100] azimuths.

In the following, we will concentrate our attention on the diffusion along

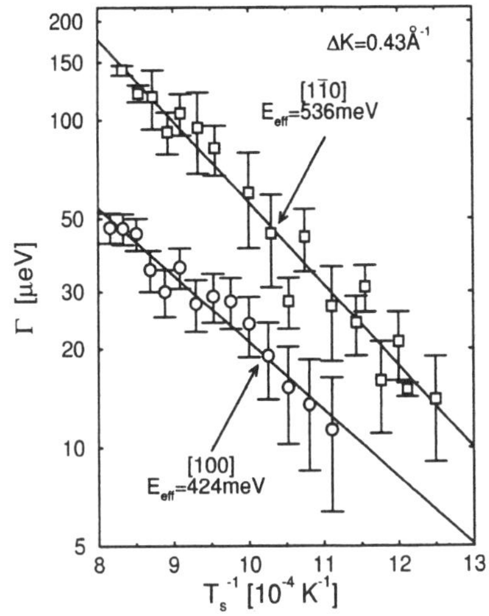


*Figure 2.* The best-fit quasielastic peak widths (log scale) on the Ni(110) surface at a parallel momentum transfer $\Delta K = 0.43$ Å$^{-1}$ and an incident energy of 11.0 meV along the [1$\bar{1}$0] and [100] azimuths are plotted as a function of the inverse temperature. From the Arrhenius plot the *effective* diffusion activation energies are determined to be $E_{\text{eff}} = 424 \pm 40$ meV and $536 \pm 40$ meV for the [100] and [1$\bar{1}$0] directions, respectively.

the [1$\bar{1}$0] azimuth because diffusion along the [100] direction perpendicular to the atom rows is estimated to proceed via exchange diffusion paths which are not open to a simple discussion appropriate to this article. Many of these effects will be discussed in a forthcoming paper [18] and are included in the Ph.D. thesis of Silvestri [19]. Along the [1$\bar{1}$0] direction, however, the best-fit QE peak widths fit very well to an Arrhenius dependence yielding an *effective* activation energy of $E_{\text{eff}} = 536 \pm 40$ meV. This value lies directly between those determined with other methods [12, 13] and agrees well with MD simulations based on embedded-atom potentials [19].

The Ni(110) surface temperatures in these QHAS experiments correspond to thermal energies in the range $k_B T_s = 77 - 103$ meV, far below the effective activation barrier of 536 meV, suggesting that diffusional jumps between favorable adsorption sites are rare. This is supported by further QHAS measurements as a function of parallel momentum transfer which yield clear evidence for single site jump frequencies of $\nu_1 = 4.2 \times 10^{10}$ Hz and $\nu_1 = 8.0 \times 10^{10}$ Hz at temperatures of 1000 K and 1100 K, respectively [1]. These frequencies represent only 0.19 and 0.36% of the jump attempt frequency of $2.2 \pm 0.1 \times 10^{13}$ Hz for the [1$\bar{1}$0] direction [1] and a considerable time (> 10 ps) is spent between successive jumps, ruling out correlated jump events. Consequently, the *effective* diffusion activation energy determined from an Arrhenius fit to the QHAS results for the Ni(110) surface along the [1$\bar{1}$0] azimuth is probably very close to the *dynamic* activation

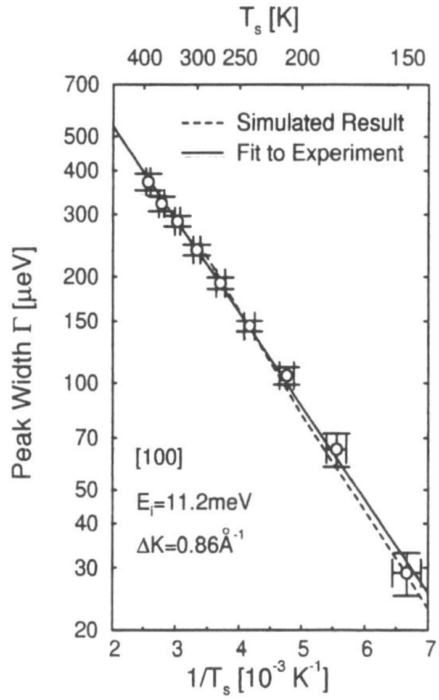


*Figure 3.* Arrhenius plot of the variation of the QHAS peak width Γ with surface temperature $T_s$ for a sodium coverage $\Theta_{Na} = 0.028$ on Cu(001), and a parallel momentum transfer $\Delta K = 0.86$ Å$^{-1}$ along the [100] azimuth. The open circles represent the best-fit QE widths following deconvolution of the HAS TOF spectra with a Lorentzian broadening. The solid line fit to the points leads to an *effective* activation energy of 52.5 ± 0.9 meV, which is lower than the *dynamic* barrier from MD simulations of 65±6 meV (dashed line for $T_s < 290$ K). Both of these values are smaller than the energy difference between the PES minima and saddle point (*adiabatic barrier*) of 75 meV determined from the MD simulations [7, 8].

energy and is confirmed by the good agreement with other measurements.

## 4. Na/Cu(001)

Sodium on Cu(001) is the system which has been studied most intensively with QHAS [7, 8, 20, 21]. These results have been carefully modeled with MD methods to yield the adiabatic PES [7, 8], the dynamical interaction with the copper substrate [22] and the interaction between sodium atoms at high coverages [21, 23]. These comparisons between the MD simulations and QHAS results provide detailed information on the relationship between the adiabatic, dynamic, and effective activation energies for this and other systems and have provided a great deal of insight into how the QHAS experiments should be interpreted. In the following discussion we will focus on the low coverage ($\Theta_{Na} \approx 0.03$) QHAS results and MD simulations because the strong interactions between Na atoms at higher coverages complicates the interpretation [21, 23].

Figure 3 shows an Arrhenius plot of the QHAS width for $\Theta_{Na} = 0.028$ on Cu(001) at a momentum transfer of $\Delta K = 0.86$ Å$^{-1}$ along the [100] azimuth. The experimental points (open circles) are fitted well over the entire temperature range by the solid line with an effective activation energy of $E_{\text{eff}} = 52.5 \pm 0.9$ meV. In fact, all of the three low coverage QHAS investigations of the diffusion of Na/Cu(001) have yielded an effective diffusion activation energy of $E_{\text{eff}} \approx 52$ meV in the temperature range $T_s = 150-390$ K [7, 8, 20]. The corresponding thermal energies are $k_B T_s = 12.9 - 34.4$ meV, which are a substantial fraction of the effective activation energy, indicating that sequential successful jump attempts may be correlated. Put another way, the time between successful jumps at 390 K is only $\approx 4.8$ ps which corresponds to about 7 oscillations of the sodium atoms in the potential minima [24].

Molecular dynamic simulations of the QHAS measurements indicate that the friction coefficient $\eta$ describing the rate of energy transfer between the sodium atom and the Cu(001) substrate is $1/\eta = 4.5$ ps [7, 8], almost the same as the time between successful jumps. Thus, it is clear that when the thermal energy becomes a significant fraction of the diffusion activation energy, the Arrhenius law breaks down and the barrier determined is too low. It is important to note that similar effects are expected with other methods whenever the thermal energy is a significant proportion of the activation energy.

An upper limit of $k_B T_s < E_{\text{ad}}/3$, where $E_{\text{ad}} = 75$ meV is the adiabatic activation energy used to model the QHAS results in the MD simulations, for application of the Arrhenius law has been suggested [8]. The dashed line in Fig. 3 shows MD simulations of the QE width in the temperature range 150-290 K, corresponding to $k_B T_s < E_{\text{ad}}/3$ and which pass through the experimental data points, with a *dynamic* activation energy of $E_{\text{dyn}} = 65 \pm 6$ meV. $E_{\text{dyn}}$ is 24% higher than $E_{\text{eff}}$ highlighting the deviation from the Arrhenius behavior at higher temperatures. Note that $E_{\text{dyn}}$ is still less, by 13%, than the adiabatic activation energy of 75 meV due to the effect of the parallel vibrational mode of the Na atoms in the potential minima, as discussed in the Introduction.

## 5. Na/Pt(111)

The final example which will be considered here is that of the diffusion of sodium on the Pt(111) surface [9]. Adsorption of alkali metals on the fcc(111) transition metal surfaces is particularly interesting because of the unexpected top-site adsorption positions which have been determined by LEED as opposed to the hollow-site observed on the fcc(001) surfaces [25]. Figure 4 shows an Arrhenius plot of the diffusion rate of a low coverage $\Theta_{\text{Na}} = 0.043$ of sodium atoms on the Pt(111) surface along the $[11\bar{2}]$ azimuth in the temperature range 100-350 K. The diffusion rates have been

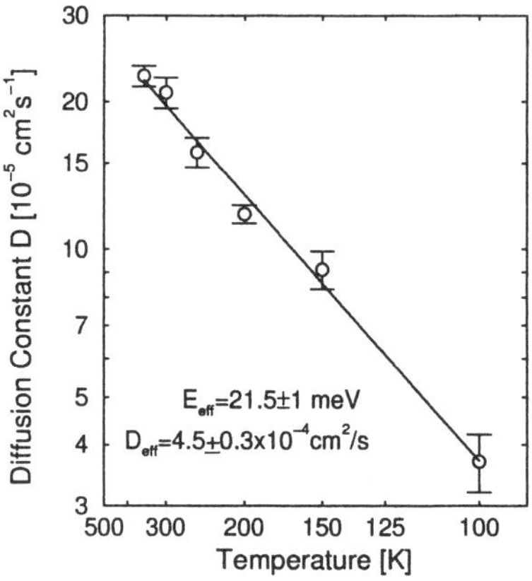


*Figure 4.* Arrhenius plot of the macroscopic diffusion rate $D$ of $\Theta_{\mathrm{Na}} = 0.043$ sodium adsorbed on Pt(111) along the $[11\bar{2}]$ azimuth determined from QHAS measurements close to specular $\Delta K < 0.3$ Å$^{-1}$. The diffusion rate increases from $D = 3.7 \times 10^{-5}$ cm$^2$/sec to $2.3 \times 10^{-4}$ cm$^2$/sec in the temperature range 100-350 K. The best fit (solid line) to the diffusion rates (open circles) yields an effective activation energy of $E_{\mathrm{eff}} = 21.5 \pm 1$ meV.

obtained by fitting the QE widths over a range of parallel momentum transfers close to specular $|\Delta K| < 0.3$ Å$^{-1}$, corresponding to diffusion over longer distances $l > 20$ Å, with the quadratic $\Delta K$ relationship adapted from QNS investigations of macroscopic diffusion [2].

Fitting the diffusion rates for the temperatures from 100 to 350 K with the Arrhenius relationship produces an effective activation energy of only $21.5 \pm 1$ meV, less than half that for Na/Cu(001) in the same temperature range. This activation energy is clearly much lower than the dynamic or adiabatic barriers, a fact which is supported by the observation of jump-like behavior in the momentum transfer dependence at 300 K [9], which points towards an activation energy of at least $> k_B T_s$ ($> 26$ meV). Such rapid diffusion certainly indicates a very low diffusion barrier, lower than for Na/Cu(001). An *estimate* of the dynamic barrier can be made by considering the lowest temperature measurement only, i.e., 100 K ($k_B T_s = 8.6$ meV) and assuming the diffusion prefactor $D_0 = 6 \times 10^{-3}$ cm$^2$/sec obtained for Na/Cu(001) from the MD simulations [7, 8]. This estimate, which is only very approximate, since the number of multiple jumps and jump lengths are different for the two systems, provides a much higher activation energy of 44 meV. Even this value is only about half of the Na/Cu(001) adiabatic barrier strongly indicating that the Na/Pt(111) lateral PES is very flat. It is interesting to note, however, that the parallel vibrational frequency of the Na atoms in the potential minima is $\hbar\omega = 5.3$ meV at 50 K [9], not much lower than for Na/Cu(001). Molecular dynamics simulations of the vibrations and diffusional motions of Na/Pt(111) are necessary to determine the adiabatic activation energy.

## 6. Conclusions

Quasielastic helium atom scattering measurements from three different diffusion systems have been discussed with particular emphasis on the interpretation of the activation energies. Since for Ni/Ni(110) the thermal energies are much lower than the activation energies, the effective activation barrier extracted by fitting the QHAS data is the same as the dynamic barrier. For Na/Cu(001) extensive MD simulations have shown that when the highest temperature corresponds to $k_BT_s \sim E_{\text{ad}}/2$, the effective activation barrier is significantly lower then the dynamic barrier and much lower then the adiabatic energy difference from the potential well minima to the saddle, or crossing points between sites by 30% [7, 8]. For Na/Pt(111) the effective activation energy appears to be only about half of the dynamic barrier which is probably lower than the adiabatic barrier.

The discussion presented here clearly indicates that care must be taken when dealing with diffusion parameters measured in a range where the thermal energy is a significant fraction of the activation barrier. However, together with suitable molecular dynamic simulations, QHAS can provide much more information on the microscopic dynamics governing diffusion than available with any other method. Thus, it is a desirable complementary probe for the study of surface diffusion processes. Indeed, the discussion presented here further substantiates the much lower diffusion barriers coming from the QHAS experiments compared to those measured by conventional techniques. As discussed in our earlier papers [1, 3], the QHAS values are confirmed by PEEM experiments [26] and appear to be due to the impairing effect of defects on the standard macroscopic measurements. Defects are expected to have a negligible effect on QHAS because of the shorter distances probed, and are avoided in PEEM [26].

## References

1. Graham, A.P., Silvestri, W. and Toennies, J.P. (1997) Elementary Processes of Surface Diffusion Studied by Quasielastic Helium Atom Scattering, in M. Tringides (ed.), *Proceedings of the NATO ASI Conference on Surface Diffusion: Atomistic and Collective Processes*, Plenum Press, New York, **B360**, pp. 565-580.
2. Bée, M. (1988) *Quasielastic Neutron Scattering*, IOP Publishing, Bristol.
3. Graham, A.P. and Toennies, J.P. (1999) Determination of the lateral potential energy surface of single adsorbed atoms and molecules on single crystal surfaces using helium atom scattering, *Surf. Sci.* **427-8**, 1-10.
4. Frenken, J.W.M., Hinch, B.J., Toennies, J.P. and Wöll, Ch. (1990) Anisotropic diffusion at a melting surface studied with He-atom scattering, *Phys. Rev.* **B41**, 938-946.
5. Ge, Q. and King, D.A. (1999) Surface Diffusion Potential Energy Surfaces from First Principles: CO Chemisorbed on Pt{110} *J. Chem. Phys.* **111**, 9461.
6. Frenken, J.W.M. and Hinch, B.J. (1992) Quasielastic Helium Scattering Studies of Adatom Diffusion on Surfaces, in E. Hulpke (ed.), *Helium Atom Scattering from Surfaces, Springer Series in Surface Sciences* **27**, Springer Verlag, Berlin, pp. 287-313.

7. Graham, A.P., Hofmann, F., Toennies, J.P., Chen, L.Y. and Ying, S.C. (1997) Determination of the Na/Cu(001) Potential Energy Surface from Helium Atom Scattering Studies of the Surface Dynamics, *Phys. Rev. Lett.* **78**, 3900-3903.
8. Graham, A.P., Hofmann, F., Toennies, J.P., Chen, L.Y. and Ying, S.C. (1997) Experimental and Theoretical Investigation of the Dynamics of Sodium Atoms on a Copper(001) Surface, *Phys. Rev.* **B56**, 10567-10578.
9. Graham, A.P. and Toennies, J.P. (2000) The macroscopic diffusion and low frequency vibrations of sodium on Pt(111) investigated with helium atom scattering, *J. Phys. Chem.*, submitted.
10. Toennies, J.P. (1991) Experimental Determination of Surface Phonons by Helium Atom and Electron Energy Loss Spectroscopy, in W. Kress and F.W. de Wette (eds.), *Surface Phonons, Springer Series in Surface Sciences*, Springer Verlag, Berlin, **21**, pp. 111-166.
11. Maiya, P.S. and Blakely, J.M. (1967) Surface Self-Diffusion and Surface Energy of Nickel, *J. Appl. Phys.* **38**, 698-704.
12. Bonzel, H.P. and Latta, E.E. (1978) Surface self-diffusion on Ni(110): Temperature dependence and directional anisotropy, *Surf. Sci.* **76**, 275-295.
13. Tung, R.T. and Graham, W.R. (1980) Single atom self-diffusion on nickel surfaces, *Surf. Sci.* **97**, 73-87.
14. Cao, Y. and Conrad, E.H. (1990) Approach to Thermal Roughening of Ni(110): A Study by High-Resolution Low-Energy Electron Diffraction, *Phys. Rev. Lett.* **64**, 447-450.
15. Chen, E.T., Barnett, R.N. and Landman, U. (1990) Surface melting of Ni(110), *Phys. Rev.* **B41**, 439-450.
16. Kern, K. (1994) Thermal roughening of surfaces: experimental aspects, in D.A. King and D.P. Woodruff (eds.) *Phase Transitions and Adsorbate Restructuring at Metal Surfaces, The Chemical Physics of Solid Surfaces*, Elsevier, Amsterdam, **7**, pp. 291-340.
17. Silvestri, W., Graham, A.P. and Toennies, J.P. (1998) Adatom Formation on the Ni(110) Surface, *Phys. Rev. Lett.* **81**, 1034-1037.
18. Bertino, M.F., Graham, A.P., Hofmann, F., Silvestri, W. and Toennies, J.P., in preparation.
19. Silvestri, W. (1998) Theory and Experiments on Surface Diffusion, Ph.D. Thesis, University of Göttingen.
20. Ellis, J. and Toennies, J.P. (1993) Observation of Jump Diffusion of Isolated Sodium Atoms on a Cu(001) Surface by Helium Atom Scattering, *Phys. Rev. Lett.* **70**, 2118-2121.
21. Ellis, J., Graham, A.P., Hofmann, F. and Toennies, J.P. (2000) The Coverage Dependence of the Microscopic Diffusion of Na Atoms on the Cu(001) Surface: A Combined Helium Atom Scattering Experiment and Molecular Dynamics Study, *Phys. Rev. B*, submitted.
22. Ellis, J. and Toennies, J.P. (1994) A molecular dynamics simulation of the diffusion of sodium on a Cu(001) surface, *Surf. Sci.* **317**, 99-108.
23. Cucchetti, A. and Ying, S.C. (1999) Diffusion of Na atoms on a Cu(001) surface, *Phys. Rev.* **B60**, 11110-11117.
24. The parallel vibrational frequency of the sodium atoms in the potential minima is $h\nu = 6$ meV [7, 8, 20], corresponding to a vibrational period of 0.69 ps.
25. Diehl, R.D. and McGrath, R. (1996) Structural studies of alkali metal adsorption and coadsorption on metal surfaces, *Surf. Sci. Rep.* **23**, 43-172.
26. Ondřejček, M., Stenzel, W., Conrad, H., Cháb, V., Chvoj, Z., Engel, W., and Bradshaw, A.M. (1993) Surface diffusion of K on Pd{111} studied by photoemission electron microscopy, *Chem. Phys. Lett.* **215**, 528-534.

# EXPERIMENTAL ASPECTS OF METAL PARTICLE DIFFUSION ON A SILICON SURFACE

JIRI SLEZAK, PAVEL SCHUKRINOV, ALEXEI SAVCHENKOV,
PINGO MUTOMBO and VLADIMIR CHAB
*Institute of Physics, Academy of Sciences of the Czech Republic,*
*Cukrovarnicka 10, CZ 162 53, Prague, Czech Republic*

## 1. Introduction

Silicon-based technology dominates the chip industry, even if new materials are continuously appearing, showing better parameters for particular applications. The reason for this is that silicon seems to be the best option for most uses in the field of electronics, and this fact has provoked an enormous interest on the part of scientists for more than 50 years. The VLSI process has focused attention on two technologically important points to solve: reduction of the thermal load during the production of metal contacts and Schottky barriers, and the development of nano- or quasi-one-dimensional (1D) wires connecting the particular elements. Both these tasks are associated with the well-known physical phenomenon of surface diffusion, which is the limit to further miniaturization [1]. Surprisingly, there are not many experimental data available for the diffusion of metals on silicon surfaces at *low* coverage [2].

Surface diffusion is a well-defined problem, which describes either mass transport or density fluctuation around the equilibrium in an adsorbate layer [2,3]. A theoretical description based on a thermally-activated random walk process is appropriate in systems where the hard ball model applies, i.e., in a system where strong interactions, chemical reactions, and phase transitions in the adsorbate layer do not govern particle motion. This is often satisfied at a very low coverage on metal surfaces and to a lower extent on semiconductors. Increasing the concentration of adparticles (to a value less than one monolayer), the additional interaction among particles complicates the picture. The most widely accepted interpretation of the concentration-dependence of the diffusion coefficient is derived from modifications of heat transfer or Green-Kubo equations [4,5]. In order to simplify the many body problem, the interaction between nearest neighboring adparticles is usually considered. However, surface processes taking place in most real systems (such as phase transitions, alloying with substrate atoms or adsorbate-induced surface reconstruction) do not allow us to express the diffusion coefficient D in sufficiently precise terms. It is generally believed that the hopping rate of adparticles and the free energy of the system are two competing factors, which determine the concentration dependence of the diffusion coefficient. From present-day theories it follows that the equilibrium is characterized by isotropy in the concentration of adparticles, single atoms, or clusters.



*M.C. Tringides and Z. Chvoj (eds.),*
*Collective Diffusion on Surfaces: Correlation Effects and Adatom Interactions,* 11–22.

Experimental data on the diffusion coefficient are commonly divided into two groups according to the lateral resolution of an applied method. Methods in the first group, which do not allow observation of the transport of an isolated adparticle, lead to the determination of the *macroscopic* form of D by measuring the time development of the concentration profiles on a surface. If the lateral resolution is good enough to resolve some morphological surface features, the influence of steps, imperfections and defects can be studied. Experimental works using this *macroscopic* resolution show that the macroscopic D is usually a combination of diffusion coefficients corresponding to less perfect parts of a crystal surface. To evaluate the time evolution of a measured profile, solution to the second Fick`s law is used in the form of the error function. In this way, the activation energy of the diffusion and the exponential prefactor are frequently misinterpreted because the particular contributions cannot be resolved. Methods taken from the second group are qualitatively different: Working with a lateral resolution allowing us to follow the diffusion path of a single adparticle, the hopping rate can be estimated. If the experiment is performed at different temperatures, the true activation energy can be found. Using a reasonable field of view, e.g., in Scanning Tunneling Microscopy (STM), the motion of many particles can be observed simultaneously. However, particle-particle interaction cannot be easily evaluated. This interaction is commonly omitted and attention is mostly devoted to the formation of very small clusters.

In this paper we will present several aspects of the equilibrium state in metal/semiconductor systems. Low coverage of Au, Cu and Pb atoms (< 0.1 monolayer) on a Si(111)-7x7 surface were studied with variable temperature STM (VT STM). The statistical ensembles of "as deposited" atoms were sampled on surface areas and were compared to the similar regions of the surface after annealing at different temperatures. The initial non-equilibrium distribution of atoms on the surface does not tend to become spatially isotropic. This contradicts the predictions of classical diffusion theories. It implies that a stronger emphasis on the chemical nature of adsorbates in theoretical modeling is desirable.

## 2. Experimental

*Microscopic* experiments were carried out in a VT STM multi-chamber (VT STM Omicron, Germany) equipped with LEED-AES facilities, sample transfer, ion sputtering, mass spectrometer, heating/L-He-cooling capabilities and an UHV evaporator of Au, Cu (Mo crucible used) and Pb (C crucible used) - water cooled Focus EFM3. The Si(111) samples used took the form of strips having the following dimensions: 9 mm x 2 mm x 0.3 mm. The clean Si(111)-(7x7) surface was prepared in a conventional way by flashing the sample up to ~ 1500 K. This was followed by a subsequent slow cooling to room temperature (RT). Heating was achieved by passing a direct current through the sample. About 0.1 monolayer (ML) and less of the metals (1 ML is defined as $7.84\times10^{14}$ atoms/cm$^2$) were evaporated separately on the clean Si(111)-(7x7) surface at RT. The temperature of annealing was obtained by a direct current provided by a stabilized HC1100 power supply (Omicron). The value of the current was chosen in such a way that at the reached temperature the changes in the position of the metal atoms were detectable within a reasonable time interval. Images were taken in a time range of about ~ 3 min/image. Pressure in the vacuum system did not exceed $10^{-10}$ mbar, even during layer deposition. Using the same sample holder and current settings, the temperature of the sample surface was measured outside the experimental chamber, (i.e., separately in air) with a precise thermometer.

STM images presented throughout the paper are basically raw data, processed with a simple linear background subtraction only. We wished to avoid subjective suppression or enhancement of measured facts.

*Macroscopic* experiments on Au/Si(111) were performed using the multi-purpose SPELEEM [6] located at the ELETTRA synchrotron radiation light source in Trieste, Italy. This instrument can be used alternately for PEEM (photoemission electron microscopy employing a mercury short-arc lamp), XPEEM (soft X-ray photoemission electron microscopy exploiting synchrotron radiation), LEEM (low energy electron microscopy), LEED (low energy electron diffraction), and MEM (mirror electron microscopy) imaging modes.

Experimental details are described in Ref. [7]. For the sake of clarity, we sum up basic facts. Sample preparation was similar respecting a different design of the sample holder in the SPELEEM: p-type silicon wafers were cut to 1x1 $cm^2$ squares and mounted on a sample cartridge with a Mo cap. The initial in situ cleaning consisted of short flashes up to 1500 K following outgassing at lower temperatures, usually overnight. Heating was achieved by electron bombardment of the Si wafer from the back.

In the SPELEEM preparation chamber, gold was deposited from a UHV evaporator (EFM3 Focus, Omicron Ltd.) with a tungsten spiral supporting a gold droplet. The evaporation rate was calibrated with LEEM and LEED prior to this, since the completion of the Au-(5x1) and √3-Au structures are clearly visible in LEEM. A movable shutter was positioned about 1 mm above the sample plane. During evaporation it touched the cap of the cartridge (not the sample) to eliminate mechanical vibrations. It also enabled one half of the sample to be covered with a gold layer. The sample temperature was measured with a W-Re thermocouple spot-welded to a small tantalum plate touching the Si sample.

## 2. Results

### 2.1. Au-MACROSCOPIC RESULTS

In order to demonstrate the different experimental views of an identical system - Au/Si(111)-7x7, we will summarize briefly our published results obtained with SPELEEM [7] and compare them to the STM data.

The SPELEEM experiments were used for the experimental determination of D on a macroscopic scale. The initial step-like profile was prepared by evaporation through the mask at a substrate temperature of about 750 K. On the covered part of the surface the concentration of the gold was ~ 1ML. In order to obtain quantitative data across the step, we determined an intensity profile along the oblong region from A to B (marked in Fig. 1) averaged over the shorter side of the rectangle. The transition zone exhibits a concentration increase over ~ 10 μm. As the sample temperature during evaporation was clearly above the formation limit of the various reconstructions, we utilized the LEEM modus of the SPELEEM to obtain locally resolved structural information from the transition zone. The sequence demonstrates (see Fig. 2) that starting with the (7x7) LEED pattern on the uncovered Si(111)-7x7 part, regions **b**) and **c**) display the lowest coverage Au-(5x1) reconstruction. Region **d**) shows the transition from Au-(5x1) to the (√3x√3)R30° structuring (evident from the superposition of either LEED pattern). The ß√3-pattern characteristic for the monolayer gold coverage is found in region **e**) although the ordering is not as distinct as in the other regions, as is apparent from the more diffuse diffraction spots. Summarizing, the border region of the initial state represents a well ordered transition zone where the gold concentration decreases continuously from 1 ML Au layer to the clean reconstructed Si(111)-7x7.

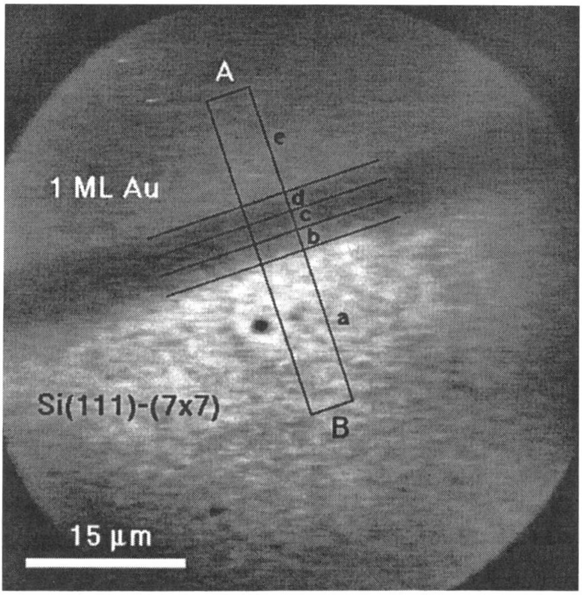


Fig. 1: A PEEM image of the profile taken immediately after Au deposition at 750 K. Parallel selections denoted *a-e* refer to the areas with different LEED patterns observed (see Fig. 2).

The onset of Au diffusion across the border can be determined by using the XPEEM imaging mode of the SPELEEM. By excitation of the characteristic photoemission lines of the element to be monitored with elevated photon energies and application of an energy filter, one can image the substrate in the light of these photoelectrons, i.e., chemically resolved. Normalized intensity profiles obtained from the sequence of XPEEM images are plotted in Fig. 3a. The respective intersections yield 2 points: the highest (**A**) and the lowest (**B**) Au concentration. These two points are considered to define the diffusion zone, i.e., the transition zone between the gold covered regions and the clean Si(111)-7x7. The positions of points **A** and **B** are plotted in Fig. 3b as a function of time with the border positions of the first profile serving as time zero. It is evident that the propagation velocity of the border diminishes with time.

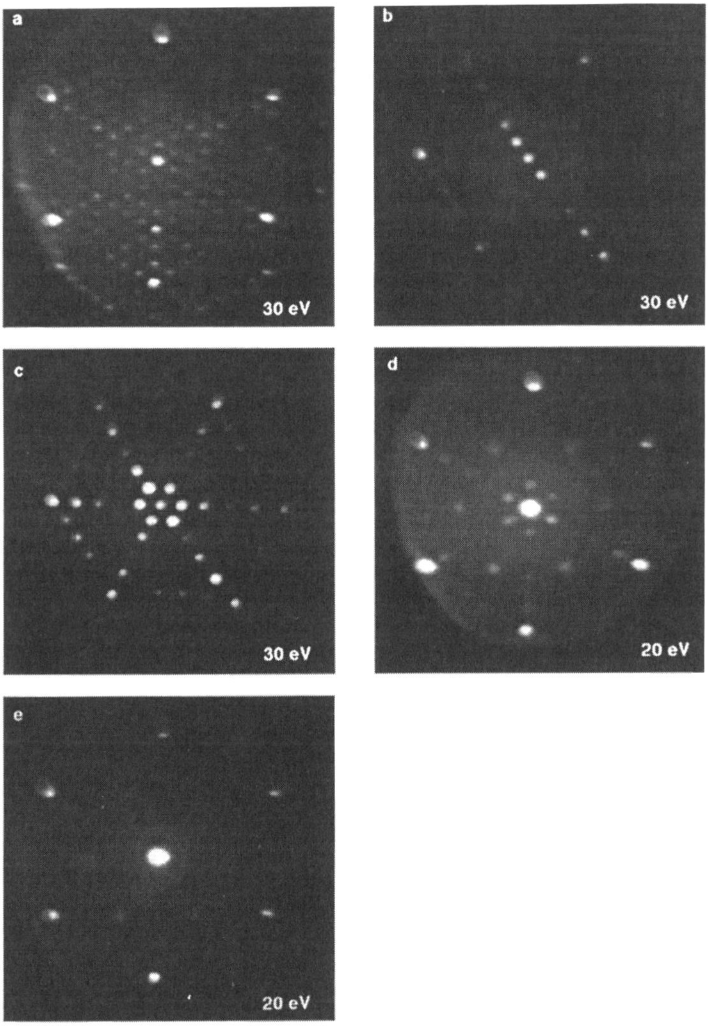


Fig. 2: LEED patterns of selected areas a-e in Fig. 1. The energies of the primary electron beam are given in the lower right-hand corners.

The separation between the two characteristic points (**A** and **B**) increases, however. To summarize, the movement of the diffusion zone can be described as a combination of two effects: i) the evolution of a characteristic S-shape concentration profile which is ii) moved as a whole in the direction of lower concentration.

In order to trace in more detail the expansion of the Au-(5x1) reconstructed area, and in particular the step-like borderline, we used the SPELEEM in the MEM imaging mode (Fig. 4). Since the incoming electrons in MEM have energies just below

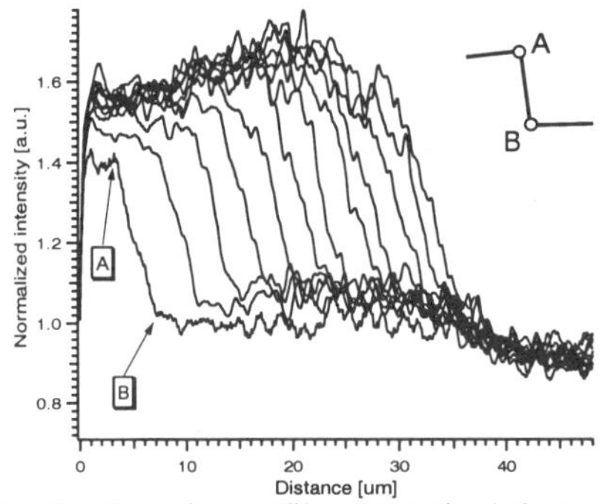


Fig. 3a: Intensity profiles determined from the sequence of XPEEM images. Each profile was fitted with three straight lines. The cross sections of the lines define the point A (the highest Au concentration) and the point B (the lowest Au concentration).

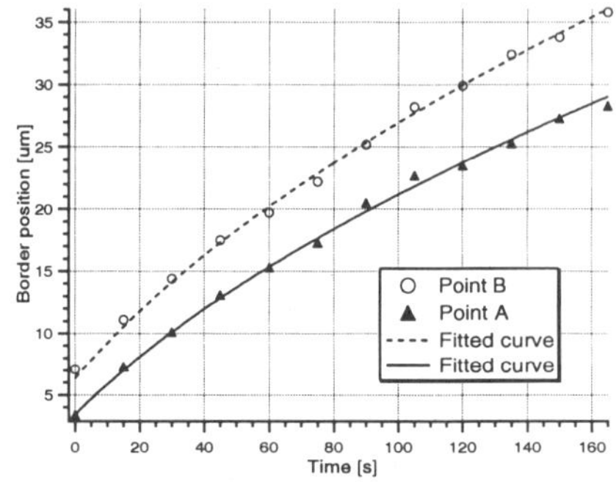


Fig. 3b: Border position as a function of time. The two sets of points correspond to the points A and B.

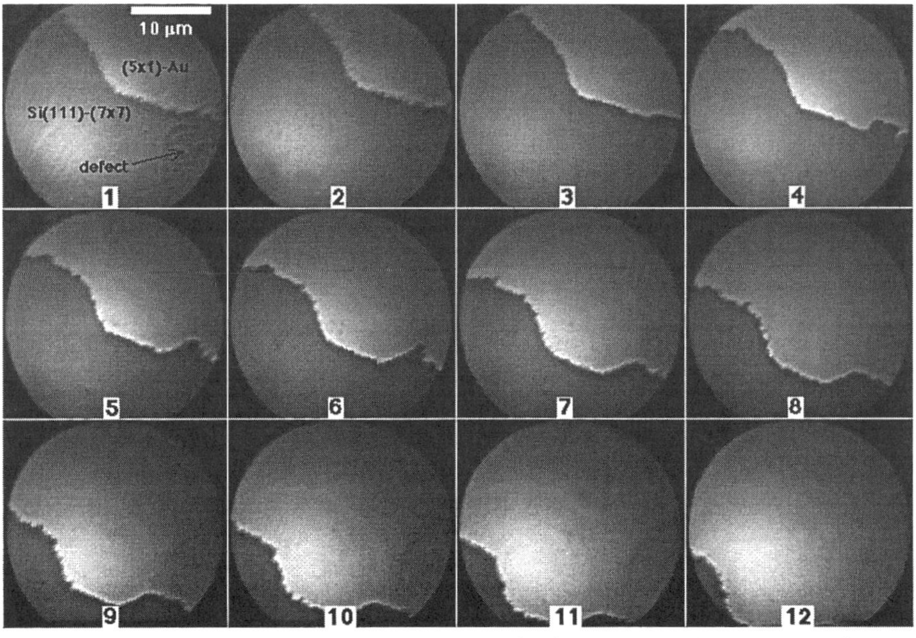


Fig. 4: Selected sequence of images in MEM mode taken during diffusion. The filed of view is 27 µm in each image, and the time interval between neighboring images is 30 seconds. Diffusion starts in the third image.

the work function threshold, they are reflected at the repulsive potential of the substrate, and the contrast of a MEM image arises from the corrugation of the surface potential. The MEM mode is thus ideally suited to detect inhomogeneities on a surface, like defects or potential steps, with high sensitivity and resolution. After preparation of an Au layer diffusion was activated until the Au-(5x1) area had sufficiently developed to be clearly separated from the √3-Au area. Diffusion resumed after the sample was once again heated to 970 K, essentially starting with image 3 and proceeding in the sequence 4 to 12 acquired every 30 seconds. In the published structure model the unit cell consists of either 4 or 5 atoms, giving rise to a coverage of $\Theta$=0.4 ML or $\Theta$=0.5 ML, respectively.

## 2.2. Au-STM

The deposited layers of Au (< 0.1 ML) were shortly flashed (< 5s) up to a temperature ~ 900 K, i.e., close to the value taken when the diffusion process started, as shown by the SPELEEM experiments. The Au coverage used in experiments was well below the interval where the saturated Au-(5x1) reconstruction was observed with LEED. The mixture of Si(111)-7x7 and Au-(5x1) was, however, identified [8]. Figure 5a shows the

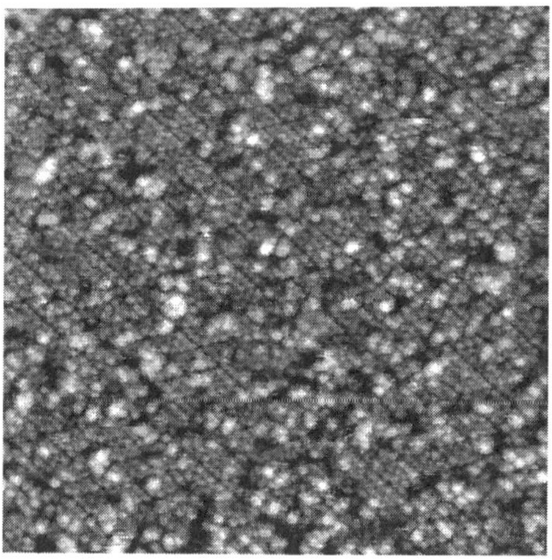

Fig. 5a: Filled-state image after Au deposition. Sample bias $V_S$ = -2.8V, $I_T$ = 0.3 nA, field of view 50x50 nm$^2$.

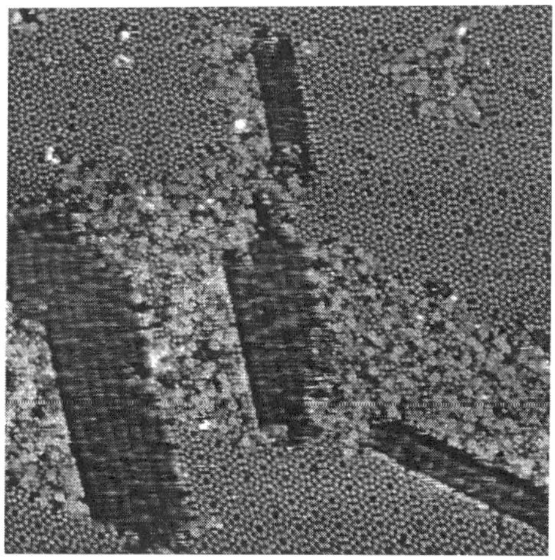

Fig. 5b: Empty-state image after annealing. Sample bias $V_S$ = +2V, $I_T$ = 0.77 nA, field of view 65x65 nm$^2$.

Si surface after Au deposition at RT. The distribution of the atoms is random and the partial erosion of the 7x7 unit cells is evident. Comparison of the empty- and filled-state images reveals different intensities of particular atoms in eroded parts, which demonstrates their different chemical nature. A short flash to the temperature taken at the beginning of macroscopic diffusion produces substantial changes in the atomic arrangement on the surface (see Fig. 5b). Three different domains of the Au-(5x1) phase (rotated by 60° with respect to each other) are formed. These are surrounded by disordered regions. The rest of the surface area is filled with Si(111)-7x7 reconstruction. Repeated flashing does not significantly change the surface stage. Observing the area on a larger scale, the distribution of the islands is seen to be homogenous mixture of three domains without any preferential orientation in respect to the surface (as is seen in Fig. 5b).

2.3. Cu-STM

Copper layers were similarly deposited at RT with a concentration less than 0.1 ML. Figure 6a demonstrates the initial stage of a layer after the deposition. The 7x7 reconstruction partially survives and the images show two prevailing features: First, the dark "vacancies" observed in *both* polarities regardless of applied voltage which we ascribe to *true* vacancies. Second, in some halves of 7x7 cells there are triangular clusters with altered appearance and intensities in empty- and filled-state images. The appearance of these *virtual* vacancies is interpreted as being due to the presence of single or more Cu atoms [9]. Generally, it can be concluded that the layer is formed by a mixture of clusters containing a maximum of three atoms in half of the 7x7 unit cell.

Fig. 6a: Filled-state image after Cu deposition. Sample bias $V_S$ = -2V, $I_T$ = 0.5 nA, field of view 75x75 nm$^2$.

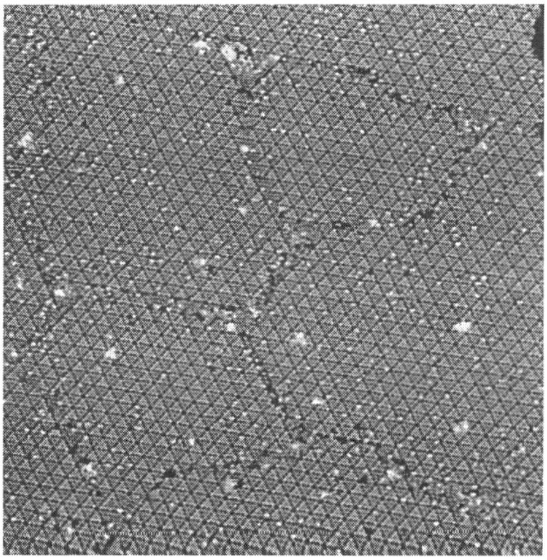

Fig. 6b: Filled-state image after annealing. Sample bias $V_S$ = -2V, $I_T$ = 0.2 nA, field of view 90x90 nm$^2$.

The layers were subsequently thermally treated. The temperature was increased up to ~ 700 K and held at this point for a few minutes (see Fig. 6b). At first sight the silicon surface appears more ordered, showing the ordered 7x7 terraces (as happens in the case of a clean surface). The only difference is that the terraces are divided by boundaries into a mosaic-like structure of 7x7 domains. The positions of the cells in two neighbor domains are shifted in respect to each other. Measuring a series of images as a function of tunneling voltage, we identified copper atoms in two positions. First, single atoms are located in halves of the 7x7 unit cells where they substitute one adatom, preferentially the corner adatoms. Similarly as with the deposited layer, they exhibit a "vacancy" character. Second, the rest of the copper atoms are found in the boundaries between the domains, preferentially at the crossroads point of three boundaries forming small triangular regions. The atoms inside these areas are arranged in a variety of local surroundings created by randomly distributed Cu and Si atoms. Even if the annealing continues at similar temperatures, no further changes in the morphology of the layers are observed.

2.4. Pb-STM

Evaporation of a small amount of Pb (< 0.01 ML) produces a layer with a characteristic STM image. The underlying Si(111)-7x7 substrate does not show any perturbation caused by the presence of Pb atoms. At RT, isolated Pb atoms move quickly in one-half of the 7x7 unit cell, appearing in STM images as a fuzzy triangle (Fig. 7a). If at least two Pb atoms are located in the same half of the 7x7 unit cell, they can be clearly distinguished as "well behaved balls" slowing down their motion. A slightly increased temperature (~ 310 K and more) initiates the diffusion: single Pb atoms may jump into

Fig. 7a: Filled-state image after Pb deposition. Sample bias $V_S$= -1.8V, $I_T$=0.88 nA, field of view 50x50 nm$^2$.

Fig. 7b: Filled-state image after annealing. Sample bias $V_S$= -1.8V, $I_T$= 0.65 nA, field of view 50x50 nm$^2$.

neighboring 7x7 cells, in this way traveling until they meet another single Pb atom, Pb cluster, or defect. The jumping rate exhibits an activation character and the diffusion coefficient can be found by measuring its temperature dependence [10]. The diffusion process ends when all single atoms form small clusters located in halves of the 7x7 unit cells with a maximum occupation of four atoms (Fig. 7b). The clusters are very stable up to the temperature of Pb desorption. We have not observed the clusters to disintegrate or move outside their "home" unit cells.

Sampling an area of 10 x 10 $\mu m^2$ of an evaporated layer with images having a frame of 90 x 90 nm$^2$, the difference in the Pb concentration was found. The value changes ± 20% among the randomly chosen frames. The same kind of statistical evaluation was carried out for annealed layers with the finished clustering of Pb atoms. No tendency was observed to equalize the concentration (counting the total number of atoms in a frame) and the estimated deviation from the previously calculated average value fits the same range.

## 3. Discussion

Three particular metals are used to demonstrate on a microscopic scale the complexity of the diffusion process on surfaces of elemental semiconductors. The choice of the metals depended on the strength of their interaction with the substrate: Lead is known for its negligible solubility in the Si bulk. A well-defined Pb/Si(111) interface is therefore formed. Stronger interaction between gold and silicon atoms results in their mixing in the surface layer, but without substantially penetrating into the Si bulk. The last case of Cu/Si(111) is distinguished by the strongest interaction, with forming silicides even close to RT.

Experimental results show that lead does not form any compound or solution with silicon, and that a minimum influence of the substrate can be expected (for morphological and topographical factors only). A single Pb atom at RT occupies one half of the 7x7 unit cell. Above a certain temperature threshold, Pb atoms start moving outside the half of the 7x7 unit cell. The diffusion can easily be modeled within the lattice gas model, omitting the diffusion of a particle inside this unit. Random walk modeling of the evolution of the number of monomers and clusters within lattice gas and non-interacting particles approximation fits well with the experimental data taken on a frame of 90 x 90 $nm^2$. On the other hand, the existing non-equilibrium distribution of Pb atoms on the frame of 10 x 10 $\mu m^2$ should force atoms to diffuse to less populated areas within the applied model. Missing evidence to support such a tendency suggests an idea of more local balance than on the mesoscopic scale. The most significant feature of the process is the trapping of two or more Pb atoms inside one-half of the 7x7 unit cell where they originally met. This indicates a strong, minimally short range Pb-Pb interaction, which could overwhelm the pure thermodynamical requirements imposed on the equilibrium.

Gold on a Si(111)-7x7 surface results in a more complicated system showing a few phases in the concentration range up to 1 ML. Au atoms do not possess a stable substitutional site in 7x7 reconstruction and therefore Au-(5x1) appears to be able to alter its gold content depending on temperature. Analysis of our mesoscopic data on the diffusion in the presence of the concentration gradient was concentrated mostly on the spreading of this phase, together with the behavior of the diffusion zone. The early stages of diffusion demonstrate that the movement of the concentration at points A and B obeys $\sqrt{Dt}$ behavior (see Fig. 3b) and the process initiates at temperatures above 950 K. The different velocity of both points opens up the possibility of spreading the Au-(5x1) reconstruction, which is in fact the alloy of gold and silicon. Keeping the temperature constant, the necessary gradient ruling diffusion over the Au-(5x1) region has to be created by other mechanisms, as, for instance, the difference in concentration at the opposite ends of this region. This should work in the initial stages of the profile evolution. On the other hand, the identical gradient is stretched over a longer and longer distance (tenths of μm) and the expansion of Au-(5x1) proceeds at constant velocity. Supposing the release of Au atoms from the $\sqrt{3}$–Au source to also be of constant rate, this concept will appear questionable in the sense of more local view of variance in the concentration observed for a Pb/Si(111) system.

The fact that no diffusion was detected macroscopically at temperatures below ~950 K strongly contradicts the microscopic findings. The short flash to the temperature (at least 100 K less than the former value) produces a very stable mixture of three phases which does not change under continuing heating at this temperature. The transition from a disordered homogenous Au layer on top of the eroded 7x7 substrate to the Au-(5x1) phase (Au-Si alloy), Si(111)-7x7 Si phase (pure Si) and a disordered Au phase (unknown mixture of Au and Si) must induce a massive rearrangement of nearly all the atoms on the surface.

The intuitive idea of *massive atomic rearrangement* was even more emphatically demonstrated in the Cu/Si system. The short annealing of the initial Cu layer leads to formations of 7x7 domains with scattered single Cu atoms substituting for the regular Si atoms. In the filled-state images it can be observed that the interaction of a Cu atom with the Si surrounding extends the dimer wall between two neighboring 7x7 unit cells. The increasing concentration of Cu atoms in a particular place causes the formation of a boundary region between 7x7 domains which possess a 7x7 reconstruction shifted with respect to each other. We conclude that macroscopic rearrangement of the surface atoms is necessary to obtain the resulting picture of "slipped" 7x7 domains.

Experiments with the Au/Si and Cu/Si systems exhibit a similar feature, which was shown by temperature-dependent measurements. Up to a certain threshold the metal atoms are well localized in the 7x7 unit cells, very rarely jumping inside or into the neighboring cell. No adsorbate-induced reconstruction, alloying, or chemical reactions were observed in this low temperature region. When the temperature increases above the threshold, the stage of the surface is rapidly changed. This fact suggests the thermally-activated character of the process. If a chemical reaction happened forming a silicide, then the reaction could be of either endo- or exothermic character. The former would freeze the atom positions because the energy can be found in the phonon bath only. However, the latter case may release significant energy, which is again transferred to the phonon bath of the substrate. This is the case of the $Cu_3Si$ silicide that is formed at a relatively low temperature (~ 470 K). Significant energy release of about 0.1 eV per 1 reacted Cu atom [11] accompanies the reaction. This *local* energy gain produces a *local* increase in surface temperature (~ 1200 K in an ideal case). It is obvious that such an effect could incite processes such as diffusion and/or other reactions involving a *macroscopic* amount of atoms. The high entropy of alloying between Au and Si atoms similarly leads also to an exothermic process, which could locally increase the temperature. This could produce a rapid outburst of energy stimulating the rearrangement of the substrate. Such an effect is known from bulk reactions, e.g., explosive crystallization [12,13].

## 4. Conclusions

The microscopic view of the diffusion of metal adsorbates on a Si surface shows the presence of additional effects, which strongly influence the equilibrium distribution of atoms as is predicted in the classical theory of diffusion. The exothermal process taking place between a Si surface and adsorbed metal atoms was suggested as a possible candidate for an explanation.

## 5. Acknowledgments

This investigation was supported by the Grant Agency of the Czech Academy of Sciences (Grant No.A1010718), the Grant Agency of the Czech Republic (Grant No. 202/99/P001 and No. 202/98/K002).

### 6. References

[1] Yasunaga, H. (1991) *Surf. Sci.* **242**, 171; Yamanaka, A. and Yagi, K. (1991) *Surf. Sci.* **242**, 181.

[2] Seebauer, E.G. and Allen, C.E. (1995) *Prog. Surf. Sci.* **49**, 265.

[3] A.G.Naumovets, A.G. and Yu.S.Vedula, Yu.S. (1985) *Surf. Sci. Rep.* **4**, 365.

[4] Snabl, M., Ondrejcek, M., Chab, V., Chvoj, Z., Stenzel, W., Conrad, H. and Bradshaw, A.M. (1998) *J. Chem. Phys.* **108**, 4212.

[5] Crank, J. (1975) *The Mathematics of Diffusion*, 2nd edition, Oxford University Press, London.

[6] Schmidt, Th., Heun, S., Slezak, J., Diaz, J., Prince, K.C., Lilienkamp, G. and Bauer, E. (1998) *Surf. Rev. Lett.* **5**, 1247.

[7] Slezak, J., Ondrejcek, M., Chvoj, Z., Chab, V. (2000) *Phys. Rev. B* **61**, 16121.

[8] Hasegawa, T., Hosoki, S. (1996) *Phys. Rev. B* **54**, 10300; Hasegawa, T., Hosoki, S., Yagi, K. (1996) *Surf. Sci.* **355**, L295.

[9] Tosch, St., Neddermeyer, H. (1989) *Surf. Sci.* **211/212**, 133; Atwell, A.R., Zuo, J.-K. (1998) *Surf. Sci.* **417**, 261; Ikeda, T., Watanabe, T., Itoh, H., Ichinokawa, T. (1996) *Surf. Rev. Lett.* **3**(3), 1377.

[10] Slezak, J., Chab, V., Choj, Z., Mutombo, P. (2000) *JVST B* **18**(3), 1151; Gomez-Rodriguez, J.M., Veuillen, J.-Y., Cinti, R.C. (1997) *Surf. Rev. Lett.* **4**, 335.

[11] Chromik, R.R., Neils, W.K., Cotts, E.J. (1999) *J. Appl. Phys.* **86**, 4273.

[12] Hurle, D.T.J. (editor) (1993) *Handbook of Crystal Growth*, Elsevier Science Publishers, Amsterdam.

[13] Lengsfeld, P., Nickel, N.H. and Fuhs, W. (2000) *Appl. Phys. Lett.* **76**, 1680; Cerny, R., Prikryl, P. (1998) *Phys. Rev. B* **57**, 194.

# SURFACE DYNAMICS OF STEPPED Si(001) STUDIED BY TEMPORAL LEED SPECTROSCOPY

M. KAMMLER AND M. HORN VON HÖGEN
*Fachbereich Physik, Universität Essen, 45117 Essen, Germany*

N. VOSS AND M. TRINGIDES
*Ames Laboratory, Iowa State University, Ames, IA 50011, USA*

AND

A. MENZEL AND E.H. CONRAD
*School of Physics, Georgia Institute of Technology, Atlanta, GA 30332, USA*

**Abstract.** We present the first demonstration of Temporal LEED Spectroscopy's ability to measure surface dynamics as shown by a study of step dynamics on vicinal Si(001). While we measure activation energies consistent with other work, the observed microscopic mechanism driving the fluctuations is inconsistent with previous claims. We exclude evaporation/condensation of atoms from the steps as the rate limiting kinetics in the temperature range of this experiment. Possible reasons for these differences are discussed.

## 1. Introduction

Knowing the dynamics of surfaces is critical to understanding how crystals grow, how surface structures form or phase separate, and how chemical reactions proceed. Unfortunately, our ability to measure surface kinetics is severely limited. STM has been used to study the kinetics of single atoms, but these studies are restricted to video acquisition rates [1]. Preliminary STM studies based on tunneling current fluctuations seem promising but they still must cope with field induced kinetics. [2] FEM has also been used to study surface kinetics, again for multiple atoms, but the systems that can be studied are restricted to materials suitable for the field emitting tips [3]. Recently LEEM has been used to study kinetics at much higher temperatures and closer to equilibrium [4], but length scales are limited to $\gtrsim 100$Å. We have developed a new method for measuring surface kinetics. It is the reciprocal space analog of LEEM with the advantages of being able to measure dynamics over larger length scales with time scales as low as a few $\mu$sec. The technique is called Temporal LEED Spectroscopy, TLS [5,6].

A TLS experiment consists of measuring LEED diffraction intensity as a function of time for a momentum transfer vector $\boldsymbol{q}$, i.e., $I(\boldsymbol{q},t)$. Fluctuations in the diffraction intensity are due to fluctuations in the surface structure related to a length scale proportional to $q^{-1}$. By choosing $\boldsymbol{q}$ properly, the diffraction can be made sensitive to the dynamics of steps, structural phases, grain boundaries, etc. For pulse counting, a discrete-time series, $I_{\boldsymbol{q}}[n]$, of $I(\boldsymbol{q},t)$ is generated by summing the number of pulses from the detector over both the



*M.C. Tringides and Z. Chvoj (eds.),*
*Collective Diffusion on Surfaces: Correlation Effects and Adatom Interactions,* 23–32.

gate period, $t_g$, and instrument resolution, $\Delta \boldsymbol{q}$. In principle $t_g$ can be $< 1\mu$sec. Because the collection time, $t_c$, is finite, $I_{\boldsymbol{q}}[n]$ is only defined for $0 \leq n < n_T$, where $n_T = t_c/t_g$. In the following discussions the index $\boldsymbol{q}$ will be dropped, although it is still implied.

¿From the time series $I[n]$ the autocorrelation function, $\hat{G}[m]$, is then generated:

$$\hat{G}[m] = \langle \delta I[n] \delta I[n+m] \rangle \,, \tag{1a}$$

where $\delta I[n] = I[n] - \langle I \rangle$, and $\langle I \rangle$ is the average number of counts per gate-time. The brackets, $\langle \ldots \rangle$, represent a time average over the collection time, $t_c$. It is important to note that Eq. (1a) is used as an estimator for the continuous-time autocorrelation function,

$$G(\tau) = \langle\!\langle \delta I(\boldsymbol{q}, t) \delta I(\boldsymbol{q}, t+\tau) \rangle\!\rangle, \tag{1b}$$

where $\langle\!\langle \ldots \rangle\!\rangle$ represent an ensemble average or its expectation value.

## 2. Theory of Temporal LEED Spectroscopy

As a specific example of how TLS is used to extract surface dynamic information, we discuss the kinetics of step fluctuation on vicinal Si(001). We review the relevant theory of step motion; showing how kinetic pathways, activation energies, etc., can be extracted from TLS. We also highlight an number of important experimental considerations.

### 2.1. EXPERIMENTAL CONSIDERATIONS

Since the diffraction intensities are integrated over the diffractometer's resolution width, the TLS signal is only coherent over a length, $\zeta = 2\pi/\Delta q$. On the other hand, the incident beam diameter, $D$, is much larger than $\zeta$, making the diffraction intensity an incoherent sum over $M$ incoherent domains; $M = (D/\zeta)^2$ (for high resolution LEED, $M \sim 10^5$). While it appears the incoherent averaging will severely reduce the TLS signal, the situation is not as bad as it seems. This is because the TLS signal averaged over $M$ domains is [5]:

$$\frac{G(\tau)}{\langle\!\langle I \rangle\!\rangle} = \delta(\tau) + \frac{\langle\!\langle I \rangle\!\rangle}{M} \frac{\langle\!\langle \delta I_{p,i}(t) \delta I_{p,i}(t+\tau) \rangle\!\rangle}{\langle\!\langle I_p \rangle\!\rangle^2}. \tag{2}$$

The first term in Eq. (2) is from statistical noise. The second term in Eq. (2) is the autocorrelation function from a single domain, defined as $G_p(\tau) \equiv \langle\!\langle \delta I_{p,i}(t) \delta I_{p,i}(t+\tau) \rangle\!\rangle$ (averaged over $M$ domains). Since $M \propto D^2$, $\langle I \rangle / M$ is proportional to the current density from the LEED gun. Therefore, if the incident current density is large enough so that $G_p(\tau)$ is above the statistical noise in the experiment, $G_p(\tau)$ will be measurable [6].

There are other factors that must be considered in a TLS experiment. First, $\Delta q$ of the instrument must be small. If it is too big, the TLS signal will be an average over a range of $\boldsymbol{q}$'s. Since relaxation times depend on the $q$ of the disturbance [7], $\boldsymbol{q}$-integration causes time constants measured by $\hat{G}[m]$ to be convoluted with those from structures not intended to be probed. In general, we need $\zeta > \xi$, where $\xi$ is the length over which structural fluctuations are correlated. Since in a high resolution LEED $\zeta \gtrsim 10^3$Å, $\boldsymbol{q}$-averaging is not a problem.

The second factor concerns the choice of $t_g$ and $t_c$. If the gate time is long compared to the decay time of $G(\tau)$, $\tau_{1/2}$, the autocorrelation estimate, $\hat{G}[m]$, will be significantly broadened compared with $G(\tau)$ [8]. However, decreasing $t_g$ makes $\langle I \rangle$ smaller and decreases the signal to noise ratio in $\hat{G}[m]$ (see Eq. (2)). A compromise can usually be found.

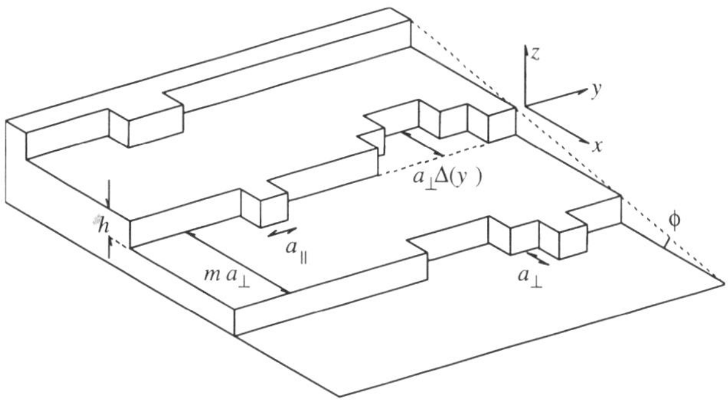


*Figure 1.* Schematic of a stepped surface with average terrace width $L = ma_\perp$.

The collection time is a more serious matter. If $t_c \sim \tau_{1/2}$, $G(\tau)$ becomes seriously distorted [9,10]. This is because estimating ensemble averages (Eq. (1b)) by time averages (Eq. (1a)) introduces a bias that cannot be corrected or minimized by standard techniques. An expression for the bias can be found if $G(\tau)$ is known. The reciprocal problem, however, is far more difficult. To find $G(\tau)$ based on the estimate $\hat{G}[m]$, one must guarantee that the bias is relatively small. Within limits, this can be done by increasing the ratio between total collection time relative to the decay time of $G(\tau)$ (a rule of thumb would be $t_c/\tau_{1/2} > 10$). Unfortunately experimental conditions, such as gun current drift, limit $t_c$.

Drifts cause additional problems because $G(\tau)$ is only defined for stationary signals (i.e., $\langle I \rangle$ = constant) [11]. Even relatively small drifts or fluctuations will introduce new distortions in the measured $\hat{G}[m]$. Background-subtraction and data filter-designs can help, but constitute a considerable challenge. To some extent problems concerning non-stationary signals can be reduced under the assumption that the signal is piecewise stationary. Then, $\hat{G}[m]$ can be calculated from an ensemble average of many shorter data-series assuming $t_c \gg \tau_{1/2}$. Note that each series must be weighted by $1/\langle I \rangle^2$ (see Eq. (2)).

## 2.2. TLS FROM STEPPED SURFACES

Figure 1 shows a vicinal surface cut at an angle $\phi$ from a low index plane. At T=0K the surface consists of ordered straight steps of height $h$ separated by terraces $L = ma_\perp$ wide ($a_\perp$ is the distance between rows on the terrace and $m$ is an integer). At elevated temperatures kinks form on the step edges (minimum length $a_{\|}$) causing them to meander with an amplitude that depends on the kink energy, $\epsilon$, and the step-step interaction [12].

The diffraction is sensitive to the position and shape of the steps if the components of $\boldsymbol{q}$ normal to the steps, $q_x$ and perpendicular to the terraces, $q_z$, correspond to an out-of-phase condition (i.e., $q_x = \pi/ma_\perp$ and $hq_z = (1-n)\pi$, where $n$ is an integer). Then the amplitude scattered from adjacent terraces is 180° out of phase [13] and $G(\tau)$ can be written analytically assuming independent steps (a slight modification of the Gruber-Mullins model) [14]. Writing the displacement at a position $y$ on the $j^{th}$ step (relative to the $T = 0$K position) as $a_\perp \Delta_j(y,t)$ for some time $t$, the second term in Eq. (2) becomes [6]:

$$\frac{G_p(\tau)}{\langle I_p \rangle^2} = \frac{4}{N_x \langle \widetilde{x} \rangle^2} \left\{ \langle\langle \widetilde{x}(t)\widetilde{x}(t+\tau) \rangle\rangle - \langle\langle \widetilde{x}(t) \rangle\rangle^2 \right\}, \tag{3}$$

where

$$\widetilde{x}_j(t) = \sum_{y}^{\zeta = a_{\|} N_y} \cos\left(\pi \Delta_j(y,t)/m\right). \tag{4}$$

Note that the TLS signal is proportional to $1/N_x = (\zeta/a_\perp)^{-1}$, the number of independent steps sampled by the coherent part of the beam. To evaluate Eq. (3), we need to calculate the appropriate averages of Eq. (4). This requires a model for the dynamics of $\Delta_j(y,t)$.

Several continuum models have been applied to STM, LEEM, and REM studies of step dynamics [4, 7, 15–19]. Common to all these models is a stochastic driving force causing local fluctuation in the step's position and a restoring force that tries to return the step to its equilibrium position. The driving force is related to the step mobility $\Gamma$, while the restoring force is parameterized by the step-edge stiffness, $\widetilde{\beta}$, and a local constraining potential, $V(x,L) \approx 2U(L) + c(L)x^2$ (where $U(x)$ is the potential between steps separated by $x$ and $c(L) = \partial^2 U/\partial x^2|_L$) [20]. $\widetilde{\beta}$ measures the steps' tendency to remain straight and depends on the kink energy through the step diffusivity, $b^2(T)$ [15, 21, 22].

$$\widetilde{\beta}(T) = \frac{a_{\|} kT}{b^2(T)}. \tag{5}$$

Both $\widetilde{\beta}$ and $c$ govern two equilibrium properties of the step: the mean squared step displacement, $w_\infty^2$, and the correlation length along the step, $\xi$. In a continuum model [6, 15, 20]:

$$w_\infty^2(T) = kT\left(\frac{1}{8c\widetilde{\beta}}\right)^{1/2} \quad \text{and} \quad \xi(T) = \left(\frac{\widetilde{\beta}}{2c}\right)^{1/2}. \tag{6}$$

We will consider three possible kinetics models for step motion: Step-edge Diffusion (SD), Terraces Diffusion (TD) and Evaporation/Condensation (EC). Evaporation/Condensation kinetics are important when diffusion along the step edges is slow and terraces diffusion is fast. Then, step fluctuations are limited by evaporation/condensation of terrace atoms to and from the step edges [15]. In this case $\Gamma$ is related to the time an atom takes to evaporate from the step [7, 15]. If both the EC and TD processes are slow, then the kinetics will be governed by diffusion along the step edges. In this case $\Gamma$ is related to the hopping time of atoms between adjacent step edge sites. Finally, if the SD process is slow and the EC and TD processes are fast, the fluctuations are determined by the diffusion time of an atom or vacancy across the terrace. Then, $\Gamma$ is determined by both the diffusion coefficient, $D_s$, and the concentration, $C_o$, of vacancies and adatoms on the terraces [16].

All three of these models can be cast in the form of linear Langevin Equations [7,15,16]. ¿From these Langevin equations the appropriate averages in Eqs. (3) and (4) can be found and the TLS signal can be calculated [6]. Based on numerical calculations the long time behavior of $G(\tau)$ can be summarized for SD, TD, and EC kinetics in a single equation. For $\tau > \tau_{1/2}$, the correlation function decays as [6, 23]:

$$G(\tau) \propto \exp\left[-A\left(\tau/\tau_o\right)^{1/n}\right], \tag{7}$$

where $A$ and $n$ are constants that depend on the microscopic process. The time constant, $\tau_o$, is the characteristic decay time of the correlation function proportional to $\Gamma$ and $c$ [6]. For the EC model $n = 1$. When evaporation is fast, the TD model gives $n = 3$. On the other hand, when the kinetics are limited by step-edge diffusion, the SD model gives $n = 4$. Therefore, the functional form of the TLS signal gives information on the type of kinetics driving the step fluctuation. In addition to $G(\tau)$'s form, its decay time gives information on the microscopic time scale and thus the microscopic energetics. Scaling relations between $\tau_{1/2}$ and $\tau_o$ have been produced [6]. These relationships depend on both the kinetic model and on the ratio between the instrument transfer width and the correlation length, $\zeta/\xi$. The reason for the latter is that, if $\zeta/\xi < 1$, the instrument only samples small wavelength fluctuations. Since the relaxation time for a fluctuation increases with wavelength [7], the measured decay time is faster than if $\zeta/\xi > 1$, even though the kinetics are the same.

## 3. Kinetics of stepped Si(001)

The bulk symmetry and the $2 \times 1$ phase of Si(001) cause adjacent terraces on Si(001) (miscut towards the [110]) to alternate between dimers parallel and perpendicular to the steps. Steps with dimers on the upper terrace parallel or perpendicular to the step edge are labeled $S_A$ and $S_B$ steps, respectively. The $S_A$ step are known to be relatively smooth while the $S_B$ steps are much more rough [19]. As we will see, the TLS data will be much more sensitive to $S_B$ step motion. This is verified by analyzing the diffraction peak widths.

Our Si(001) sample was miscut with its surface normal tilted $\sim 1°$ towards the [110] direction producing $\langle L \rangle = 65$Å terraces. Samples were cleaned and annealed using standard techniques [24]. Data were taken using a SPA-LEED system with an external electron gun operating at 300eV (incidence angle was $\sim 67°$). The system transfer width was $\zeta \sim 400$Å and 1600Å perpendicular and parallel to the steps, respectively. For time series data, $\boldsymbol{q}$ was fixed at the out-of-phase condition: $q_z = 6.93$Å$^{-1}$ and $q_x = 0.0455$Å$^{-1}$. For all data the gate period was $t_g = 2$ms. A single time series consisted of $n_T = 9000$ points for a total collection time, $t_c = 18$s. At each temperature $\hat{G}[m]$ were generated and ensemble averaged over 200 individual correlation functions.

### 3.1. STATIC ENERGIES

Before discussing the step kinetics, it is important to mention how static energies are obtained. As the amplitude of the step fluctuations increases with temperature, the decreased correlation along the steps causes the diffraction peaks to broaden in the direction parallel to the step edges. That is, the widths of these peaks are proportional to $\xi^{-1}$ [22]. Figure 2 shows diffraction profiles perpendicular and parallel to the step edges at an out-of-phase condition ( $q$ is given in units of % of the Brilluoin Zone $2\pi/a$ : $a = 3.84$Å). In Fig. 2(a) the peak splitting perpendicular to the steps is shown where the peak separation is $\Delta q_x = 2\pi/L$. Figure 2(b) is a profile parallel to the steps taken through the peak marked by the dashed line in Fig. 2(a), i.e., $q_x = 2.78$. The solid line in Fig. 2(b) is a two component fit; a broad Lorentzian due to the larger fluctuating $B$ steps and a narrow Gaussian for the straighter $A$ steps. The dashed line in Fig. 2(b) is a single Lorentzian fit showing the poor quality of the fit without two compents. The widths of each component were measured as a function of temperature to calculate $\xi^{-1}$ for both the $A$ and $B$ steps [22]. Using the measured correlation lengths from the line shapes and Eqs. (5) and (6), the ratio of the step diffusivity to the interaction energy can be computed. The temperature

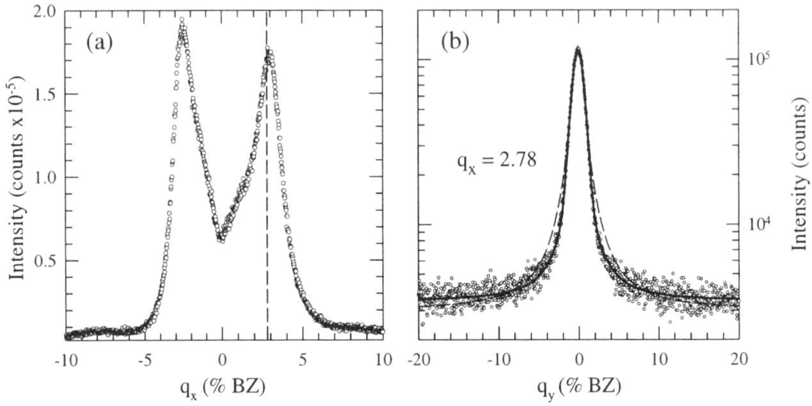


*Figure 2.* (a) Intensity profile perpendicular to the steps for $T = 1023$K showing peak splitting. Dashed line marks the value of $q_x$ where profile in (b) was taken. (b) Intensity profile parallel to the steps. (○) data. Solid line is a two component fit as described in text. Dashed line is a single Lorentzian fit.

dependence of this ratio is determined by $b^2$. In the TSK plus corner energy model [22]:

$$b^2 = a_\perp^2 \frac{2}{\sinh^2(\epsilon/2kT)} \left\{1 + \tanh(\epsilon/2kT)\left[\exp(\epsilon_c/2kT) - 1\right]\right\}^{-1}, \tag{8}$$

where $\epsilon$ and $\epsilon_c$ are the kink and corner energy, respectively. Fitting the temperature dependence of $\xi$, gives both the kink energies and $c$. For Si(001) A and B steps fluctuate but with different amplitudes [19, 25]. Again, assuming that adjacent steps are independent, the diffraction lineshapes will consist of two components: a sharp term whose width is related to the straighter A-steps and a broad component related to the coarser B steps. Figure 3(a) shows the experimental $\xi$'s for both steps as a function of temperature derived from diffraction profiles parallel to the steps. We have fit $\xi$ for both A and B steps to the experimental values using a $\chi^2$ routine to Eqs. (5), (6), and (8). We assume that the step interaction potential is $U(x) = C_1 + C_2 \ln(x/a)$ [26,27] (so from the definition above Eq. 8, $c = C_2/L^2$) [20], and that $\epsilon_c$ and $c$ are the same for both steps. With these assumptions we find $cL^2 = 1.2 \pm 0.2$meV/Å, $\epsilon_A = 65 \pm 6$meV, $\epsilon_B = 38 \pm 5$meV, and $\epsilon_c = 80 \pm 20$meV. The fit is plotted in Fig. 3(a) and is rather insensitive to $\epsilon_c$, although our fitted value is fortuitously equal to previous measurements [19]. Our kink energies are within error bars of previous STM and LEEM measurements [19, 28]. The step-step interaction is also within error of those derived from strained Si(001) experiments [27].

Using the experimental values of $\xi$ and the fitted value of $c$, we can calculate the step-edge stiffness using Eq. (6). The values of $\widetilde{\beta}$ are plotted in Fig. 3(b) along with the theoretical values from the TSK model using the kink energies given above. We have also plotted $\widetilde{\beta}$ from a slightly higher temperature regime as determined by Bartelt *et al.* (using LEEM) [28]. Our A step results are in good agreement. Our B step stiffness is about 50% higher at 800K, but our extrapolated $\widetilde{\beta}$ approaches the LEEM values above 1200K.

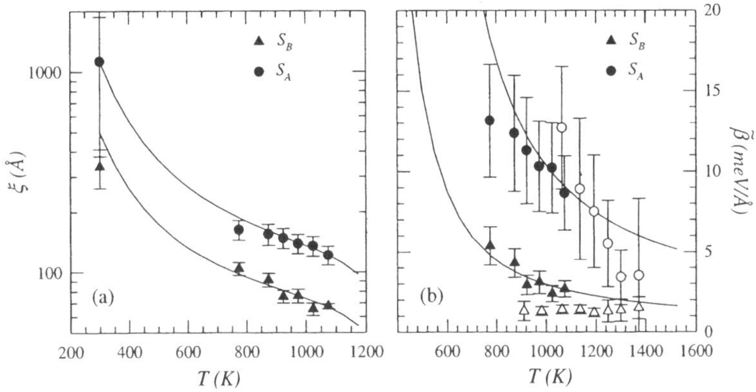


*Figure 3.* The correlation length (a) and step stiffness (b) versus temperature. Solid symbols are those measured from this work, open symbols are those measured in Ref. [28]. Solid lines are best fit values based on a TSK+corner model for $b^2$ and step-step interaction $c$.

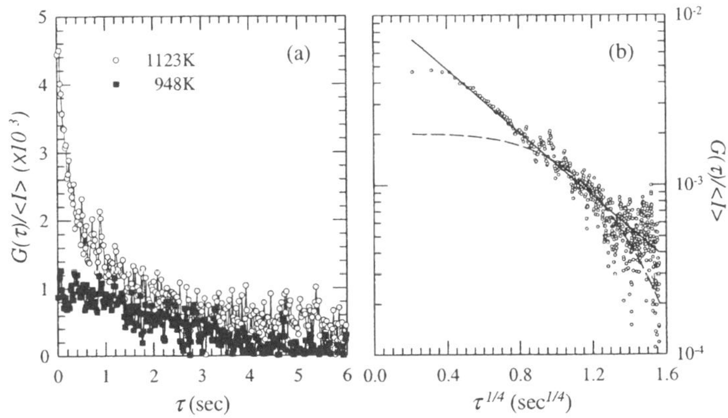


*Figure 4.* (a) The normalized autocorrelation function $G_p(\tau)/\langle I\rangle$ versus time for an out-of-phase condition. (○) $T = 1123$K. (■) T=948K. (b) A plot of $\log(G(\tau)/\langle I\rangle)$ vs $\tau^{1/4}$ for $T = 1123$K. The heavy solid line is the best fit line for the SD model. The dashed line is the best fit line for the EC model.

### 3.2. STEP KINETICS

Figure 4(a) shows the normalized autocorrelation function, $G(\tau)/\langle I\rangle$, for out-of-phase scattering at two different temperatures. The statistical noise term at $\tau = 0$ (the delta function in Eq. (2)) has been removed, since it contains no information about the kinetics [5]. Also, the calculated background, due to finite collection time effects, has been added to $G(\tau)/\langle I\rangle$ [9]. The temperature-dependence of $G(\tau)$ is obvious in Fig. 4(a); both increasing in strength and decaying faster with increasing temperature. This is consistent with faster kinetics and larger step fluctuations at higher temperatures.

To determine the rate limiting kinetics, fits were made to the experimental $G(\tau)$ using Eq. (7). The best fit value for the exponent $n$ for all the data was $n = 3.9 \pm 0.5$. This is close to $n = 4$ suggesting SD kinetics. An example fit is shown in Fig. 4(b). The data are plotted on a log plot versus $\tau^{1/4}$ to emphasize the poor fit to an exponential decay ($n = 1$).

The fact that the data cannot be fit to $n = 1$ excludes evaporation/condensation as the rate limiting kinetics (at least for 940K $< T <$ 1130K). On the other hand, the $\chi^2$ values for an $n = 4$ fit were only slightly better than those for $n = 3$. This means that while the SD model is a better representation of the data, we cannot rule out the TD model.

To determine activation energies and prefactors for the microscopic process, note that the decay times, $\tau_{1/2}$, of the autocorrelation functions represent a macroscopic time associated with distorting the step shape. Relating $\tau_{1/2}$ to the microscopic process driving the step fluctuations depends on the rate limiting kinetics [6]. Since we have ruled out EC kinetics, we will restrict our discussion to SD and TD kinetics. Note that the majority of the TLS signal must be coming from the rough B step fluctuations. This is because the magnitude of the TLS signal is proportional to $w^4$. Since $w^2 \propto 1/\xi$ (see Eq. (6)), it follows that $G(\tau)/\langle I\rangle^2 \propto \xi^{-2}$ [6]. Using the values of $\xi$ from Fig. 3(a), we can estimate that the B steps contribute 3–4 times more signal than the A steps. In subsequent discussions we will refer to activation energies and prefactors as being those from B steps.

The scaling relation between $\tau_{1/2}$ and the microscopic times can be determined numerically [6]. While in general the scaling depends on the ratio $\zeta/\xi$, in our temperature regime $\zeta/\xi > 1$. This makes the scaling independent of $\zeta$. In this limit there are good analytical approximations for the scaling. For SD and TD kinetics, it can be shown that [29]:

$$\tau_{1/2} \approx 1.35 \left(\frac{\xi}{a_{\parallel}}\right)^2 \left[\frac{kT}{2ca_{\parallel}a_{\perp}^2}\tau_s\right], \text{ for SD kinetics} \tag{9a}$$

$$\tau_{1/2} \approx 0.054 \left(\frac{\xi}{a_{\parallel}}\right) \left[\frac{kT}{2ca_{\parallel}a_{\perp}^2}\frac{1}{D_S C_o}\right], \text{ for TD kinetics,} \tag{9b}$$

where $\tau_s$ is approximately the time for an atom to hop to an adjacent site. $D_s$ and $C_o$ are the diffusion coefficient and equilibrium concentration of the diffusing species on the terraces, respectively. Using the experimental values of $\tau_{1/2}$ from $G(\tau)$ and $c$ and $\xi$ obtained from the profile fits, we can calculate either $\tau_s$ for SD kinetics or $1/D_sC_o$ for TD kinetics from Eqs. (9a) or (9b). Arrhenius plots of $\tau_s$ and $1/D_sC_o$ are shown in Fig. 5. If the hopping time for step-edge diffusion is $\tau_s = (1/\nu_s)\exp(\Delta E_s/kT)$, where $\nu_s$ is the attempt frequency and $\Delta E_s$ is the activation energy, we find $\Delta E_s = 0.8 \pm 0.2$eV and $\nu_s \sim 1 \times 10^{10}\text{s}^{-1}$. If we assume TD kinetics, then $1/D_sC_o = (1/\nu_T)\exp(\Delta E_T/kT)$ and $\Delta E_T$ is the sum of a diffusion barrier and a formation energy with $\nu_T$ being the attempt frequency for diffusion. For TD kinetics we find $\Delta E_T = 1.0 \pm 0.2$eV and $\nu_T \sim 3 \times 10^8\text{s}^{-1}$. The reason $\Delta E$ depends on the model is due to the temperature dependence of $\xi$ in Eqs. (9a) and (9b).

The low $\Delta E$'s we measure favor step-edge diffusion over terrace diffusion. For TD the activation energy should be due to both a diffusion and a formation barrier: $\Delta E_T = \Delta E_D + \Delta E_F$. For Si(001) the diffusion barrier is known to be ~0.67eV [30]. This implies that $\Delta E_F \sim 0.1 - 0.3$eV, well below the theoretical dimer formation energy of 1.6eV [31].

Our measured activation energy is consistent with earlier STM results [32]. Pearson *et al.* found $\Delta E = 0.97 \pm 0.12$eV. LEEM studies have also investigated the step kinetics of this surface but in a higher temperature regime. [4] They find $\Delta E = 1.45$eV, higher than our value. Two important differences must be mentioned between this work and the previous studies. First, the STM studies measure a much lower prefactor ($\nu \sim 10^4$ to $10^6$) [32] than we find. The reason for the low value of $\nu$ is not known, but since LEEM studies [4] measure frequencies similar to ours, we suspect flaws in the STM analysis.

The second difference is in the type of kinetics we find responsible for the step motion. The STM studies of Pearson *et al.* [32] suggest "kink diffusion" (defined as any process

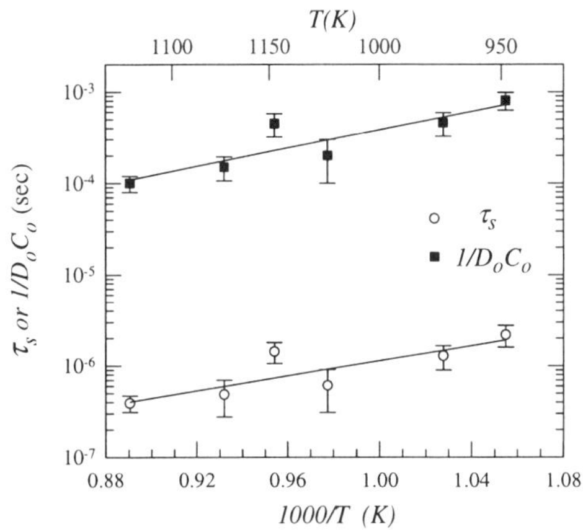


*Figure 5.* An Arrhenius plot of the microscopic time constants for Step-edge Diffusion, $\tau_s$, (○) and Terrace Diffusion, $1/D_sC_o$ (■).

causing a kink edge to move) as the responsible microscopic mechanism. However, they could not distinguish between SD and TD as the source of the kink movement because they can only determine if a site is filled and not where the atom filling that site arrived from. This is because the kinetic process leading to site filling is too fast to be resolved in the image acquisition time. The LEEM results make a much stronger statement [4]. From their Fourier analyzed time constant they find that $\tau(q) \propto q^{-2}$; consistent with evaporation/condensation of atoms from the step edges. This is clearly inconsistent with our results as demonstrated in Fig. 4(b). This apparent difference may be resolved by noting that the Arrhenius behavior of the step mobility reported in Fig. 10 of Ref. [4] shows evidence of breaking down for temperatures below 1100K. This suggests that there is a crossover to different kinetics below this temperature. The low temperature data in the Arrhenius plot of Ref. [4] gives a $\Delta E \sim 0.9$eV, consistent with both our results and the STM results [32]. Assuming this to be true, we can conclude that the dynamics of steps on this surface are rich and may have various crossover regimes at different temperatures.

## 4. Conclusions

We have discussed how Temporal LEED Spectroscopy can be used to measure the dynamics of surfaces. As a specific example of the application of this technique, we have studied the kinetics of stepped Si(001). The kink formation energies and step-step interaction we have measured are in good agreement with previous work. These measurements, however, raise new questions concerning the kinetics of this surface. From the TLS autocorrelation functions we find that step-edge diffusion or terrace diffusion must be responsible for the step kinetics. Our low activation energy, however, favors step-edge diffusion because it would imply an unreasonably low adatom or dimer formation energy ($\Delta E_F < 0.3$eV).

While our activation energies are consistent with STM results, there are differences that need to be reconciled. Our attempt frequencies are much higher than STM results [32], by nearly 4 orders of magnitude. Whether or not this is due to a fundamental difference between STM and TLS (e.g., sampling rates) or due to differences in the temperature ranges investigated ($T < 700$K for the STM studies) must be addressed. In addition, LEEM studies find that evaporation/condensation kinetics are responsible for the step fluctuations for $T > 1000$K. We instead find that most likely step-edge diffusion (or possibly terrace diffusion) dominate the kinetics below 1123K. It is possible there is a crossover to evaporation/condensation at higher temperatures, but such a supposition cannot be supported at this time. It is obvious that more work needs to be done to resolve these issues.

## Acknowledgments

Ames Laboratory is operated for the US Department of Energy by Iowa State University under Contract No. W-7405-Eng-82 (M. Tringides). This work was supported in part by the NSF under grant No. DMR-9211249 (E. Conrad).

## References

1. I.S. Hwang, S.K. Theiss and J. Golovchenko, *Science* **265**, 490 (1994).
2. M.L. Lozano and M.C. Tringides, *Europhys. Letters* **30**, 537 (1995).
3. R. Gomer (1997) in, M.C. Tringides (ed.), *Surface Diffusion Atomic and Collective Processes*, Proc. NATO ASI, Plenum Press, New York, pp. 427–442.
4. N.C. Bartelt and R.M. Tromp, *Phys. Rev.* **B54**, 11731 (1996).
5. E.H. Conrad, A. Menzel, S. Kiriukhin and M.C. Tringides, *Phys. Rev. Lett.* **81**, 3175 (1998).
6. A. Menzel, K. Wiesenfeld, E.H. Conrad and M.C. Tringides, *Phys. Rev.* **B61**, 2997 (2000).
7. N.C. Bartelt, J.L. Goldberg, T.L. Einstein, E.D. Williams, J.C. Heyraud and J.J. Mètois, *Phys. Rev.* **B48**, 15453 (1993).
8. M.I. Larsson, M.C. Tringides, H. Pfnür, H. Frischat, K. Budde, M. Kammler, and M. Henzler, *Surf. Sci. Lett.* **411**, L789 (1998).
9. A. Menzel, E.H. Conrad, and M.C. Tringides (to be published).
10. A. Gelb (ed.) (1984), *Applied Optimal Estimation*, M.I.T. Press, Cambridge
11. A.V. Oppenheim and R.W. Schafer (1999), *Discrete-Time Signal Processing*, Prentice Hall, Upper Saddle River; E.O. Schultz-DuBois and I. Rehberg, *Appl. Phys.* **24**, 323 (1981).
12. E.D. Williams and N.C. Bartelt, *Science* **251**, 393, (1991).
13. E.H. Conrad (1996), in, *Physical Structures*, W.N. Unertl (ed.), *Handbook of Surface Physics* Vol 1, Elsevier, Amsterdam, pp. 271-360.
14. E.E. Gruber and W.W. Mullins, *J. Phys. Chem. Solids* **28**, 875 (1967).
15. N.C. Bartelt, J.L. Goldberg, T.L. Einstein and E.D. Williams, *Surf. Sci.* **273**, 252 (1992).
16. N.C. Bartelt, T.L. Einstein and E.D. Williams, *Surf. Sci.* **312**, 411 (1994).
17. A. Pimpinelli, J. Villain, D.E. Wolf, I. Elkinani, G. Uimin, J.C. Heyraud and J.J. Mètois, *Surf. Sci.* **295**, 143 (1993).
18. J.B. Hannon, N.C. Bartelt, B.S. Swartzentruber, J.C. Hamilton, and G.L. Kellogg, *Phys. Rev. Lett.* **79**, 4226 (1997).
19. B.S. Swartzentruber, Y.-W Mo, R. Kariotis, M.G. Lagally, and M.B. Webb, *Phys. Rev. Lett.* **65**, 1913 (1990).
20. N.C. Bartelt, T.L. Einstein and E.D. Williams, *Surf. Sci. Lett.* **240**, L591 (1991).
21. P. Nozières (1991), in, *Solids Far From Equilibrium*, C. Godrèche (ed.), Cambridge University Press, Cambridge, p. 1.
22. N.C. Bartelt, T.L. Einstein and E.D. Williams, *Surf. Sci.***276**, 308 (1992).
23. While the $\exp(\tau/\tau_o)^{1/3}$ for $G(\tau)$ has not been derived explicitly for the TD model, the form is justified by analogy to the SD model. That is, the time constant for a fluctuation in the SD model is; $\tau(q) \propto q^{-4}$ and in the TD model $\tau(q) \propto q^{-3}$ (see Refs. [7] and [16]).
24. B.S. Swartzentruber, Y.-W Mo, M.B. Webb, and M.G. Lagally, *J. Vac. Sci. Technol.* **A7**, 2901 (1989).
25. B.S. Swartzentruber, N. Kitamura, M.G. Lagally, and M.B. Webb, *Phys. Rev.* **B47**, 13432 (1993).
26. O.L. Alerhand, David Vanderbilt, Robert D. Meade and J.D. Joannopoulos, *Phys. Rev. Lett.* **61**, 1973, (1998).
27. M.B. Webb, K.K. Men, B.S. Swartzentruber, R. Kariotis and M.G. Lagally, *Surf. Sci.* **242**, 23, (1991).
28. N.C. Bartelt, R.M. Tromp and E.D. Williams, *Phys. Rev. Lett.* **73**, 1656 (1994).
29. M. Kammler, M. Horn von Högen, N. Voss, M.C. Tringides, A. Menzel and E.H. Conrad, (to be published).
30. Y.-W Mo, J. Kleiner, M.B. Webb and M.G. Lagally, *Phys. Rev. Lett.* **66**, 1998 (1991); *Surf. Sci.* **268**, 275 (1992).
31. A. Rockett, *Surf. Sci.* **312**, 201 (1994).
32. C. Pearson, B. Borovsky, M. Krueger, R. Curtis, and E. Ganz, *Phys. Rev. Lett.* **74**, 2710, (1995).

# ELECTRIC CURRENT INDUCED ASYMMETRY OF SURFACE DIFFUSION - INSTABILITY OF VICINAL CRYSTAL SURFACES

S. S. STOYANOV[1], J. J. MÉTOIS[2], V. TONCHEV[1]

[1] *Institute of Physical Chemistry, Bulgarian Academy of Sciences, 1113 Sofia, Bulgaria*

[2] *CRMC2-CNRS, Campus de Luminy, 13288 Marseille Cedex, France*

**Abstract**

We measured the width $L_b$ of a bunch, consisting of a fixed number of steps, as a function of the temperature of the Si crystal, heated by direct electric current. We carried out these measurements in three different regimes - during crystal-vapor equilibrium, during sublimation, and during crystal growth. The central result of this work is the experimentally-determined temperature-dependence g(T) of the step interaction coefficient in the expression $f(\rho) = f(0) + \kappa\rho + g\rho^3$ for the surface free energy (per unit projected area) of a vicinal surface of Si(111) with a density of steps ρ. We evaluated g(T) in the high temperature interval on the basis of the derived expression $\frac{g(T)}{g(T_0)} = \frac{L_b^3(T)}{L_b^3(T_0)} \frac{I(T)}{I(T_0)}$ and the measured values of $L_b$ and I(T) during crystal-vapour equilibrium (I(T) is the electric current, flowing through the Si crystal to keep it at a temperature T). The experiments on the temperature-dependence of the bunch width $L_b$ under non-equilibrium conditions (growth of the Si crystal) manifest a sharp peak (at T = $1230^0$ C), which indicates a significant change of the structure of the Si(111) surface.

## 1. Introduction

Bunching of steps at vicinal surfaces during sublimation and growth of crystals received recently a lot of attention as a promising way to grow nanostructures as well as a valuable tool to study the interactions between the steps. The possibility to use the step bunches, induced by direct current heating of the Si crystal, as a tool to study the step-step interaction is the subject of this paper. The physical ground is simple - the steps in the bunch keep a certain distance from each other because the step-step repulsion



*M.C. Tringides and Z. Chvoj (eds.),*
*Collective Diffusion on Surfaces: Correlation Effects and Adatom Interactions,* 33–45.

balances the tendency to further compression of the bunch. According to the theoretical results [1,2] the distance between the steps in the bunch depends on the step-step repulsion magnitude A(T), which takes into account of both entropic and energetic repulsions between neighboring steps. Under crystal-vapor equilibrium, A(T) is equal to g(T) - the coefficient in front of $\rho^3$ in the expression for the surface free energy per unit projected area [3]

$$f(\rho) = f(0) + \kappa\rho + g\rho^3, \tag{1}$$

where $\rho$ is the density of steps.

Our aim is to determine the temperature-dependence of the coefficient g(T) from relevant experiments on step bunching, induced by direct electric current heating. The step interaction coefficient g(T) is involved in the theoretical expression for the width $L_b$ of a bunch of steps. This circumstance provides a ground to study the temperature-dependence of g(T) by studying the temperature-dependence of $L_b$ of a bunch consisting of a fixed number of steps. We explored this possibility by reflection electron microscopy (REM) observations of electric current induced bunches of steps at Si(111) under equilibrium conditions (as an indication for reaching equilibrium at the observed Si surface, we used the vanishing of the rate of motion of the single steps). Working under equilibrium conditions justifies the use of thermodynamic quantities and relations and, on the other hand, it makes very easy the task to keep the number of steps in the bunch constant. It is essential to note, however, that the system under consideration is not in a true thermodynamic equilibrium, because of the electric force applied to the adatoms and causing their drift on the crystal surface.

We studied also the temperature dependence of the width $L_b$ of the current induced bunches of steps under non-equilibrium conditions, i.e., during sublimation as well as during growth. In these cases the step-step repulsion magnitude A(T) is not equal to the thermodynamic quantity g(T) and an interpretation of the results is given on a qualitative basis. Nevertheless, the experiments on the step bunching during sublimation and growth are helpful in understanding the step dynamics at (111) Si surface.

## 2. Experimental Set-up and Results

We used an ultra high vacuum electron microscope which allows in situ experiments [5]. The pressure around the sample was in the $10^{-9}$ torr range. The samples (18 x 2 x 0.3 $mm^3$) were cut out of an n-doped silicon (111) wafer. Their longer sides were parallel to <112>. The specimen holder, described in detail in previous papers, was a special design [5]. In fact, two samples were mounted on this specimen holder, facing each other at a very short distance (about 100μm). They could be heated independently by passing two different electric currents through them (DC current with different directions). Due to the short distance between the samples, the mutual heating between them had to be taken into account for a temperature calibration versus the electric current. The temperature was measured by optical pyrometry (monochromatic, $\lambda$ = 0.65μm).

During the in-situ experiments, one of the samples, named substrate, was observed and the other, named buffer, was used to supply atoms to the substrate. In this way, the sublimation, the growth, and the equilibrium conditions of the substrate can be achieved. The reflection electron microscopy (REM) technique was used to observe the substrate [7]. The electron beam passing between the two samples came under grazing incidence on the substrate. This mode of observation allows visualization of the monoatomic steps along the direction very close to <110>. It is reasonable to note that the foreshortening of the REM images was about 35 (see Fig.1).

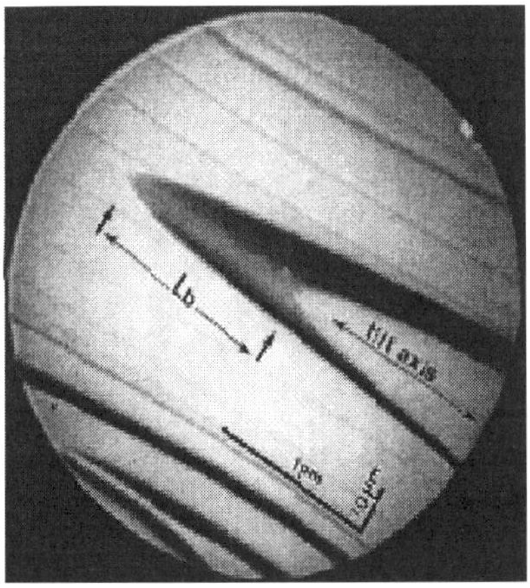


*Figure 1.* REM image of a bunch of steps on Si (111): we measure the width of the bunch along its larger slope in the direction of the tilt axis shown on the image. The usual disorientation of the bunches observed is no more than 1°, so it is not necessary to take into account the difference between the real width of the (111) terraces in the perpendicular direction of steps and the width along the larger slope of the bunch.

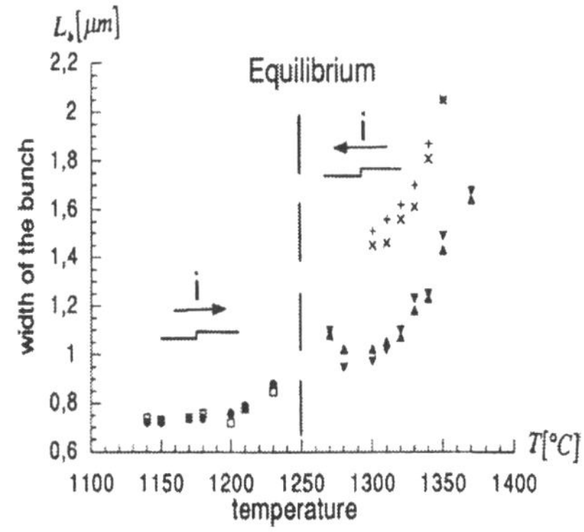


*Figure 2.* Width of 4 selected bunches under crystal–vapor equilibrium (2 bunches are observed below 1250°C (at step-up current), and 2 other bunches are subject to measurements above 1250°C (at step-down current)). The temperature-dependence $L_b(T)$ above 1250°C was scanned 2 times for each bunch (with some annealing time between the two scans) and the data are presented with different symbols.

This technique is particularly suitable for studies of the temperature-dependence of the width of a bunch of steps at (111)Si surface, because it is possible to observe (and measure the width of) an appropriately selected bunch at different temperatures in the entire temperature interval, where step bunching is induced by an electric current with a given direction. During crystal-vapor equilibrium and during sublimation, the current with step-up direction induces step bunching in the range of temperatures between 1100 and 1250°C (middle temperature interval), whereas above 1250°C as far as 1350°C (high temperature interval) the step-down direction of the current is necessary to get a bunching of steps. This circumstance prevented us from measuring the width of the selected bunch in the whole temperature range between 1100 and 1350°C. Starting from the lower limit (1100°C) and increasing the temperature (by increasing the electric current through the Si crystal) we observed debunching in the vicinity of 1250°C. To study the temperature-dependence of the bunch width above 1250°C, we reversed the current direction from step-up to step-down. When we changed the direction of the current, however, the electron beam was deviated (so that we got an REM image of another part of the crystal surface) and it was practically impossible to find again and measure the width of the bunch, observed in the middle temperature interval. That is why the data for the bunch width during crystal-vapor equilibrium (as well as during sublimation) refer to different bunches (one bunch in the middle temperature interval and another bunch in the high temperature interval) usually consisting of a different number of steps (see Figs. 2 and 3 ). During growth, however, the step-down direction of the current induces bunching of steps in the entire range (1100 - 1350°C) of the temperatures and this circumstance provides the unique possibility to observe and measure the width $L_b$ of the same bunch at all temperatures (see Fig. 4).

Making use of the experimental technique, described above, we measured the width of selected bunches at different temperatures. The obtained results for the cases of crystal-vapor equilibrium, sublimation, and growth are shown in Figs. 2, 3 and 4.

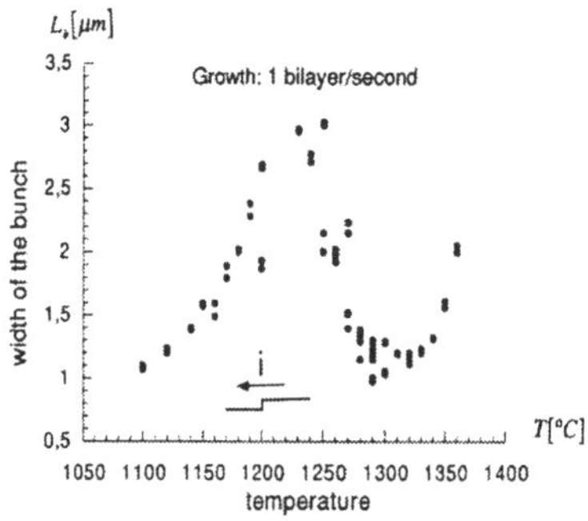


*Figure 3.* Width of bunches in the case of sublimation (the heating current directions are indicated on the figure).

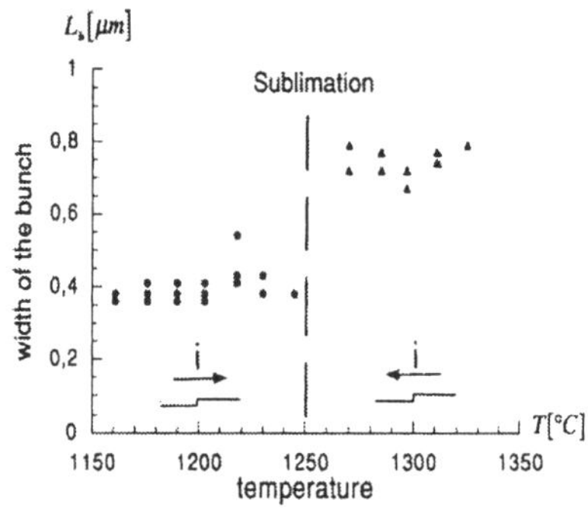


*Figure 4.* Width of a bunch during growth (the growth rate is maintained at around 1 bilayer per second). The current heating direction remains the same for the two temperature domains (below and above 1250°C) so that we can observe one and the same bunch in both domains. The sharp peak of the width $L_b$ at around 1230°C indicates a transition-like change of the Si surface structure.

## 3. Interpretation of the Experimental Results

It is convenient to start with the interpretation of the results obtained in the high temperature range (1250 - 1350$^0$C), where the current-induced bunching of steps is satisfactorily described by the generalized Burton, Cabrera, Frank (BCF) model [1]. An attempt for an interpretation of the data obtained in the medium temperature interval 1150 - 1250$^0$C will be given in the second subsection.

## 3.1. TEMPERATURE-DEPENDENCE OF THE WIDTH OF A BUNCH OF STEPS IN THE INTERVAL 1250 – 1350$^0$C

We shall consider first the case of crystal-vapor equilibrium. These experiments produce very reliable data, since it is easy to keep the number of steps in the bunch constant for a long time (the steps do not attach and/or detach from the bunch -- they only fluctuate around fixed positions), so that the bunch achieves its quasi-equilibrium shape.

### *3.1.1. Bunch of steps under crystal - vapor equilibrium.*

Here, we provide a theoretical consideration of the step dynamics under crystal-vapor equilibrium in a presence of electromigration of the adatoms (the theoretical results published in Ref. [1] refer to sublimation in ultra-high vacuum). The processes of atom migration (in a presence of an external force F), desorption and deposition during crystal-vapor equilibrium are described by the following equation

$$D_s \frac{d^2 n_s}{dx^2} - \frac{D_s F}{kT}\frac{dn_s}{dx} - \frac{n_s}{\tau_s} + R_e = 0, \tag{2}$$

where $D_s$ is the coefficient of surface diffusion, $n_s$ is the concentration of mobile atoms, adsorbed on the surface, $\tau_s$ is the life time of adatoms, $R_e$ is the equilibrium value of the deposition rate of atoms to the crystal surface, x is an axis, perpendicular to the step edges (a system of parallel steps with straight edges is under consideration), k is the Boltzmann constant, and T is the absolute temperature. The exchange of atoms between the crystal phase and the dilute layer of adatoms takes place at the steps and determines the boundary conditions for the Eq. (2). They have the form [1]:

$$D_s\left(-\frac{dn_s}{dx} + \frac{Fn_s}{kT}\right) = -\frac{\beta}{a_\perp a_{//}} \frac{n_s - n_s^e(x_i)}{n_s^e} \quad \text{at } x = x_i$$

$$D_s\left(-\frac{dn_s}{dx} + \frac{Fn_s}{kT}\right) = \frac{\beta}{a_\perp a_{//}} \frac{n_s - n_s^e(x_{i+1})}{n_s^e} \quad \text{at } x = x_{i+1}, \tag{3}$$

where $\beta$ is the step kinetic coefficient. This quantity is defined by the expression for the velocity of the step motion

$$v = \beta\,[n_s(x_i) - n_s^e(x_i)]/\,n_s^e, \tag{4}$$

where $n_s^e$ $(x_i)$ is the equilibrium value of the adatom concentration in the vicinity of the step, situated at $x_i$ (the value of $n_s^e$ $(x_i)$ depends on the distances to the (i-1)-th and (i+1)-th step). It is essential to note that $n_s^e$ in the denominator of Eqs. (3) and (4) is the equilibrium value of the adatom concentration at a vicinal surface with equidistant steps (i.e., zero curvature) and, therefore, it does not depend on x.

Solving the diffusion problem (Eq. (2) with the boundary conditions (3)), one obtains an expression for the adatom concentration $n_s$ (x) on the terrace between the i-th and (i+1)-th step. Substituting $n_s$ ($x_i$) into Eq.(4) one obtains the contribution of the terrace between the i-th and (i+1)- th step to the velocity of the i-th step. Accounting in the same way for the contribution of the terrace between the i - th and (i -1) - th step, one can write an expression for the velocity $dx_i / dt$ of the i-th step as a function of the widths of the lower and upper terraces. In other words, one can write a set of equations, describing the motion of a system of straight steps (which interact with each other via energetic and entropic repulsion) under the conditions of crystal-vapor equilibrium. These equations contain 4 dimensionless parameters - $l / \lambda_s$, $\gamma = \beta\lambda_s / D_s a_\perp a_{//} n_s^e$, Fl/2kT, and $l_0 / l$, where $\lambda_s = \sqrt{D_s \tau_s}$ is the mean diffusion distance, characterizing the migration of the adatoms on the crystal surface, the quantity $l_0$ is related to the step interaction coefficient g(T) through the expression

$$l_0 = \left( \frac{2a_\perp a_{//} g(T)}{kT} \right)^{1/3} \tag{5}$$

and l is the interstep distance in absence of any bunches (l is determined by the miscut angle of the Si wafer).

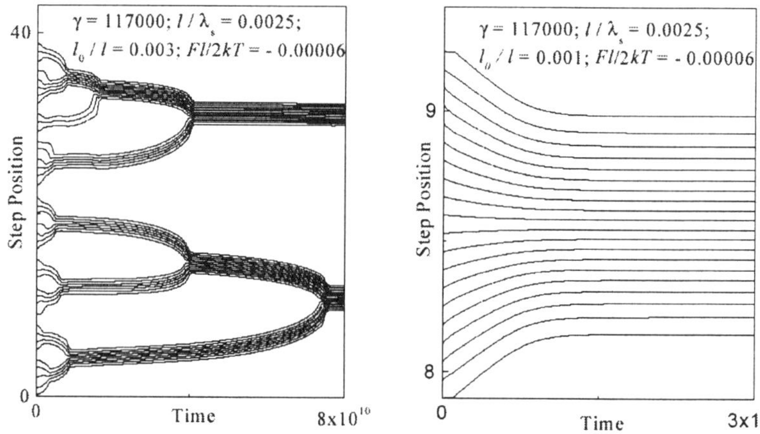


*Figure 5.* The step trajectories, shown in this figure, are obtained by numerical integration of the equations of step motion under crystal-vapor equilibrium. The values of the parameters, involved in the equations of step motion, are given inside the figure. In its initial configuration the steps are randomly deviated from their regular positions (a), the steps form a bunch, which is expanded with respect to its equilibrium configuration (b).

The numerical integration of the equations of step motion starts from an initial configuration of steps randomly deviated from their equidistant positions. As seen in Fig. 5a, the steps form initially small bunches, which later coalesce to form larger bunches of steps. The numbers at the time axis refer to the dimensionless time $\tau = \beta t / l$, so they are not instructive. It is interesting, however, to briefly discuss the values (given at the top side of Fig. 5a) of the parameters $l / \lambda_s$ , $\gamma$, $Fl / 2kT$ and $l_0 / l$, involved in the numerical integration of the equations of step motion. The value $l / \lambda_s = 0.0025$ means $\lambda_s = 12$ µm when $l = 30$ nm , and $\lambda_s = 48$ µm at $l = 120$ nm. These values of $\lambda_s$ are in agreement with the observations on step spacing at Si(111) during sublimation [7]. The value $\gamma = 117000$ corresponds to $\lambda_s = 35$µm, since the condition for "non-transparency" of the steps $\beta/D_s a_\perp a_{//} n_s^e = 1$ reduces the expression $\gamma = \beta\lambda_s/ D_s a_\perp a_{//} n_s^e$ to $\gamma = \lambda_s/ a_{//}$ and, therefore, $\lambda_s = 117000\ a_{//}$ (the atomic site length is $a_{//} = 0.3$nm ). The value $l_0 / l = 0.003$ is realistic for a Si surface with average interstep distance $l = 120$ nm (it corresponds to a step interaction coefficient $g \approx 0.04$ eVnm). For vicinal surfaces of a larger miscut angle, the interstep distance $l$ is smaller and the value of $l_0 / l$ (corresponding to $g \approx 0.04$ eVnm) is larger. Finally, the value $Fl / 2kT = -0.00006$ corresponds to an effective electric charge $z_{eff} = 0.26|e|$ (see Ref. [8]) at a temperature T = 1250$^0$C and an average interstep distance $l = 120$ nm (the electric field, applied to the Si wafer in the typical experiments is around 5V/cm).

Since our REM observations provide direct data for the width $L_b$ of a bunch of steps, it is of interest to study the dependence of $L_b$ on the step interaction coefficient g(T). For this purpose we integrated the equations of step motion of a system of 20 steps (with periodic boundary conditions) for a set of values of the parameter $l_0 / l$, starting the integration at each value of $l_0 / l$ from an initial step configuration, corresponding to a bunch of 20 steps. Figure 5b shows the step trajectories, obtained in an integration run with values of the parameters shown at the top side of the figure. As can be seen, it is easy to determine the equilibrium width $L_b$ of the bunch at each value of the parameter $l_0 / l$, while the other 3 parameters have fixed values. The obtained results are plotted in

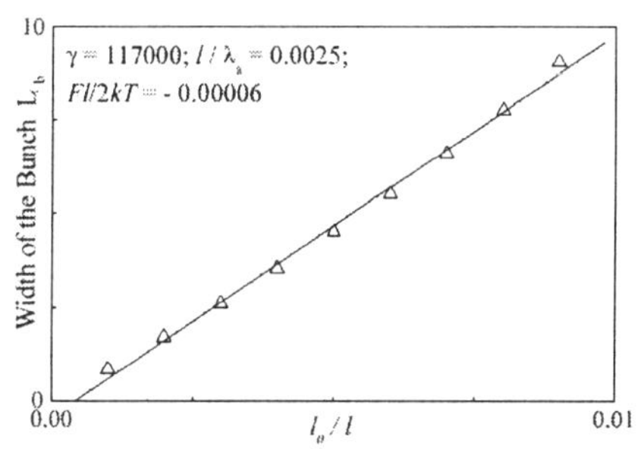


*Figure 6.* The width $L_b$ of a bunch of 20 steps under equilibrium plotted versus the step-step repulsion parameter $l_0 / l$

Fig. 6. In a similar way we determined how the equilibrium width $_b$ of a bunch of steps depends on each of the other 3 parameters: $l / \lambda_s$, $\gamma = \beta\lambda_s/ D_s a_\perp a_{//} n_s^e$ and $Fl/2kT$.

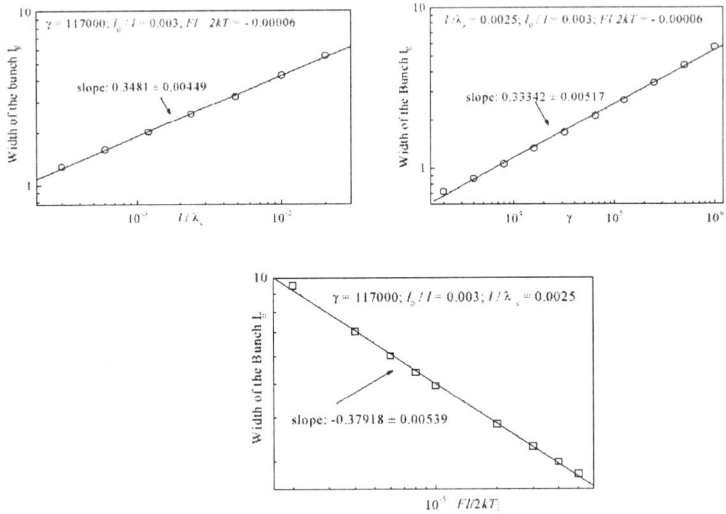


*Figure 7.* The dependence of the equilibrium value of $L_b$ on the parameters: $l/\lambda_s$ (a), $\gamma=\beta\lambda_s/D_s a_\perp a_{//} n_s^e$ (b) and $Fl/2kT$ (c), involved in the equations of step motion

The results are shown in Fig. 7 and can be presented as scaling relations $L_b \sim \left(\frac{l}{\lambda_s}\right)^{1/3}$, $L_b \sim \left(\frac{\beta\lambda_s}{D_s a_\perp a_{//} n_s^e}\right)^{1/3}$ and $L_b \sim \left(\frac{Fl}{kT}\right)^{-1/3}$

On the other hand the continuum model equation, governing the shape of the crystal surface during crystal-vapor equilibrium, gives the following expression for the width of the bunch

$$L_b = N^{1/3}\left(\frac{18a_{//}g(T)}{|F|}\right)^{1/3} B \tag{6}$$

where $N$ is the number of the steps in the bunch and B is a constant. Equation (6) is in agreement with the results obtained by numerical integration of the equations of step motion $Lb \sim l_0 \sim [g(T)]^{1/3}$ (see Fig. 6 and Eq. (5)) and $L_b \sim F^{-1/3}$ (see Fig. 7c).

Since we are interested in the temperature-dependence of g(T), it is convenient to rewrite the Eq. (6) into the following normalized form

$$\frac{g(T)}{g(T_0)} = \frac{L_b^3(T)}{L_b^3(T_0)}\frac{F(T)}{F(T_0)}, \tag{7}$$

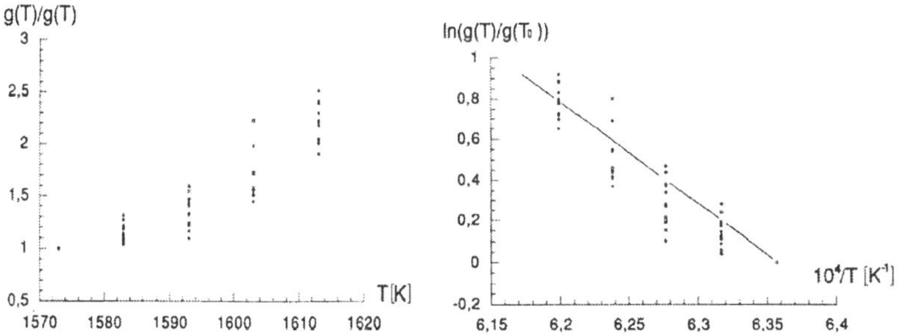


*Figure 8.* The temperature dependence of the ratio $g(T)/g(T_0)$, where $T_0$ is chosen to be $1300^0C$ because the bunch width $L_b$ displays a sharp increase above this temperature (a): an Arrenius plot of the same temperature dependence results an activation energy of $4.2 \pm 0.4$ eV (b). The data presented were obtained in independent measurements of the temperature dependence of the width of 12 different bunches.

where $T_0$ is a temperature selected in the interval 1250 - $1350^0C$, whereas F(T) reflects the fact that the higher temperature of the Si crystal is achieved by passing a stronger current through the Si crystal and, therefore a larger electromigration force is applied on the adatoms (Ichikawa and Doi [9] proved that F is proportional to the current through the crystal).

Finally, we substitute our experimental data into the expression:

$$\frac{g(T)}{g(T_0)} = \frac{L_b^3(T)}{L_b^3(T_0)} \frac{I(T)}{I(T_0)}, \tag{8}$$

obtained by substituting $F(T) / F(T_o) = I(T) / I(T_0)$ (where I(T) is the electric current, flowing through the Si crystal to keep it at a temperature T) into Eq. (7). The results obtained in independent measurements of 12 different bunches are shown in Fig. 8 (we selected $T_o = 1300^0C$ because the sharp increase of $L_b$ starts at this temperature). To get an idea about how strong is the obtained temperature dependence of the step interaction coefficient g(T), we presented the data as an Arrenius plot (see Fig. 8b), which results in an activation energy $E_g$=4.2 ± 0.4 eV.

### *3.1.2 Bunch of steps under sublimation conditions*

Our REM observations of step bunches during sublimation definitely show a constant value of the bunch width $L_b$ in the interval 1270 - $1330^0$ C (our attempts to measure $L_b$ at $1340^0$ C were not successful because the frequencies of attachment and detachment of steps to the bunch are very high and one cannot be sure that the number of steps in the bunch remains constant during the observation time). The temperature-dependence (in the interval 1270 - $1330^0$ C) of the magnitude A of step interaction during sublimation is also given by Eq. (8) with $L_b$ =const. (the coefficient g(T) is replaced by A(T)). The increase of A(T) with the temperature is given, therefore, by the increase of the electric current, heating the Si crystal (I(T) in Eq. (8)). Making use of our data for the temperature calibration versus the heating current during sublimation, we obtain $A(1330^0C) / A(1270^0C) = 3.25 / 2.8$.

The temperature-independent value of $L_b$ measured in our sublimation experiments (see Fig.3) in the interval 1270 - 1330$^0$C is in contrast with the sharp increase of $L_b$, documented under crystal-vapor equilibrium at $T > 1300^0$ C (see Fig. 2). We are not in a position to offer a clear quantitative explanation of the considerable difference between the temperature dependencies of $L_b$ observed under sublimation and under equilibrium conditions. We could only mention a few qualitative arguments in an attempt to understand this experimental fact. It is reasonable to note this basic difference is related to the non-equilibrium structure of the steps during sublimation. Pimpinelli et al. [10] claim the large amplitude step fluctuations to be kinetically hindered (these fluctuations have no time to appear) when the time to sublimate one monolayer is shorter than the time needed to build a meander of large size. Although the physical ground for the observed step smoothing is not yet clear, the absence of large amplitude fluctuations of the step shape decreases the entropic repulsion, since the no crossing constraint between the steps is not essential - the steps are rather straight (due to kinetic reasons) and do not "feel" their neighbous (the steps "feel" their neighbors mainly because of the large amplitude fluctuations). As a result, the repulsion between the steps during sublimation is mainly energetic (elastic or dipolar interactions) which does not depend so strongly on the temperature as the entropic repulsion (i.e., A(T) is rather different from g(T)).

*3.1.3 Bunch of steps during growth of Si (111) surface*

In our experiments we selected the buffer temperature to provide a constant growth rate of the substrate - about 1 bilayer per second in the whole interval 1100 - 1350$^0$C (for each temperature of the substrate we adjusted the buffer temperature to keep the growth rate of 1 bilayer per second). The measured values of the bunch width $L_b$ are given in Fig. 4 (in the whole temperature interval 1100 - 1350$^0$C the current, heating the Si crystal, has a step-down direction). Focusing our attention to the temperature interval 1250 - 1350$^0$ C we notice the temperature behavior of the bunch width $L_b$ to be very similar to the behavior observed in the experiments during crystal-vapor equilibrium (see Fig. 2). This similarity is not surprising because the desorption rate estimated in our experiments (we measured the time one step needs to travel the average interstep distance) is 2.68 bilayers/s at 1300$^0$C and 7.07 bilayers/s at 1340$^0$ C. This means that the growth rate of 1 bilayer per second is realized under relatively small supersaturation, i.e., the crystal surface is under near to equilibrium conditions at 1300$^0$C and even closer to equilibrium at 1340$^0$C.

## 3.2 TEMPERATURE-DEPENDENCE OF THE WIDTH OF A BUNCH OF STEPS IN THE INTERVAL 1100 – 1250$^0$C

We made an attempt to interpret the experimental results in this temperature interval on the basis of the continuum model of a crystal surface with "transparent" steps [2] (the "transparency" of the steps means that the adatoms, migrating on a given terrace easily jump to the neighbouring terraces, in contrast with the BCF model, where adatoms arriving to a step join the crystal phase by attaching to a kink position). We extended the model of "transparent" steps (formulated in Ref. [2] for the case of sublimation in vacuum) by taking into account the deposition rate R (in this way the model can describe the step bunching during growth, equilibrium and sublimation of the crystal. In the framework of the extended model, we obtained the following expression for the width of a bunch of "transparent" steps

$$L_b = \sigma\sqrt{a_{//}a_{\perp}}N^{1/4}\left[\frac{\frac{A}{F}}{\beta\frac{n_s^e - R\tau_s}{n_s^e}}\right]^{1/4}\lambda_s^{1/4}\left(\frac{n_s^e}{\tau_s}\right)^{1/4}. \tag{9}$$

As seen, Eq. (9) predicts debunching ($L_b \to \infty$) under crystal-vapor equilibrium, i.e., when $R = Rc = n_s^e / \tau_s$. This theoretical result is in a clear contradiction with our experimental data (see Fig. 2), showing a finite width of a bunch under crystal-vapor equilibrium in the temperature interval 1140 -1230$^0$C, where the model of "transparent" steps was believed to describe the mechanism of step bunching. In our experiments, we observed bunches of steps under crystal-vapor equilibrium even after 12 hours of heating with a step-up electric current. The absence of any tendency to a relaxation of the step bunches under crystal-vapor equilibrium is a fundamental difficulty in the use of the continuum model[2] of a bunch of "transparent" steps in the interpretation of our observations and measurements of $L_b$. Obviously, this model fails to explain a basic experimental fact - the existence of step bunches under crystal-vapor equilibrium.

## 4. Discussion and Conclusions

We used REM observations of step bunches on (111)Si surface as a tool to study the temperature-dependence of the step interaction coefficient g(T) in the expression $f(\rho) = f(0) + \kappa\rho + g\rho^3$ for the surface free energy (per unit projected area) of a vicinal surface with a density of steps $\rho$. The novel experimental technique allowed us to study the temperature-dependence of the width $L_b$ of a bunch of fixed number of steps under crystal-vapor equilibrium. The experimentally obtained data for $L_b$ were substituted into the Eq. (8) and the results are shown in Fig. 8, where data obtained for the temperature-dependence of the width $L_b(T)$ of 12 different bunches are plotted together to get statistically reliable dependence of g on T. As seen in Fig. 8b, the data for g(T) fit an Arrenius plot with an activation energy 4.2 ± 0.4 eV.

The step interaction coefficient g(T) is a fundamental quantity in the thermodynamics of crystal surfaces. Our data for the temperature dependence of g(T) are restricted to the interval from 1300 to 1340$^0$C, where the step bunching is induced by electric current with step-down direction. We observed debunching at temperatures higher than 1350$^0$C. That is why we use only the data for $L_b$ obtained between 1300 and 1340$^0$C. As seen in Fig. 8 the temperature-dependence of the step interaction coefficient g(T) is rather strong. It is considerably stronger than the g(T) dependence obtained by Akutsu and Akutsu [11] on the basis of statistical mechanical calculations of the two-dimensional honeycomb lattice gas model.

The basic disadvantage of the technology for determining the temperature-dependence g(T) of the step interaction coefficient, developed in this paper, is related to the existence of transition temperatures, where the electric current direction should be changed to keep step bunches at the Si surface. Our technology does not work in the vicinity of the transition temperatures, where the increase of the bunch width $L_b$ is due to the small efficiency of the step bunching mechanism. As a result, we can use the technology for determining g(T) by measuring $L_b(T)$ in relatively narrow temperature intervals, situated away from the transition temperatures. On the other hand, revealing the physical nature of the transition temperatures is a challenging problem and the experiments on step bunching could help to understand it.

The most interesting result of the experiments under non-equilibrium conditions is the relatively sharp peak in the temperature-dependence of the bunch width $L_b$ during the growth of the Si crystal (see Fig.4). It is essential to note that $L_b$ displays its maximum value at a temperature around 1230$^0$C, which is close to the temperature (1200$^0$C) of the minimum of the step spacing in the sublimation experiments of Homma et al. [7] (steps were formed by nucleation of macrovacancies on the Si(111) surface). Both the maximum of $L_b$ (in Fig. 4) and the minimum of the step spacing (reported in [7]) indicate a change of the Si(111) surface structure. Having in mind the very high adatom concentration (higher than 2.8% of the density of adsorption sites) proven to exist at 1250$^0$C (see Ref. [8]), we are convinced to support the opinion of Homma et al. [7] that the surface structure change under discussion is an incomplete surface melting at Si(111).

## 5. References

1. Stoyanov, S. and Tonchev, V. (1998), *Physical Review B* **58**, 1590.

2. Stoyanov, S. (1997), *Surface Science* **370**, 345; *ibid* (1998) **416**, 200.

3. Jeong , H.-C. and Williams, E. (1999), *Surface Science Reports* **34**, 171.

4. Métois, J. J. , Nitsche, S., and Heyraud, J.C. (1989) *Ultramicroscopy* **27**, 349.

5. Heyraud, J. C., Métois, J. J. and Bermond, J. M. (1999), *Surface Science* **425**, 48.

6. Bermond, J. M., Métois, J. J., Heyraud, J. C. and Alfonso, C. (1995), *Surface Science* **331-333**, 855.

7. Homma, Y., Hibino, H., Ogino, T. and Aizawa, N. (1997), *Physical Review B* **55**, R10237.

8. Fujita, K., Ichikawa, M. and Stoyanov, S (1999), *Physical Review B* **60**, 16006.

9. Ichikawa, M. and Doi, T. (1992), *Applied Physics Letters* **60**, 1082.

10. Pimpinelli, A., Elkinani, I., Karma, A., Misbah, C. and Villain, J. (1994), *J. Physics: Condensed Matter* **6**, 2661.

11. Akutsu, N. and Akutsu, Y. (1999), *J.Physics: Condensed Matter* **11**, 6635.

# MEMORY EFFECTS AND MEMORY FUNCTIONS IN SURFACE DIFFUSION

T. HJELT, E. KUUSELA, J. M. LAHTINEN AND T. ALA-NISSILA*
*Helsinki Institute of Physics and Laboratory of Physics, Helsinki University of Technology, P.O. Box 1100, FIN-02015 HUT, Espoo, Finland*

I. VATTULAINEN
*Department of Chemistry, Technical University of Denmark, Building 207, DK-2800 Lyngby, Denmark*

AND

S. C. YING
**Department of Physics, Brown University, Providence, R.I. 02912-1843, U.S.A.*

**Abstract.**

We consider the diffusive dynamics of particles in various model systems with strong interactions. We study the temporal dependence of the single-particle velocity autocorrelation function $\phi(t)$, and its corresponding memory function. We find $\phi(t)$ to decay non-exponentially and in most cases follow a power-law $\phi(t) \sim t^{-x}$ at intermediate times $t$, while at long times there is a crossover to an exponential decay. We characterize the possible values of the decay exponent $x$, and show that $x$ correlates with interaction and ordering effects. In many cases, the memory function follows behavior similar to that of $\phi(t)$. These results suggest that $\phi(t)$ can be used to obtain information about the ordering of the system and about the nature of predominant interactions between adparticles using experimental techniques such as scanning tunneling microscopy, in which $\phi(t)$ can be measured in terms of discrete adparticle displacements. Finally, our studies suggest that the decay of velocity correlations in collective diffusion is qualitatively similar to the case of tracer diffusion.



*M.C. Tringides and Z. Chvoj (eds.),*
*Collective Diffusion on Surfaces: Correlation Effects and Adatom Interactions,* 47–57.

## 1. Introduction

Surface diffusion plays an important role in various technologically important processes such as surface growth [1] and chemical reactions [2]. The quantity that characterizes the motion of individual particles is the tracer diffusion coefficient [3]

$$D_T = \lim_{t\to\infty} \frac{1}{2dt} \langle |\vec{r}(t) - \vec{r}(0)|^2 \rangle, \tag{1}$$

where $d = 2$ is the dimensionality of the system and $\vec{r}(t)$ is the position of the tagged particle at time $t$. Equivalently, $D_T$ can be defined in terms of the velocity autocorrelation function $\phi(t) \equiv \langle \vec{v}(t) \cdot \vec{v}(0) \rangle$ through the Kubo-Green formula [4]

$$D_T = \frac{1}{d} \int_0^\infty \phi(t)\, dt, \tag{2}$$

where $\vec{v}(t)$ represents the velocity of the tagged particle. Another important quantity in describing surface diffusion is the collective (chemical) diffusion coefficient $D_C$, which describes the decay rate of collective density fluctuations in equilibrium. $D_C$ can also be conveniently expressed via the Kubo-Green formula [4]

$$D_C = \frac{1}{k_B T\, \theta\, \kappa_T\, N\, d} \int_0^\infty \psi(t)\, dt, \tag{3}$$

where $\theta$ is the coverage of the particles that form an adlayer on the surface, $N$ is the total number of particles, and $\kappa_T$ is the compressibility. The main quantity of interest here in Eq. (3) is the flux correlation function $\psi(t) \equiv \langle \vec{J}(t) \cdot \vec{J}(0) \rangle$, where $\vec{J}(t) = \sum_{i=1}^{N} \vec{v}_i(t)$ is the flux of all diffusing adparticles $i = 1, \ldots, N$ in terms of their velocities $\vec{v}_i(t)$. It plays a role similar to the single-particle velocity correlation function $\phi(t)$ in that the dynamics is wholly described by the flux correlation function $\psi(t)$. It is important to note that for interacting systems at higher coverages, $D_C$ can be very different from $D_T$, since $D_C$ is only indirectly related to the motion of individual tagged particles [3, 5].

As can be seen from the definitions above, the correlation functions $\phi(t)$ and $\psi(t)$ are the fundamental quantities from which the diffusion coefficients arise. In surface diffusion, an exponential decay of $\phi(t)$ is observed, if adatom diffusion is considered to be the Brownian motion of independent particles in a periodic potential, with uncorrelated collisions with the substrate excitations [4-7]. However, in real systems the decay of $\phi(t)$ depends on the adatom-substrate coupling and may show behavior very different from pure exponential decay [7, 8].

In the case of interacting many-particle systems, memory effects are even more subtle. In this paper we wish to review our recent works [9-19] on such memory effects, and the time-dependence of the correlation functions [$\phi(t)$ and $\psi(t)$] and the associated memory functions [4, 20] in particular. Besides fundamental theoretical interest, this issue turns out to have important implications to surface diffusion experiments. To this end, we consider a variety of model systems and study them in detail to find what the factors are that determine the behavior of the correlation functions. We show that within the lattice-gas approximation, the memory effects can be traced back to correlations between successive jumps of the adatoms. Such correlations are largest within the ordered phases and close to first and second order phase transition boundaries. Additional studies of models of flexible chains and hard particles on surfaces, and diffusive dynamics under steady-state sedimentation reveal that memory effects in dissipative systems in general manifest themselves in an intermediate time algebraic decay of the relevant correlation functions. The corresponding values of the decay exponents can be related to the nature of interactions and degree of spatial ordering in the systems.

## 2. Memory Expansion Method

We have recently introduced an efficient way to compute $D_T$ and $D_C$, which is called the "memory expansion" method (MEM) [11]. It can be used to obtain information of velocity correlations from particle displacement data in simulations and in experiments. This is done by defining the generalized velocity autocorrelation function as

$$\phi_g(t) = \langle \vec{v}_g(t_m + t) \cdot \vec{v}_g(t_m) \rangle, \tag{4}$$

where $\vec{v}_g(t_m) \equiv \delta\vec{R}(t_m)/\tau_0$ is the generalized particle velocity at time $t_m = m\tau_0$, with $\tau_0$ denoting a (fixed) discrete time step. $\delta\vec{R}(t_m) \equiv \vec{r}(t_m) - \vec{r}(t_{m-1})$ is the change in the position of the tagged particle between two consecutive observations at times $t_m$ and $t_{m-1}$. For lattice-gas models, the terms $\phi_g(t > 0)$ characterize dynamical correlations (memory effects) between consecutive displacements of a given particle. The case where memory effects are completely neglected constitutes the Dynamical Mean Field (DMF) approximation [9, 13].

For collective diffusion, the MEM proceeds in a similar manner. It is formulated in terms of the generalized flux correlation functions

$$\psi_g(t) = \langle \vec{J}_g(t_m + t) \cdot \vec{J}_g(t_m) \rangle, \tag{5}$$

where the generalized flux $\vec{J}_g(t_m) \equiv \delta\vec{R}_{\rm cm}(t_m)/\tau_0$, and $\delta\vec{R}_{\rm cm}(t_m) = \vec{R}_{\rm cm}(t_m) - \vec{R}_{\rm cm}(t_{m-1})$ is the center-of-mass displacement of $N$ particles. Again here, the terms $\psi_g(t > 0)$ characterize dynamical correlations (memory effects) between consecutive displacements of the center-of-mass of the adlayer.

## 3. Tracer Diffusion

### 3.1. RESULTS FOR THE MODEL ADSORPTION SYSTEM O/W(110)

We first consider the diffusion of oxygen adatoms on a W(110) surface in a lattice-gas model with Transition Dynamics Algorithm [21] Monte Carlo dynamics. The oxygen-oxygen interaction Hamiltonian contains pair interactions and also a contribution from three-body interactions that are important at large coverages. These lead to the appearance of a number of ordered phases and coexistence regions such that the resulting phase diagram is in close agreement with the experimental observations [22-24] for the O/W(110) system. Details of this model system and the Hamiltonian parameters can be found in Refs. [10, 14].

In the case of O/W(110), the exact separation of dynamical correlations allows a detailed study of the contribution of memory effects to diffusion. A detailed comparison between the diffusion coefficients and their DMF counterparts $T = 590$ K and 465 K reveals [14] that directional correlations between successive jumps of the adatoms are responsible for the memory effects. Such memory effects are strongest at low temperatures and close to the ideal coverages corresponding to ordered phases, *i.e.*, around $\theta = 1/2$ and $\theta = 3/4$ in the model of O/W(110). In all finite-coverage cases studied, we have found that the velocity autocorrelation function $\phi_g(t)$ can be well fitted to the form [17, 18]

$$\phi_g(t) = \frac{\phi_g(0)}{1 + At^x}, \tag{6}$$

which leads to an *algebraic decay* $\phi_g(t) \sim t^{-x}$ for $At^x \gg 1$. In a similar fashion, we have analyzed the power-law behavior of $\psi_g(t)$.

In Fig. 1, we show typical correlation functions for different coverages showing the algebraic decay, and Fig. 2 shows the effective exponent $x$ as a function of temperature [18]. At high temperatures where interaction effects become less important, $x$ approaches the value of two in accordance with the Langmuir gas limit [15]. Within the ordered phases, however, the value of $x$ depends on the details of ordering in a nontrivial way [18]. Within disordered phase, ordering plays no significant role and then the nature of interactions seems to be a sufficient criterion to determine the value of $x$: For repulsive interactions $x \simeq 2$ (or larger), while attractive

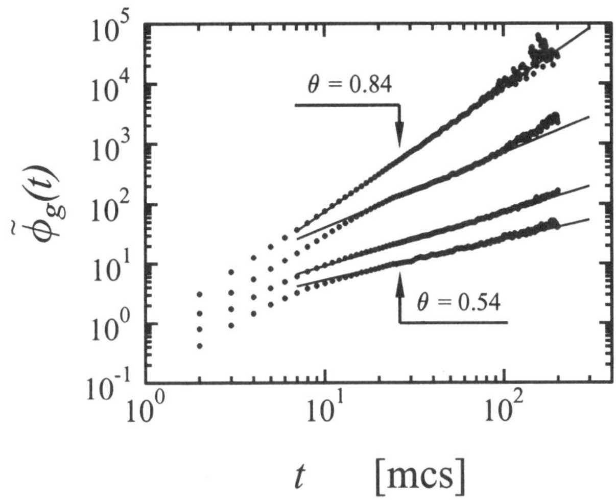


*Figure 1.* Typical results for $\tilde{\phi}_g(t) \equiv [\phi_g(t)/\phi_g(0)]^{-1} - 1$ at $T = 465$ K in the O/W(110) model system. At later times the decay of $\phi_g(t)$ starts to cross over to an exponential form. Results are shown for four coverages from top to bottom: $\theta = 0.84$, 0.74, 0.59, and 0.54 respectively. Full lines are fits to the data. Some curves have been shifted for clarity.

interactions lead to $x < 2$. In ordered phases, however, the same idea is not sufficient for a proper description of $x$. Namely, if one considers the case at high coverage and at low temperature, then the system is dominated by repulsive interactions, but nevertheless $x < 2$ within the ordered phases and in the vicinity of phase boundaries. Therefore, within the ordered phases, the behavior of $x$ is governed by ordering effects that determine the local structure in which adatoms diffuse, while the nature of interactions is less important.

### 3.2. MODEL POLYMER SYSTEM

The second model we have considered is for flexible, chainlike molecules on flat surfaces. The chains are modeled by the 2D fluctuating-bond model with Monte Carlo dynamics [25, 26], in which each segment excludes four nearest and next-nearest neighbor sites on a square lattice. The exclusion induces a strong entropic repulsion between the molecules even if there are no direct interactions present. The repulsion strongly influences diffusion in this system, and has been studied in detail in Ref. [27]. For the present purpose, the model has been augmented with an additional direct Lennard-Jones type of attraction between the segments of different chains. This model allows a systematic tuning of the interplay between repulsive and

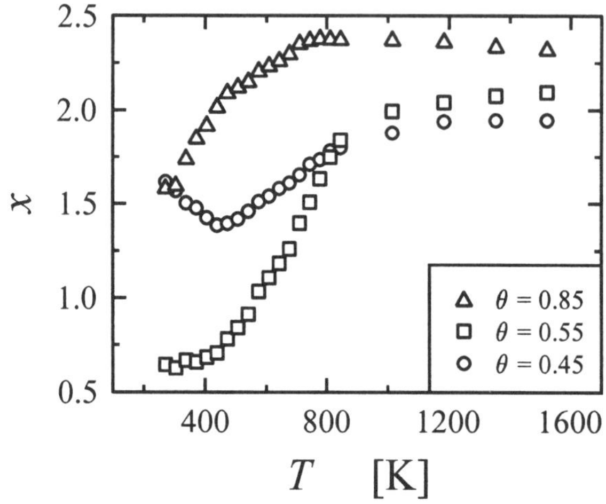


*Figure 2.* Results for the temperature dependence of the fitted decay exponent $x$ at various coverages $\theta$ in the O/W(110) model system. Error bars are smaller than the size of the symbols.

attractive interactions. Details on the model and parameters can be found in Ref. [11].

The results are in complete qualitative agreement with the results for the O/W(110) system. In all cases studied, we again found a power-law decay $\phi_g(t) \propto t^{-x}$ at intermediate times, with an exponent $x$ whose value depends on the interaction strength. At a fixed coverage $\theta = 0.25$, we found $x = 2.0 \pm 0.1$ in the athermal limit, $x = 1.7 \pm 0.2$ for a weak attractive interaction between the segments of different chains, and $x = 1.5 \pm 0.2$ for strongly attractive chains. Since there are no ordered phases under present circumstances, we can conclude that the effective value of $x$ *decreases* when attractive interactions become stronger.

We also studied how $\phi_g(t)$ depends on $\theta$. To this end, we used the model without direct interactions, in which case the entropic interchain repulsion increases with coverage [27]. We found exponents $x = 2.0 \pm 0.1$, $2.6 \pm 0.2$ and $2.6 \pm 0.4$ for $\theta = 0.25$, 0.5, and 0.75 in respective order. These results indicate that the effective $x$ may be larger than two with increasing repulsion.

### 3.3. HARD CIRCLES ON SURFACES

The third model is for 2D hard spheres (circles) on surfaces. To this end, we consider a continuum model with purely dissipative Brownian dynamics [19, 28]. The surface is taken to be completely smooth and there is no periodic

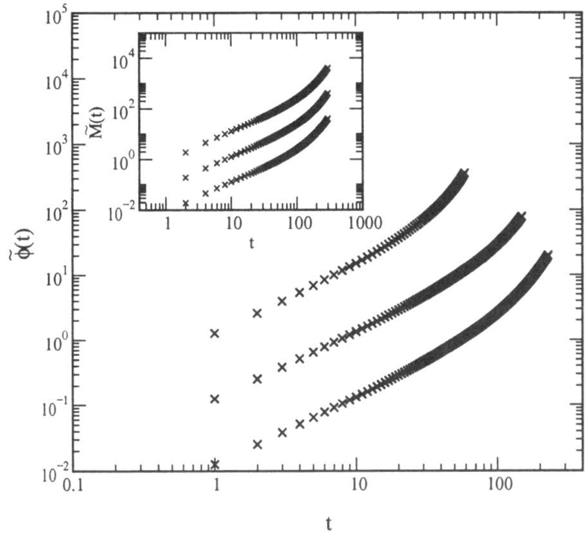


*Figure 3.* Behavior of $\tilde{\phi}_g(t) \equiv [\phi_g(t)/\phi_g(0)]^{-1} - 1$ and the corresponding memory function (inset) for 2D hard circles. From top to bottom, the curves are for areal densities $\rho = 0.986$ (in the vicinity of the freezing transition [29]), 0.329, and 0.110, respectively. Curves have been shifted for clarity.

potential. The dynamics of these particles is described by the Langevin equation:

$$\frac{d}{dt}\vec{p}(t) = -\eta\vec{p}(t) + \vec{f}(t) + \vec{F}(t), \qquad (7)$$

where $\vec{p}(t)$ is the 2D momentum of the particle, $t$ is time, and $\vec{f}(t)$ is the total interaction force with the other particles. The remaining random term $\vec{F}(t)$ is the driving force for the motion of the Brownian particles, and it obeys the standard fluctuation-dissipation relation.

The results for the velocity autocorrelation and memory functions are shown in Fig. 3. Again, these functions display an intermediate power law decay; however, in the present system the memory effects are relatively weak, and it is difficult to estimate $x$. Just fitting to the early part of $\tilde{\phi}(t)$ gives $x \simeq 1.2$ in qualitative agreement with the other models here.

## 3.4. DIFFUSIVE PARTICLE DYNAMICS UNDER STEADY-STATE SEDIMENTATION

The results presented in the previous sections suggest that the appearance of transient power-laws in the temporal dependence of $\phi_g(t)$ and the associated memory function is a universal feature of strongly interacting dissipative systems. The final example we wish to consider here concerns a case completely different from the others: a dissipative system in continuum with strong *hydrodynamic* interactions present. To this end, we consider a model of two-phase flow developed by Schwartzer [30]. In the model, the liquid is treated through a discretized Navier-Stokes equation, to which the discrete particle phase is coupled by a Stokes law. There is only a direct contact interaction between the particles, whose equations of motion are solved simultaneously with the Navier-Stokes equation. A detailed description of the model can be found in Ref. [30].

The model has only two spatial degrees of freedom, but the Stokes term is calculated as in 3D. Thus, the system can be thought of as a thin layer of suspension between frictionless walls. The system is periodic in the $xy$ plane with gravity pointing in the $-y$ direction. We focus here on the behavior of the velocity fluctuation autocorrelation functions, the corresponding memory functions, and the diffusion coefficients in the case of uniform spherical particles under steady-state sedimentation conditions. We consider the case of small but finite Reynolds numbers, $\mathrm{Re}_p \ll 1$, with Brownian motion completely neglected [17].

The solvent mediated long-range hydrodynamic interactions cause velocity fluctuations which induce diffusive behavior in the steady state even without explicit Brownian motion [17, 30]. In this case, the corresponding hydrodynamic diffusion coefficient, $D_h$, can be defined through generalizing the Green-Kubo relation as

$$D_h \equiv \frac{1}{d} \int_0^\infty \phi'(t) dt, \tag{8}$$

where $\phi'(t)$ is the particle velocity *fluctuation* autocorrelation function.

In Figs. 4 (a) and (b), we show $[\phi'_{||}(t)/\phi'_{||}(0)]^{-1} - 1$ and $[\phi'_{\perp}(t)/\phi'_{\perp}(0)]^{-1} - 1$ as a function of the scaled time $t/t_s$ on a log-log scale for several volume fractions of the particle phase. The parallel component of $\phi'(t)$ shows well-defined power law behavior over about two orders of magnitude in time, which indicates that Eq. (6) is indeed a good approximation for the times shown in the figure. The inset of Fig. 4 (a) shows results of least squares fitting to the logarithmic data. The effective exponent $x \simeq 2$ is almost independent of the volume fraction and in the same range as results for strongly repulsive dissipative systems presented above. For the perpendicular component, the effective power law is not as well-defined, and the exponent $x \simeq 2.3$. In both cases, the late-time behavior becomes eventually exponential as expected. Analysis of the corresponding memory function reveals they do not show very well-defined intermediate power law behavior within the accuracy of the present data.

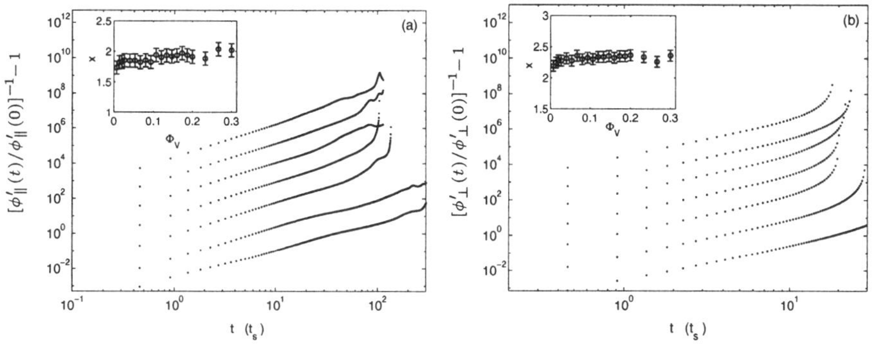

*Figure 4.* Behavior of the (a) parallel and (b) perpendicular velocity autocorrelation functions for particle volume fractions $\Phi_V$ = 0.0066, 0.0266, 0.0667, 0.133, 0.2, 0.267, and 0.367, from bottom to top ($\mathrm{Re}_p = 0.02$, and time has been normalized by the Stokes time $t_s$ [17]). Curves have been shifted for clarity. Inset shows effective exponents *vs.* $\Phi_V$.

## 4. Collective Diffusion

### 4.1. RESULTS FOR THE MODEL SYSTEM O/W(110)

We next considered the decay of velocity correlations in collective diffusion by calculating $\psi_g(t)$ numerically [18]. We focus on the case $\theta = 0.55$ and to determine $\psi_g(t)$ as accurately as possible at a few temperatures only. This choice is due to the fact that, unlike $\phi_g(t)$, $\psi_g(t)$ is a collective quantity and therefore its accurate determination requires extensive computational efforts. Furthermore, the memory effects in collective diffusion are in general much weaker than those in tracer diffusion, [11] which means that $\int_0^t \psi_g(t')\,dt'$ converges much faster than $\int_0^t \phi_g(t')\,dt'$. This implies that the time window in which a power-law decay is expected to hold is considerably narrower in case of $\psi_g(t)$.

The correlation function $\psi_g(t)$ was found to have a positive leading term, $\psi_g(0)$, and a negative tail at times $t > 0$. We find that after a very small crossover time, an effective power-law form, $\psi_g(t) \propto t^{-y}$, develops at intermediate times. The exponent $y$ was found to decrease monotonically as the temperature is lowered from high to low temperatures: we found $y = 1.23 \pm 0.05$ at $T = 676$ K and $y = 1.02 \pm 0.10$ at $T = 270$ K. Therefore, we conclude that the results for $\psi_g(t)$ are in qualitative agreement with previous results for $\phi_g(t)$ and suggest that $y$ is non-universal, and thus subject to interaction effects.

4.2. RESULTS FOR THE MODEL POLYMER SYSTEM

To complement previous studies, we carried out additional simulations for the model polymer system. We first considered the model without direct interactions and varied the coverage $\theta$. Again, there is an approximate power-law decay $\psi_g(t) \sim t^{-y}$ at relatively short times. The time window in which the power-law decay is observed is very narrow, since the memory effects in collective diffusion are very weak [11]. This, again, implies that the computational effort in studies of $\psi_g(t)$ is extensive, and makes an accurate determination of $y$ very difficult. In the present case, we found exponents $y = 1.8 \pm 0.1$ at $\theta = 0.40$ and $y = 2.0 \pm 0.1$ at $\theta = 0.65$. This behavior is in agreement with results for the decay of $\phi_g(t)$. Additionally, we studied the decay of $\psi_g(t)$ for strongly attractive chains and found $y = 1.5 \pm 0.1$ at $\theta = 0.40$, again in agreement with the behavior of $\phi_g(t)$. These results confirm that the decay of $\psi_g(t)$ follows, qualitatively, the same behavior as the decay of velocity correlations in tracer diffusion.

## 5. Discussion

In this review, we have summarized our recent works on the nature and role of memory effects in many-particle diffusion. The memory effects manifest themselves in a nontrivial temporal behavior of $\phi_g(t)$ and $\psi_g(t)$. In cases where the interparticle interactions are important, we find that both $\phi_g(t)$ and $\psi_g(t)$ display a transient non-exponential regime at intermediate times, well approximated by algebraic decays of the form $\phi_g(t) \sim t^{-x}$ and $\psi_g(t) \sim t^{-y}$. This behavior is most pronounced in $\phi_g(t)$ for tracer diffusion, where the memory effects seem to be much stronger than in collective diffusion. Based on our studies, within a disordered phase, the behavior of $x$ is related to the nature of predominant interactions, while in ordered phases $x$ is mainly governed by local ordering effects. This should be particularly useful from the experimental point of view to obtain information about the ordering and the nature of predominant interactions between the adparticles.

*Acknowledgments:* This work has in part been supported by the Academy of Finland through its Center of Excellence Program and by a grant from the European Union (I.V.).

## References

1. A.-L. Barabási and H. E. Stanley, *Fractal Concepts in Surface Growth* (Cambridge University Press, Cambridge, 1995).
2. V. P. Zhdanov and B. Kasemo, *Surf. Sci. Rep.* **20**, 111 (1994).

3. R. Gomer, *Rep. Prog. Phys.* **53**, 917 (1990).
4. J. P. Boon and S. Yip, *Molecular Hydrodynamics* (Dover, New York, 1980).
5. T. Ala-Nissila and S. C. Ying, *Prog. Surf. Sci.* **39**, 227 (1992).
6. L. Y. Chen and S. C. Ying, *Phys. Rev. B* **49**, 13838 (1994); L. Y. Chen, M. R. Baldan, and S. C. Ying, *Phys. Rev. B* **54**, 8856 (1996).
7. A. Cucchetti and S. C. Ying, *Phys. Rev. B* **54**, 3300 (1996).
8. R. Ferrando, R. Spadacini, G. E. Tommei, and G. Caratti, *Surf. Sci.* **311**, 411 (1994); F. Montalenti and R. Ferrando, *Phys. Rev. B* **59**, 5881 (1999).
9. T. Hjelt, I. Vattulainen, J. Merikoski, T. Ala-Nissila, and S. C. Ying, *Surf. Sci.* **380**, L501 (1997); *Surf. Sci.* **402–404**, 253 (1998).
10. I. Vattulainen, J. Merikoski, T. Ala-Nissila, and S. C. Ying, *Phys. Rev. B* **57**, 1896 (1998).
11. S. C. Ying, I. Vattulainen, J. Merikoski, T. Hjelt and T Ala-Nissila, *Phys. Rev. B* **58**, 2170 (1998).
12. I. Vattulainen, *Surf. Sci.* **412-413**, L911 (1998).
13. T. Hjelt, I. Vattulainen, J. Merikoski, T. Ala-Nissila, and S. C. Ying, *Surf. Sci.* **402/404**, 253 (1998).
14. I. Vattulainen, S. C. Ying, T. Ala-Nissila, and J. Merikoski, *Phys. Rev. B* **59**, 7697 (1999).
15. T. Hjelt, I. Vattulainen, T. Ala-Nissila, and S. C. Ying, *Surf. Sci.* **449**, L255 (2000).
16. T. Hjelt and I. Vattulainen, *J. Chem. Phys.* **112**, 4731 (2000).
17. E. Kuusela and T. Ala-Nissila, to be published in *Phys. Rev. E* (2001).
18. I. Vattulainen, T. Hjelt, T. Ala-Nissila, and S. C. Ying, *J. Chem. Phys.* **113**, 10284 (2000).
19. J. M. Lahtinen, T. Hjelt, T. Ala-Nissila, and Z. Chvoj, unpublished (2000).
20. N. H. March and M. P. Tosi, *Atomic Dynamics in Liquids* (Dover, New York, 1976)].
21. I. Vattulainen, S. C. Ying, T. Ala-Nissila, and J. Merikoski, *J. Chem. Phys.* **111**, 11232 (1999); I. Vattulainen, J. Merikoski, T. Ala-Nissila, and S. C. Ying, *Phys. Rev. Lett.* **80**, 5456 (1998).
22. G.-C. Wang, T.-M. Lu, and M. G. Lagally, *J. Chem. Phys.* **69**, 479 (1978).
23. C. R. Brundle and J. Q. Broughton, Chapter 3 in Volume 3A of *The Chemical Physics of Solid Surfaces and Heterogeneous Catalysis: Chemisorption Systems*, edited by D. A. King and D. P. Woodruff (Elsevier, Amsterdam, 1990).
24. K. E. Johnson, R. J. Wilson, and S. Chiang, *Phys. Rev. Lett.* **71**, 1055 (1993).
25. I. Carmesin and K. Kremer, Macromolecules **21**, 2819 (1988); *Monte Carlo and Molecular Dynamics Simulations in Polymer Science*, edited by K. Binder (Oxford, New York, 1995).
26. Here, one MCS is defined as an attempt to move each monomer of every chain in the model.
27. T. Ala-Nissila, S. Herminghaus, T. Hjelt, and P. Leiderer, *Phys. Rev. Lett.* **76**, 4003 (1996); T. Hjelt, S. Herminghaus, T. Ala-Nissila, and S. C. Ying, *Phys. Rev. E* **57**, 1864 (1998).
28. J. M. Lahtinen, T. Hjelt, and T. Ala-Nissila, *Surf. Sci.* **454-456**, 598 (2000).
29. H. Löwen, *Phys. Rev. E* **53**, R29 (1996).
30. S. Schwartzer, *Phys. Rev. E* **52**, 6461 (1996).

# DRIVEN DIFFUSION IN A MODEL OF THE O/W(110) SYSTEM

M. A. ZAŁUSKA-KOTUR
*Institut of Physics Polish Academy of Sciences*
*Al. Lotników 32/46, 02-668 Warsaw, Poland*
STANISŁAW KRUKOWSKI
*High Pressure Research Center Polish Academy of Sciences,*
*Sokolowska 29/37, 01-142 Warsaw, Poland*
AND
ŁUKASZ A. TURSKI
*Center for Theoretical Physics Polish Academy of Sciences,*
*Al. Lotnik ów 32/46, 02-668 Warsaw, Poland*

**Abstract.** We present a new model for the O/W(110) system viewed as a dilute lattice gas of particles that can be in one of three different states, two which are energetically equivalent and one of higher potential energy. The phase diagram consists of $(2 \times 1)$ and $(2 \times 2)$ phases as expected. No three-body interactions were needed to reproduce the features of experimental phase diagram. We also study diffusion of the system by analysis of step profile evolution. The possible mechanism and schemes of particle migration and corresponding diffusion processes are discussed.

## 1. Introduction

Lattice gas model investigations, analytical as well as by computer simulations, continue to be an important method in study of surface dynamical problems. Simple models with the smallest possible number of constants seem to be quite adequate to describe static properties of adsorbed systems [1-8]. Analysis of dynamical processes using lattice gas models require careful definition of transition rates matrix that decides about gas particle individual jumps. Freedom in construction of such a matrix is quite substantial. Mathematically, the only condition that must be fulfilled is a detailed balance condition that guarantees that the dynamical process leads to the



*M.C. Tringides and Z. Chvoj (eds.),*
*Collective Diffusion on Surfaces: Correlation Effects and Adatom Interactions,* 59–69.

local equilibrium between two states. Usually global equilibrium state is reached in such a way. It is well known that barriers for individual jumps may depend on interactions with neighbors and in most cases should be defined taking this into account [1-14]. It is very difficult to find convincing construction of the jump rates dependence on configuration. The collective dynamic processes close and far from equilibrium unquestionably depend on details of the transition matrix. In this work we discuss the dependence of the collective dynamics on the choice of individual jumps, using a novel model of $O/W(110)$ system[15].

Oxygen on tungsten surface is one of the most frequently studied model adsorption systems. The existence of different, coverage-dependent phases allows for the study of various phenomena: phase structure and coexistence, condensation, kinetics of ordering as well as equilibrium and far from equilibrium diffusion processes [1-24].

The lattice-gas model with properly chosen interaction constants is believed to describe properly phase transitions exhibited by the O/W(110) system. There exist several propositions for set of interaction constants. The basis of lattice models constructed to describe O/W(110) is a centered rectangular lattice-gas with competing interactions [5-8]. In general, there are four possible adsorption sites at the tungsten surface [19, 20]. An existence of site-exchange domain superstructures observed by use of low energy electron diffraction [18, 19] and scanning tunneling microscopy [20] suggests that only one of them, triply-coordinated site, is the most probable for oxygen adatoms. There are two degenerate positions where the adatom has three neighboring tungsten atoms and oxygen can occupy one of them. We have performed quantum mechanical calculations in DFT (density functional theory) formulation to determine the energy of the adsorbed single O atom on W(110) surface. Calculations confirm that the state of lowest energy is a triply-coordinated position[15]. There are two such minima separated by a barrier of about 0.2 eV. This barrier is relatively small as compared with 0.5 eV, which we find for the lowest path between two neighboring sites of the adsorption lattice. A relatively small barrier between two degenerate states suggests that interparticle interactions can modify this two-well picture. For some specific cases, it is possible that the center position, between two wells, can be the minimum of energy. On the basis of this picture, we have constructed a new lattice gas model, analysis of which shows existence of experimentally observed phases. Contrary to previous works, our model requires no three particle interactions[15].

The oxygen on a tungsten layer has two basic ordered phases and several coexistence regions. At the low density part of the phase diagram there is transformation from disordered to $(2\times1)$ striped phase. This transformation is believed to be of second order at higher temperatures and becomes the first order with wide coexistence region when the system is cooled. For higher densities oxygen forms (2 $\times$ 2) squared structure, which has its perfect ordering density at surface coverage, $\rho = 3/4$. The first (2 $\times$1) stripped phase is anisotropic. In its basic dynamical formulation, where barriers do not depend on interactions, diffusion of such $O/W(110)$ phase is much faster along lines than in the vertical direction [23, 24]. We show, however, this is not the only choice of transition rates, but the other choices can lead to completely different results.

We compare also these results with those studied by us earlier, the hexagonal lattice gas model with repulsive short-range forces [13]. Two different barrier choices have been studied resulting in different diffusion coefficient density dependence.

## 2. The Lattice Model of O/W(110) System

Two triply-coordinated positions at the Wolfram surface have minimal energy and are equivalent for the single oxygen atom. These two close local energy minima are separated by a small barrier. The adatom occupies one of these positions. Double occupancy is forbidden. Thus, atoms occupying one adsorption site can be at one of two states. The barrier for jumping between these states is around three times smaller than the barrier between two neighboring sites. Both states of one adsorption sites have the same energy for isolated adatoms. Interaction with other adatoms change this energy. Interaction depends generally on the relative distance; hence, it is different for each of two positions. In some specific situations it can cause the energy of interacting adatom at its triply-coordinated position to be higher than the barrier between two states. We assume that such situations can happen, and that for some configurations of neighbors of a given adatom, its lowest energy position is a central between both triply-coordinated states. We add such position to our model (see Fig. 1). At low densities particles do not occupy this state, because of the local energy $V$, which is equal to the barrier for a jump between two triply-coordinated minima. We define interaction constants in such a way that for dense systems some adatoms due to interaction energy have their local minima at central positions.

The construction of the model is the following: adsorbed atoms at each site can be at one of three possible states $A, B, C$, or the site can be empty. In such a way we have a three-state diluted lattice gas or a four-state lattice model. Let variables $n^A, n^B, n^C = 0, 1$ denote the occupation of corresponding states. The Hamiltonian can be written in the following way

$$H = -J \sum_{nn} (n_i^A - n_i^B)(n_j^A - n_j^B) + J \sum_{nnn} (n_i^A + n_i^B)(n_j^A + n_j^B) \qquad (1)$$

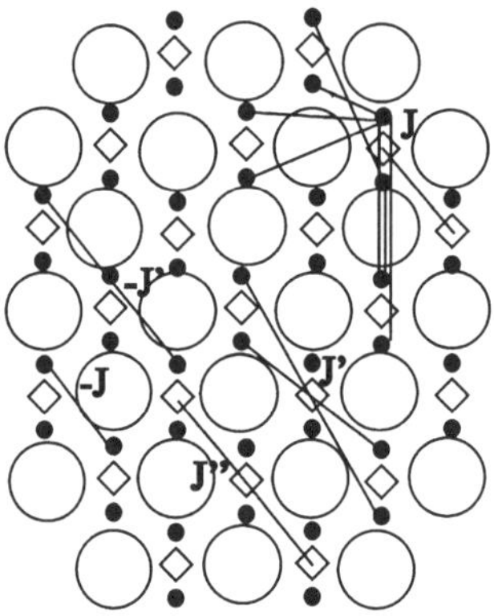


*Figure 1.* Surface W(110)with plotted adsorption sites and interactions. Circles are plotted for a triply-coordinated position and squares for the center point.

$$+V\sum_i n_i^C - J'\sum_{3n}(n_i^A - n_i^B)(n_j^A - n_j^B) + J\sum_{nn} n_i^C n_j^C \qquad (2)$$
$$-J''\sum_{3n} n_i^C n_j^C - \mu\sum_i (n_i^A + n_i^B + n_i^C),$$

where $nn$ means summing over all nearest-neighbor pairs, $nnn$ next nearest-neighbors, $3n$ third neighbors, and $\mu$ is chemical potential. First three terms in (2) describe relative interactions between equivalent positions $A$ and $B$ at the lattice site. The first and third give attractions between two atoms being at the same state, and repulsion when they are in different states. According to second term particles repel each other for all combinations of $A$, $B$ states . These interactions cause system orders in stripped $(2 \times 1)$ phase. Next, two interactions are repulsive for adatoms in the $C$ state when they are nearest-neighbors and attractive when they are third neighbors (see Fig.1). State $C$ has a higher local energy, given by potential $V$. For densities higher than 0.5, such combinations of interactions give squared $(2 \times 2)$ structures.

Four states can be easily mapped onto two-spin model, which has been widely studied in various contexts [25]. Two spin representation of the four-state model is easier to analyze and to simplify in some cases. We choose $s_i, \sigma_i = \pm 1$ as two independent, two-state variables. Now, the convention is $(s_i\sigma_i) = (+1 -1)$ and $(-1 +1)$ are two equivalent triply-coordinated minima $A, B$. Configuration $(+1, +1)$ means the particle in the central position $C$, and $(-1, -1)$ no particle at site $i$ - 0. We can express the occupation of an appropriate state by the following combinations of variables $s_i, \sigma_i$

$$n^A = \frac{1}{4}(1-\sigma_i)(1+s_i), n^B = \frac{1}{4}(1+\sigma_i)(1-s_i), \qquad (3)$$
$$n^C = \frac{1}{4}(1+s_i)(1+\sigma_i), n^0 = \frac{1}{4}(1-\sigma_i)(1-s_i).$$

For low concentrations the Hamiltonian (2) can be simplified, when we note the existence of two equivalent states that give two different domains. When system orders the symmetry is broken. One state is chosen spontaneously, like in magnetic system. To analyze the phase diagram, we can safely assume that only one of these states $A$ or $B$ is active, whereas the second one is unoccupied. Hence, the following condition can be used

$$n^B = (1 + \sigma_i)(1 - s_i) = 0. \tag{4}$$

This means that site $B$ is forced to be empty $n_B = 0$. Note that the phase diagram is kept; however, all phenomena related in any way to domain coexistence are lost. At low densities, we can also assume that $n_C = 0$. The state $C$ has a local energy potential $V > 0$, thus for lower densities the occupation of these states is highly improbable. This assumption, together with (4), gives $\sigma_i = -1$, and the simplified Hamiltonian is

$$H_0 = -J \sum_{nn} u_i u_j + J \sum_{nnn} u_i u_j - J' \sum_{3n} u_i u_j - \mu \sum_i u_i \tag{5}$$

It depends only on single variable per site $u_i = 0.5(1 + s_i)$, which has meaning of single state occupation $u_i = 0, 1$ . This Hamiltonian gives first half of the phase diagram. It describes transition to the stripped phase $(2 \times 1)$ . This is the simplest version of the classical models for O/W(110) system [5]. It has an expected structure of phase diagram, namely first order transition at lower temperatures; that transformation changes into the second order for higher T. Hamiltonian (5) can be easily extended by using more interacting constants for farther neighbors.

We present in Fig 2. a schematic plot for the phase diagram of (2). We choose the following interaction constants $J' = 0.1J, J'' = 0.4J$, and $V = J$ ($J \simeq 0.1eV$). The values are chosen in such a way that behavior of our system is close to the diluted lattice gas model with first $J_0 = -1$, second $J_0' = 0.806$ and third neighbors $J_0'' = -0.203$ [5]. Transition to $(2 \times 1)$ phase is of second order at higher temperatures and of first order at lower temperatures. It seems that transition from the $(2 \times 1)$ to $(2 \times 2)$ phase is also of second order for higher and of first order for lower temperatures. The phase diagram for densities higher than 0.5 is dominated by squared $2 \times 2$ structure, and coexistence regions of $2 \times 1$ and $2 \times 2$ , and $2 \times 2$ and $1 \times 1$ structures.

To calculate order parameters for both ordered phases, we divide all lattices into four sublattices. Comparison of density in four sublattices shows difference between striped and squared phases. In Fig. 3 we have plotted order parameters calculated as the sum of the absolute values of differences between sublattice densities $n_\alpha = 4(\sum_{i \in sub} n_i)/N$ and the global

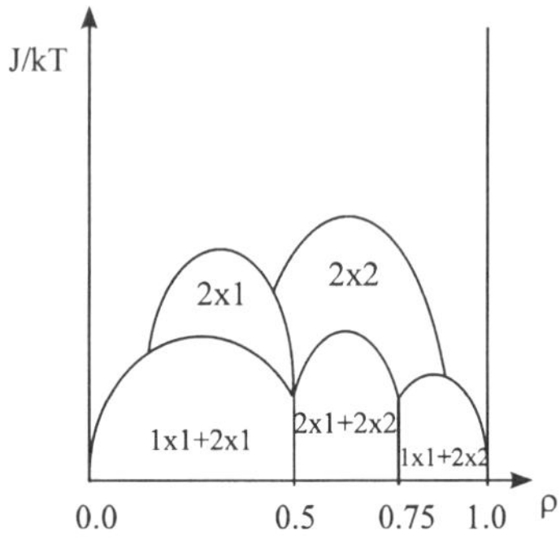


*Figure 2.* Schematic phase diagram

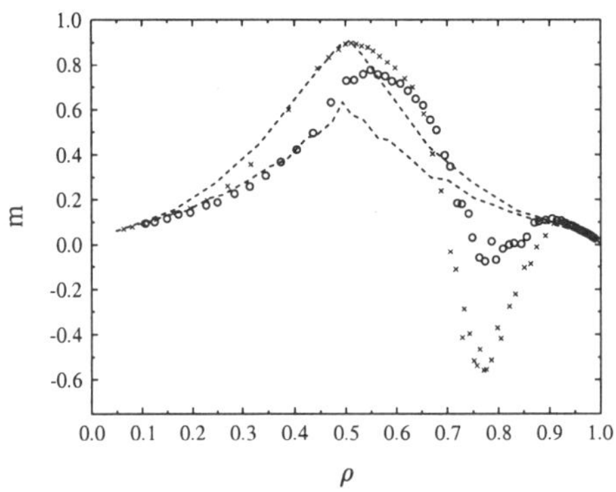


*Figure 3.* Order parameter for $T = 0.72T_c$ - upper curve, and $T = 0.9T_c$ - lower curve. In dashed lines we plot simple lattice gas model from [5] for the same temperatures.

density$\Delta n_\alpha = n_\alpha - n$, all multiplied by signs of these differences

$$m = \frac{1}{2} \prod_{\alpha=1}^{4} sign(\Delta n_\alpha) \sum_{\alpha}^{4} abs(\Delta n_\alpha). \tag{6}$$

Such an order parameter is equal to 1 for ideal ordering of the stripped phase and $-3/4$ for the ideal squared phase. We compare the order parameter $m$ given by (6) for our model, and the reference lattice gas system with defined previously set of constants in Fig 3. We can see that up to mean density $n = 0.5$, both plots for two models are close, and the difference is for higher densities, where the squared phase emerges.

## 3. Driven Diffusion Process

The model of $O/W(110)$ system, formulated above, is highly anisotropic in its $(2 \times 1)$ ordered phase. The most frequently used transition rates are zero for distances longer than this to nearest-neighbor, and for jumps between neighboring sites are defined as $W_{ij} = C \exp[\beta H(n_i)]$. They are inversely proportional to the probability the initial configuration at a given temperature. Constant $C$ does not depend on the system configuration. When such rates are used to model dynamics, the ordered $(2 \times 1)$ phase builds quickly

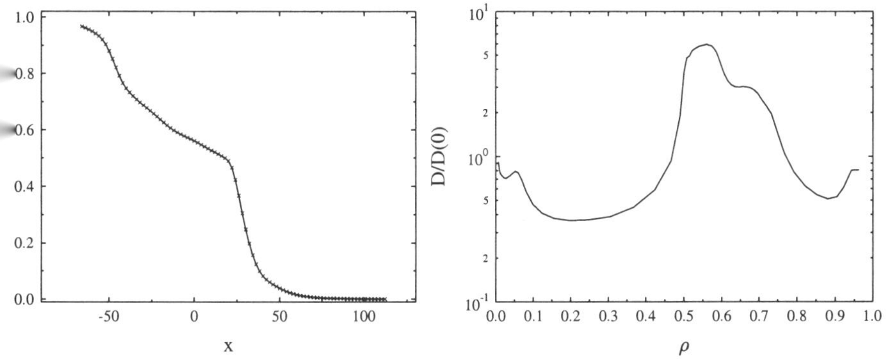


*Figure 4.* a)Density profiles and b) resulting diffusion coefficient for first type of dynamics for O/W(110) model at $T = 0.72T_c$.

along lines, whereas diffusion is very slow for vertically-oriented domains. Usually many domains are present and the system always can find a fast path. Step evolution, in the initial stage smooth, after some time, starts to build terraces around density 0.5 and, later 0.75, see Fig. 4a, [23, 24]. Such terraces mean that the diffusion coefficient has a maximum density of the best ordering. Let us now recall the Kubo-Green expression for the equilibrium diffusion coefficient, which is

$$D = \left[ \frac{< (\delta N)^2 >}{< N >} \right]^{-1} \frac{1}{4t} \left\langle \left( \sum_{i=1}^{N} \Delta r_i(t) \right)^2 \right\rangle \tag{7}$$

The first term in the square bracket is the mean squares number fluctuation in a given area, containing average $< N >$ particles, and is proportional to the two-dimensional gas compressibility. In the second term, $t$ denotes the time, which in our simulations is measured in the number of MC steps. The particle displacements $\Delta r_i(t)$ are measured in the lattice constants.

Observation that diffusion $D$ has its maximum at the perfect ordering density $\rho = 1/2$ is in agreement with the analysis of the compressibility, which has its minimum at the center of the ordered phase [1, 26]. The compressibility depends only on static properties of the system. It means that it is independent of the choice of the microscopic transition rates. The second part of the formula for the diffusion coefficient is an asymptotic value of density-density time correlation function. This quantity depends on the jump rates and can qualitatively change the behavior of the diffusion coefficient.

Jump rates $W_{ij}$ must be defined in such a way that a detailed balance condition,

$$W_{ij} \exp[-\beta H(n_i)] = W_{ji} \exp[-\beta H(n_j)], \tag{8}$$

is satisfied. This condition assures that the system has correct equilibrium properties. The detailed balance condition is satisfied by rates $W_{ij}^{DB} = \exp[\beta H(n_i)]$. The particle jumps faster when its position is that of high energy. In such a way, the system reaches its thermal equilibrium. A detailed balance is defined for two states between which particle jumps. This means that when the additional term $W_{ij}^S$, does not differ between initial and final sites, it is added to transition rates, and the equilibrium state remains unchanged. Such symmetric terms can be different for any pair of sites, and can change the collective dynamics of the entire system in significant ways. Thus, we have now jumps between individual sites given by $W_{ij} = W_{ij}^S W_{ij}^{DB}$ for neighboring sites and $W_{ij} = 0$, for all longer jumps.

Let us start with one of the simplest systems: diluted lattice gas on the hexagonal lattice interacting via repulsive forces. The difference between the diffusion coefficient in an empty system $\rho \simeq 0$ and totally occupied $\rho \simeq 1$, below critical temperature, is several orders of magnitude. The diffusion coefficient for high densities can be lowered by a well-chosen choice of symmetric terms, $W_{ij}^S$. Compare two different choices of barriers for individual jumps. The first will be equal to an interaction with two sites at both sides of the jump saddle point. Interaction is via the same constant like in the Hamiltonian. Barriers are smaller when neighboring sites are occupied, and symmetry $W_{ij}^S = W_{ji}^S$ is fulfilled. Results for diffusion coefficient are presented in Fig. 5. The upper curve grows by four orders. The main increase of the diffusion coefficient happens at $\rho = 1/3$,and then at $\rho = 2/3$. At these points the system is forced to accept presence of additional nearest neighbor. Compressibility is very low at these points, the system is very stiff. However the compressibility increases at higher densities, whereas the diffusion coefficient stays at the same level. There is no local peak of diffusion for densities $\rho = 1/3$ or $\rho = 2/3$. To study these points in a more detailed way, we can increase the individual barriers more. The interaction at the saddle point will now be 1.5 times larger than static interaction. Diffusion for this case is presented by a lower curve in Fig. 5. It can be seen that still there is no increase of the diffusion coefficient at the points of maximal ordering $\rho = 1/3, 2/3$.

We return to the model of the $O/W(110)$ system. Assumed in Section 2, interactions lead to local energies for single particle ranging from $-4J$ for four nearest-neighbors of the same $A$ or $B$ site, to $6J$ for four nearest- and four next nearest- neighbors. In spite of so large a range of local energies the diffusion coefficient following the basic version of dynamics, i.e. $W = W_{ij}^{DB}$, does not change by orders of magnitude. In this basic version, it has maxima at $\rho = 1/2$ and at $\rho = 3/4$ (see Fig. 4b)[23, 24] where the stripped phase grows. In this regime, according to Hamiltonian (2), particles have quite a high energy, and when they move between lines they jump quite fast.

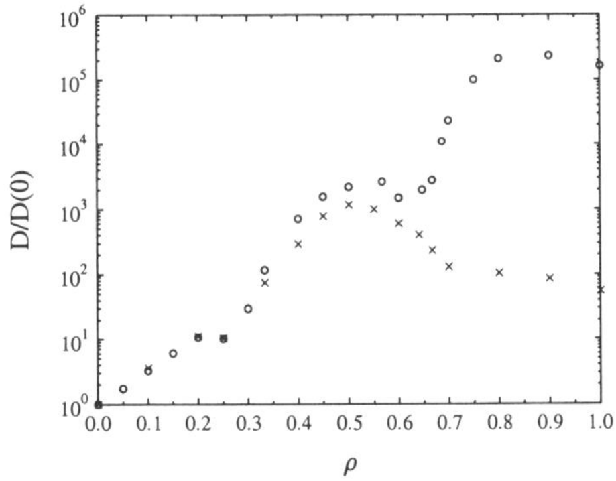


*Figure 5.* Diffusion coefficient calculated for hexagonal lattice at $T = 0.75T_c$ Upper curve - the first type of dynamics and lower -the second

When ordering lines stay vertical to the direction in which we measure the diffusion, e.g., at the top of decaying profile [23, 24] particles move rather slowly. When diffusion is measured in a direction perpendicular to lines of the stripped phase, there is no diffusion peak for $\rho = 1/2$. In a real system at thermal equilibrium, there are many domains with different orientations, thus the mean diffusion has a peak at $\rho = 1/2$.

The situation can be changed by other choices of $W_{ij}^s$. Anisotropy of jumps in $(2 \times 1)$ phase is removed by a second choice of transition rates. In this model adatom, that is in line of other adatoms, and is attracting others, has symmetric part of rates given by $W_{ij}^S = \exp(5\beta J/2)$. Jumps in and out of this state are increased by an attractive interaction at the saddle point along the transition path. At the same time particles in a state $C$ are "slowed" by choosing $W_{ij}^S = \exp(-0.7\beta J)$. In such a way both directions are dynamically equivalent. Domains can build up in the first and second directions with equal probability. In Fig. 6a we show profile for this kind of dynamics. It can be seen that it is qualitatively different from the profile obtained by the first transition rates Fig. 4a. In Fig. 6b we see the diffusion coefficient that is calculated using the Boltzmann-Matano method to profile Fig. 6a. As one can see, it has a maximum at around $\rho = 0.3$, whereas a wide minimum is present in the ordered phase $\rho \cong 0.5$. It is known that the diffusion coefficient taken from the analysis of profile decay can yet differ from equilibrium data in some details (see Ref [13, 23, 24]). More detailed studies under proper equilibrium limit of shown curves, based on harmonic profile decay [12], are continued now.

Our results shows that both collective dynamic processes discussed are different and dependent on the choice of the symmetric transition rates $W_{ij}^S$. Various choices of transition rates give the same, asymptotic equilibrium state, but can lead to qualitatively different results when the dynamics is studied [16]. This example illustrates the fact that the Hamiltonian of the lattice gas model does not determine its collective dynamical quantities,

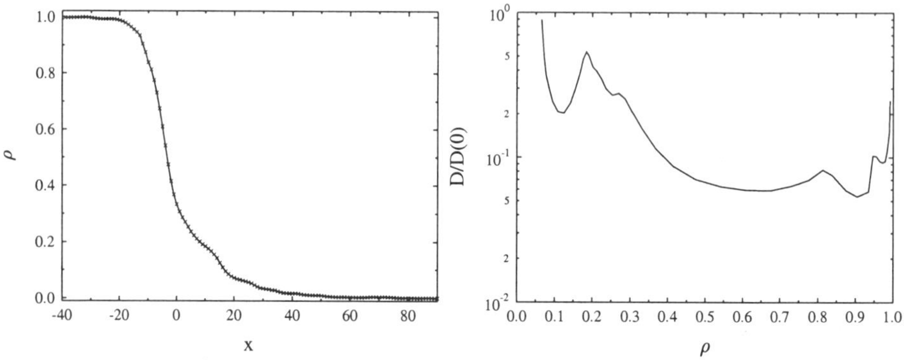


*Figure 6.* a)Density profiles and b)resulting diffusion coefficient for second type of dynamics for O/W(110) model at $T = 0.72T_c$.

even close to equilibrium values. We can change not only the global tendency of studied parameters, like in presented hexagonal system, but also have influence onto their main characteristics. A special choice of transition rates, with accordance to detailed balance, can cause diffusion maximum changes into minimum at some specific points, like density of perfect ordering, or phase transition density. More detailed study of such dependencies allow for understanding relationships between single particle transition rates and collective behavior of system.

## 4. Acknowledgment

This work was supported by KBN Grant 2 P03B 025 17.

## References

1. M.C. Tringides (Ed.) "*Surface Diffusion, Atomistic and Collective Processes* " , Plenum Press, New York 1997
2. E.Bauer in "*Structure and Dynamics of Surfaces II*" ed. W. Schommers and P. von Blackenhagen, Springer -Verlag, Berlin Heidelberg 1987
3. R.Gomer R, *Rep.Progr.Sci.*, **53**; 917(1990).
4. A.Zangwill,*Physics at surfaces.* Cambridge: Cambridge University Press, 1988
5. E.D Williams, S.L. Cunningham, and W.H. Weinberg, *J.Chem.Phys.*, **68**; 4688(1978).
6. D.Sahn, S.C. Ying, and J.M. Kosterlitz, in *The Structure of Surfaces II,* ed. by J. F. van der Veen and M.A. van Hove (Springer-Verlag, Berlin, 1988) p 410 .
7. W.Y. Ching, D.L. Huber, M.G. Lagally, and G.C. Wang, *Surf.Sci.*,**77**;550(1978).
8. P.A. Rikvold, K. Kaski, J.D. Gunton and M.C. Yalabik, *Phys.Rev.B.*, **29,** 6285(1984)
9. A. Danani, R. Fernando, E. Scalas, M. Torri and G.P. Brivio, *Chem.Phys.Lett.* **236**(1995) 533.
10. A. Danani, R. Fernando, E. Scalas and M. Torri, Int. *J.Mod.Phys.B* **11**, 2217 (1997),*Surf.Sci.* **409**(1998) 117, *Surf.Sci.* **402-404**(1998)
11. R. Ferrando and E. Scalas, *Surf.Sci.* **281** 178(1993).

12. M.A.Zaluska-Kotur, S.Krukowski,and L.A.Turski, *Surf.Sci.* **441**,320(1999)
13. M.A.Zaluska-Kotur, S. Krukowski, Z. Romanowski,and L .A.Turski, *Surf.Sci.* 457,357(2000)
14. V.P. Zhdanov, *Surf.Sci.* **257,**63, (1991).
15. M.A.Zaluska-Kotur, S. Krukowski, Z. Romanowski,and L .A.Turski, submitted to *Phys. Rev. B.*
16. C. Uebing, R. Gomer, *Surf.Sci.*, **381**; 33(1997).
17. I. Vattulainen, J. Merikoski, T. Ala-Nissila, and S. C. Ying, *Phys. Rev. B.* , **57,** 1896(1998).
18. P.K.Wu, M.C. Tringides, and M.G. Lagally, *Phys.Rev.B.* **39;** 7595(1989)
19. M. C. Tringides, *Phys.Rev.Lett.* **65**; 1372,(1990).
20. K.E. Johnson, R.J Wilson, and S Chiang, *Phys.Rev.Lett.* **71**; 1055,(1993)
21. A.T. Loburets, A. G. Naumovets, Yu. S. Vedula ,*Surf.Sci* , **399**, 297-304,(1998).
22. A.G. Naumovets, M.V. Paily, Yu S. Vedula, *Phys. Rev.Lett.*, **71**,105(1993)
23. P. Nikunen, I. Vattulainen,and T. Ala-Nissila, *Surf. Sci.Lett.* ,**447,** L162(2000)
24. ,P. Nikunen, I. Vattulainen,and T. Ala-Nissila, submitted to *J.Chem.Phys.*
25. R.J. Baxter, "*Exctly solved models in statistical mechanics*", London ,Academic Press, 1982;
26. A.V. Myshlyavtsev, A.P. Stepanov, C Uebing, and V.P. Zhdanov, *Phys.Rev.B* **52**,5977(1995).

# SURFACE DIFFUSION AND REAL-SPACE RENORMALIZATION GROUP

A. A. TARASENKO
*Institute of Physics, National Academy of Sciences of Ukraine*
*Prospect Nauki 46, 252028, Kyiv 28, Ukraine*

L. JASTRABIK
*Institute of Physics, Academy of Sciences of the Czech Republic*
*Na Slovance 2, 182 21 Praha 8, Czech Republic*

F. NIETO
*Departamento de Fisika, Universidad Nacional de San Luis,*
*CONICET, Chacabuco 917, 5700 San Luis, Argentina*

AND

C. UEBING
*Department of Physics and Astronomy, Rutgers University*
*136 Frelinghuysen Rd., Piscataway 08854-8019, USA*

## 1. Introduction

In recent years, the effects of lateral interactions on the chemical surface diffusion coefficient of adsorbed particles have been intensively investigated using many different theoretical methods applicable to critical phenomena. In fact, mean-field [1], Bethe-Peierls [2], real-space renormalization group (RSRG) [3, 4, 5] and Monte Carlo (MC) [6, 7, 8] methods have been used in order to describe the surface diffusion phenomenon. It was found that adparticle interaction strongly influences surface diffusion, especially at low temperatures and in the close vicinity of surface phase transitions.

In general, the determination of the chemical diffusion coefficient requires solution of a kinetic equation for a system of many particles. However, in the local equilibrium approximation the problem is reduced to the calculations of purely thermodynamic quantities [9, 10, 11, 12]. This approximation is used widely due the simplicity of the expression for the chemical diffusion coefficient.



*M.C. Tringides and Z. Chvoj (eds.),*
*Collective Diffusion on Surfaces: Correlation Effects and Adatom Interactions,* 71–82.

In this work we present a short review of the critical properties of different RSRG transformations used for investigations of adparticle diffusion on the honeycomb, *hc*, square, *sq* and planar triangular, *pt* lattices. The RSRG and MC approaches have been used for calculations of the thermodynamic quantities necessary for evaluation of the chemical diffusion coefficient: adsorption isotherms, the pair correlation function and the isothermal susceptibility. All these quantities have been compared to the corresponding MC data, showing a very good coincidence even for rather low temperatures.

The outline of this paper is as follows: the Hamiltonian and the derivation of the chemical diffusion coefficient are described in Sec. 2. The RSRG approach employed is described in Sec. 3. The coverage dependences of different quantities and summary of results are presented in Sec. 4.

## 2. Diffusion of adsorbed particles in a lattice gas model

In the following we will consider an idealized solid surface of definite geometry. The potential relief minima of the surface form a 2D regular array of sites with lattice constant $a$. Foreign particles, adsorbed on such a surface, are assumed to exclusively occupy these sites. If the depth of these potential minima, $\varepsilon$, is much larger than the thermal energy, $k_{\rm B}T$, the adparticles will stay within the potential minima and from time to time perform jumps to the empty neighbor sites. The duration of such jumps is much shorter than the mean time of an adparticle sojourn in a site. In this case we can define a set of occupation numbers $\{n_i\}$ according to

$$n_i = \begin{cases} 1, & \text{if site } i \text{ is occupied} \\ 0, & \text{if site } i \text{ is empty.} \end{cases} \tag{1}$$

The grand partition function, $Q$, can be written as follows

$$Q = \sum_{\{n_i\}} \exp\left[(\mu N_a - H_a)/k_{\rm B}T\right], \tag{2}$$

where $\mu$, $N_a = \sum_i n_i$ and $H_a$ denote the chemical potential, the number of adparticles and the Hamiltonian of the system, respectively. The latter is given by

$$H_a = -\varepsilon N_a + \varphi \sum_{nn} n_i n_j + \psi \sum_{nnn} n_i n_j n_k. \tag{3}$$

Here $\varphi$ is the pair interaction energy of adparticles in the nearest neighbor, *nn*, sites, $\psi$ is the energy of a three-particle interaction[1]; the summations are carried out over all lattice bonds and elementary triangles, respectively.

[1]We have considered the three-particle interaction for the *pt* lattice gas only

The occupation numbers are changed with time due to the jumps of adparticles. Here we restrict the consideration to the $nn$ uncorrelated jumps only. Any adparticle must surmount the potential barrier between the initial and the final sites. In case of interacting lattice gases the activation energies of jumps are affected by the presence of the adjacent adparticles. We assume that the lateral interactions change only the minima of the periodic potential barrier. The rate frequency for an adparticle, jumping from the $i$th site to the $j$th site can be written as follows

$$\nu_{ij} = \nu \exp\left[-\frac{1}{k_{\mathrm{B}}T}\left(\varepsilon - \varphi \sum_{k=1}^{z} n_k\right)\right]. \tag{4}$$

This simple model includes the interaction between the activated particle in the $i$th minimun with its $nns$; $z$ is the coordination number of the lattice.

Using the local equilibrium approximation (the interested reader is reffered to Refs. [2, 5] for a detailed description of this approach) and Eq. 4, one can obtain the simple expression for the chemical diffusion coefficient

$$D = D_0 \exp(\mu/k_{\mathrm{B}}T) P_{00} \theta^{-1} \partial\mu/k_{\mathrm{B}}T/\partial \ln\theta, \tag{5}$$

where $D_0 = z\nu \exp(-\varepsilon/k_{\mathrm{B}}T)a^2/4$ is the diffusion coefficient of noninteracting adparticles, $\partial\mu/k_{\mathrm{B}}T/\partial\ln\theta$ is the thermodynamical factor; $\theta$ is the surface coverage; $P_{00}$ is the correlation function of two holes in ajacent sites:

$$\theta \equiv \langle n_0 \rangle, \qquad P_{00} = 1 - 2\theta + \langle n_0 n_1 \rangle. \tag{6}$$

We introduce the free energy of the system as $F = k_{\mathrm{B}}TN^{-1}\ln Q$. Then all necessary quantities can be calculated via the first and second derivatives of the free energy over the chemical potential and interaction parameter:

$$\theta = \partial F/\partial\mu, \ \langle n_0 n_1 \rangle = -2\partial F/z\partial\varphi, \ \partial\mu/\partial\ln\theta = \theta\left(\partial^2 F/\partial\mu^2\right)^{-1}. \tag{7}$$

It is important to recall that the expression for the diffusion coefficient, Eq. 5, was derived in the hydrodynamic limit (i.e. for adparticle surface coverage inhomogeneities varying slowly in space and time) and the adparticle jump frequency is determined by Eq. 4.

## 3. Real-space renormalization group method

One must use some approximative method for calculations of the free energy. Even for the simplest models the problem remains too complex to be solved exactly. We have used the real-space renormalization group method developed by Niemeyer and van Leeuwen [13] and Nauenberg and Nienhuis [14] for investigations of the Ising spin models. It is well known that the lattice-gas model is equivalent to the Ising spin model in an external magnetic field. Using the obvious relation between spins and occupation numbers $n_i = (1+s_i)/2$, one can easily obtain the spin Hamiltonian

$$H(s) = h\sum_i s_i + p\sum_{nn} s_i s_j + t\sum_{nnn} s_i s_j s_k + Nc. \tag{8}$$

Here

$$\begin{aligned} h &= (\mu+\varepsilon-z\varphi/2-3\psi/2)/2k_{\rm B}T, \quad & p &= -(\varphi+\psi)/4k_{\rm B}T,\\ c &= (\mu+\varepsilon-z\varphi/4-\psi)/2k_{\rm B}T, & t &= -\psi/8k_{\rm B}T. \end{aligned}$$

Attraction (repulsion) between adparticles corresponds to ferromagnetic, F (antiferromagnetic, AF) interaction between spins. In the RSRG method the whole lattice is divided into blocks of $L$ sites. A block spin $S_\alpha$ is assigned to each block. All blocks together must form the same lattice as the original one but with the increased lattice constant $\sqrt{L}a$. A RSRG transformation reduces the number of spins from $N$ to $N/L$. The transformation can be described by

$$\exp\left[H(S)+g\right] = \sum_{\{s\}} P(S,s)\exp\left[H(s)\right], \tag{9}$$

where $H(S)$ is the renormalized Hamiltonian of the block spin system, $g$ is the empty-set term (independent on the spin values) and $P(S,s)$ is the weighting factor with the properties: $P(S,s)\geq 0$ and $\sum_S P(S,s)=1$. For blocks with odd $L$ the value of $S_\alpha$ is determined by the majority rule:

$$S_\alpha = sgn\sum_{i=1}^{L} s_i, \qquad \text{where} \qquad sgn(x) = \begin{cases} +1, & \text{if } x>0 \\ -1, & \text{if } x<0. \end{cases} \tag{10}$$

In order to carry out the summation in Eq. 9, one usually employs periodic boundary conditions. It is assumed that the whole lattice is given by the periodic continuation of a small cluster of blocks. The AF ordered phase consists of two (for *hc* and *sq*) or three (for *pt*) sublattices with different magnetization. Therefore, there must be at least two or three blocks in the cluster. For this case the RSRG equations for *hc* and *sq* symmetries (2 blocks) and *pt* symmetry (3 blocks) can be easily written as follows

$$\begin{aligned} &h_1(S_1+S_2)+zp_1S_1S_2+2Lg = \ln\sum P(S_{1,2},s)\exp[H(\{s\})] \ \text{ for } hc, sq,\\ &h_1(S_1+S_2+S_3)+3p_1(S_1S_2+S_2S_3+S_1S_3)+\\ &+6t_1S_1S_2S_3+3Lg = \ln\sum P(S_{1,2,3},s)\exp[H(\{s\})] \qquad \text{for } pt. \end{aligned} \tag{11}$$

The summations in the RHS of Eqs. 11 are carried out over all possible configurations of spins in the clusters, which keep constant the values of the block spins $S_{1,2,3}$. Setting definite values of the block spins, one can obtain a system of equations relating renormalized and original Hamiltonian parameters $h$, $p$ and $t$. The system describes completely the RSRG transformation. Its unstable fixed points correspond to the critical points of the Hamiltonian Eq. 8. The singular behavior of all thermodynamic quantities is controled by the RSRG exponents $y_i$, which are related to the eigenvalues $\lambda_i$ of the linearized in the critical point matrix of the RSRG transformation: $y_i = 2\ln\lambda_i/\ln L$. In the critical points all RSRG transformations have one, so-called, temperature-like $y_T$ and one field-like $y_H$ eigenvalue exponents. The exact values of the RSRG exponents are well known: $y_T = 1$ and $y_H = 15/8$ [15]. The comparison between the RSRG critical values of the interaction parameters and critical exponents with their exact values gives an estimation of the RSRG transformation accuracy.

As was shown by Nauenberg and Nienhuis [14], the free energy of the system, $F$, for any values of interaction parameters can be evaluated in the series of sequential RSRG transformations of the original Hamiltonian

$$F = k_{\rm B}T \sum_{m=0}^{\infty} L^{-m} g(h_m, p_m, t_m). \tag{12}$$

Here $h_m, p_m, t_m$ are the parameters of the $m$th RSRG transformation.

We have investigated a number of RSRG transformations (blocks up to 19 spins for $hc$, blocks up to 18 spins for $sq$ and blocks up to 13 spins for $pt$ lattices). There are some general properties of the RSRG transformations. Usually all RSRG transformations with odd $L$ have an unstable fixed point in the F region and another one in the AF region. Many of the RSRG transformations with even $L$ have only one fixed point, located in the F region. In general, the critical value of the pair interaction parameter $p_c$ approaches to its exact value if the number of sites in the block is increased. However, the accuracy of the RSRG transformations depends considerably not only on the block size $L$, but also in great extend on the symmetry of the block and composition of sites in the block, i.e. how much sites from different sublattices enter the block. There are quite similar behavior of the RSRG transformations for the $hc$ and $sq$ lattices. Good RSRG transformations have blocks, containing sites from both sublattices (so-called cell $L \times 2$ RSRG transformations) and describe almost equally well the F and AF ordering. The RSRG transformations with blocks of homogeneous composition (blocks containing spins from one sublattice only, so-called sublattice $L \times 2S$ RSRG transformations) give critical values of the pair interaction parameter with big errors.

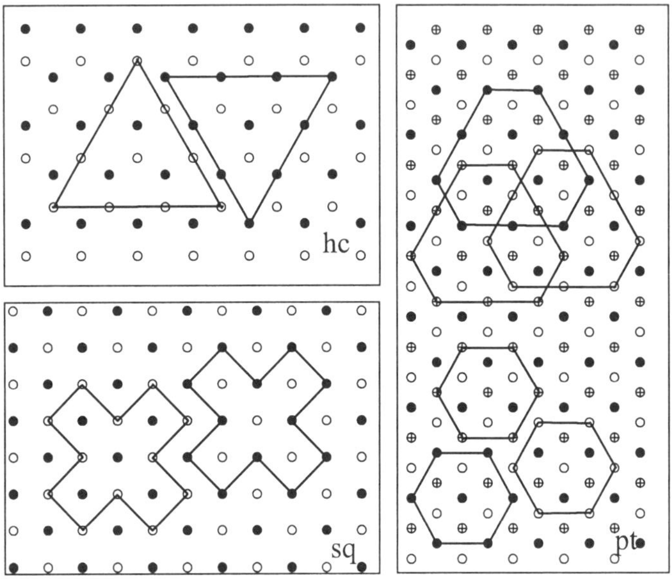


*Figure 1.* The best clusters of blocks for RSRG transformations on *hc*, *sq* and *pt* lattices.

In the case of *pt* lattice the cell $L \times 3$ RSRG transformations describe fairly well the F ordering and give poor results for AF interaction. In many cases such transformations do not show the critical AF behavior at all. A good description for AF ordering is obtained only with sublattice $L \times 3S$ RSRG transformations. These transformations describe also the critical behavior of systems with a three-particle interaction (Baxter-Wu phase transition). The accuracy of these transformations is not very sensitive to the symmetry properties of the blocks and most of them give surprisingly accurate value for zero entropy $s_0$ at half monolayer coverage ($T = 0$). The sublattice RSRG transformations describe the critical behavior in F region, but errors are much greater as compared with the cell RSRG transformations with the same $L$ analogously to the cases of *hc* and *sq* lattices. Therefore, necessary to use different $L \times 3$ and $L \times 3S$ RSRG transformation for calculations in the F and AF domains of the *pt* lattice gas. The best found RSRG transformations ($13 \times 2$ for *hc*, $17 \times 2$ for *sq*, $13 \times 3$ and $12 \times 3S$ for *pt* lattices) are shown in Fig. 1. The critical parameters of these RSRG transformation are collected in Tables 1 and 2.

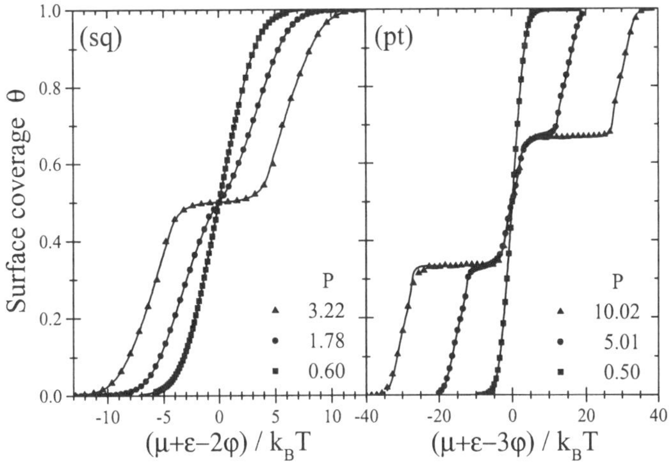


*Figure 2.* Adsorption isotherms for different temperatures. $P$ is expressed in units of $|\varphi|/k_B T$. Solid lines are obtained by the RSRG method, symbols denote the MC data.

## 4. Results and discussion

Using the most accurate RSRG transformations we have calculated chemical diffusion coefficients of adparticles on the *hc*, *sq* and *pt* lattices. We have investigated phase diagrams for pair repulsively interacting adparticles and have also calculated the temperature and coverage dependences of thermodynamic quantities of interest. In order to check the accuracy of the RSRG results we have compared the corresponding results with the MC simulation data. As the results for the *hc* and *sq* lattices are very similar, we present for the comparison some dependences for the *sq* and *pt* AF lattice gases.

The adsorption isotherms for AF interaction are plotted in Fig. 2. At high temperatures the isotherms are close to the Langmuir case ($\varphi = 0$), but as temperature is lowered, peculiarities appear at the stoichiometric coverages ($\frac{1}{2}$-M for *sq*, $\frac{1}{3}$-M and $\frac{2}{3}$-M for *pt* lattices), which develop into horizontal plateaux. The peculiarities correspond to the critical points of the second order phase transitions: the whole system is splitted into sublattices with different mean values of the occupation numbers. The plateaux show regions of existence of the ordered phases. These are the $C(2 \times 2)$ structure in case of *sq* lattice (the region of existence is centered around half coverage) and two AF ordered phases in case of *pt* lattice centered at $\frac{1}{3}$-M (one sublattice is almost filled and others are almost empty) and at $\frac{2}{3}$-M (two of sublattices are almost filled and the third is almost empty).

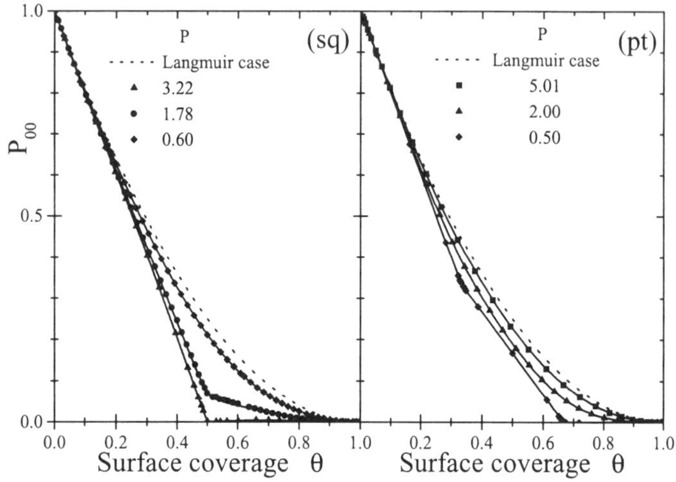


*Figure 3.* Surface coverage dependences of the correlation functions $P_{00}$ for different temperatures. Solid lines and symbols represent the RSRG and MC data, respectively.

The coincidence between the RSRG and MC data is very good over the whole range of the temperatures and coverages. In Fig. 3 we show the coverage dependences for the correlation function $P_{00}$. This quantity describes the probability of finding two empty *nn* sites. It is seen clearly that the low temperature coverage dependences of $P_{00}$ are almost linear. The slopes of the lines are changed abruptly at the stoichiometric coverages. This behavior can be explained very easily. For example, the linear dependences at small coverages $\theta \leq \frac{1}{2}$-M ($\theta \leq \frac{1}{3}$-M) is largely due to the fact that in this region adparticles are able to avoid contact with each other and, therefore, every particle adsorbing on the *sq* (*pt*) lattice deletes 4 (6) two-hole configurations. Thus, the probabilities of finding these configurations are decreased linearily with the corresponding rates. After complete formation of the ordered sublattice at $\theta \geq \frac{1}{2}$-M all two-hole configurations on *sq* lattice disappear. On *pt* lattice at $\theta \geq \frac{1}{3}$-M every additional particle has to occupy empty sites on other sublattices, deleting only 3 two-hole configurations. Therefore, the corresponding slope is two times smaller. When $\theta \geq \frac{2}{3}$-M, all two-hole configurations on *pt* lattice disappear as well. The RSRG and MC data fit very good even in the critical points.

The quantity being the most sensitive to the phase transitions is the isothermal susceptibility $\chi_T$, which is proportional to the mean square surface coverage fluctuations, the second derivative of the free energy over the chemical potential $\partial^2 F/\partial\mu^2$. The quantity enter the thermodynamic factor

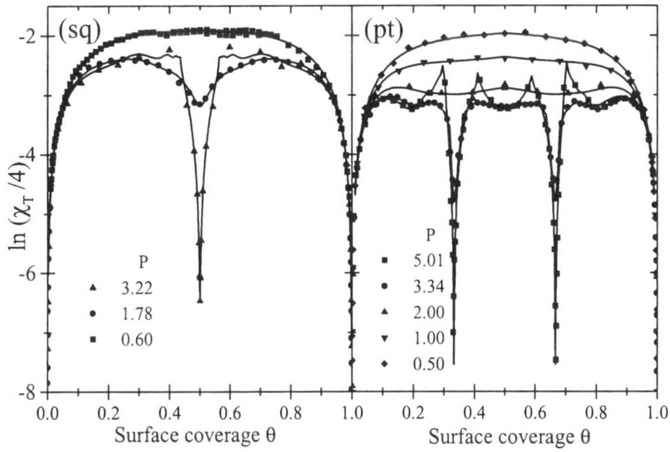


*Figure 4.* The isothermal susceptibility, $\ln \chi_T$ vs surface coverage $\theta$ for different temperatures. Solid lines and symbols represent the RSRG and MC data, respectively.

and determine the critical behaviour of the chemical diffusion coefficient. The coverage dependences of this quantity are plotted in Fig. 4. For high temperatures the coverage mean square fluctuations are close to the Langmuir case $\theta(1-\theta)$. For low temperatures the fluctuations exhibit maxima at the critical points. The coverage dependences are nonanalytic in these points. At low temperatures the fluctuations are strongly suppressed at the stoichiometric coverages due to the strong repulsion between the adparticles. Any coverage disturbance is related with displacements of adparticles from the filled sublattice(s) to the empty sublattice(s). It causes substantial increase of the free energy and is thermodynamically unfavorable. As surface coverage is not equal to the stoichiometric values, there are fluctuations of nonstoichiometric nature that can not be removed from the system by the adparticles jumps. Therefore $\chi_T$ increases when $\theta$ deviates from the stoichiometric values. There is a good coincidence between RSRG and MC data in the whole coverage range for different temperatures excluding only the close vicinities of the critical points.

The coverage dependences of the chemical diffusion coefficient are shown in Fig. 5. At low coverages the coverage dependences are similar for the both lattices. Diffusion coefficient increases with $\theta$ as the mean number of $nns$ is also increased. If the temperature is lowered, the coverage fluctuations grow and cause the reduction of the diffusion coefficient as the surface coverage approachs to the critical lines. In fact, at the critical points the dif-

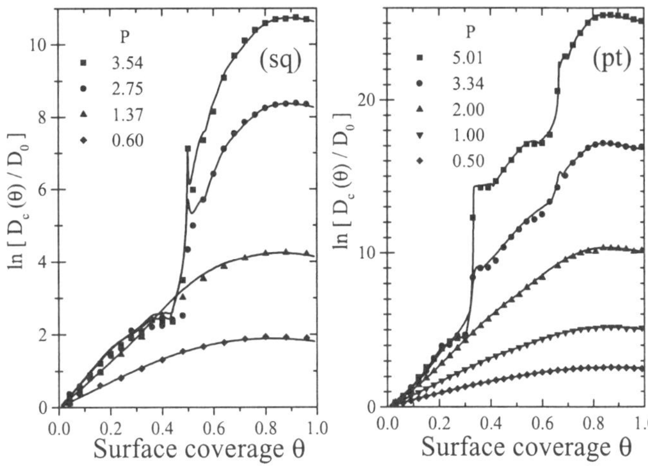


*Figure 5.* Chemical diffusion coefficient $\ln D/D_0$ vs. surface coverage $\theta$ for different temperatures. Solid lines and symbols represent the RSRG and MC data, respectively.

fusion coefficient exhibits minima, corresponding to the maxima of isothermal susceptibility. The chemical diffusion coeffcient changes abruptly at the stoichiometric coverages at low temperatures. This behaviour is clearly connected with strong changes of the diffusion activation energy due to the formation of ordered structures. When one of the sublattices on the *pt* lattice is completely occupied, the diffusion activation energy decreases by $3\varphi$. For $\theta \geq \frac{1}{3}$-M, adparticles migrate over the effective honeycomb lattice, created by two unoccupied sublattices. The coverage dependences of the diffusion coefficient for $1 \geq \theta \geq \frac{1}{3}$-M are very similar to the corresponding dependences, obtained for the *hc* and *sq* lattices in monolayer range [4, 5]. The formation of the second occupied sublattice at $\theta = \frac{2}{3}$-M is quite equivalent to the creation of the ordered $C(2 \times 2)$ structure on a *hc* or *sq* lattice and causes the same behavior of the dependences of the chemical diffusion coefficient. If $\theta \geq \frac{2}{3}$-M, two sublattices are filled completely and most of adparticles are immobile, therefore the diffusion is better described in terms of holes. The agreement between the RSRG and MC data is rather good for the whole range of coverages and temperatures. Even for low temperatures ($T < T_c$) the discrepancies between the different methods are small. However, in the vicinities of the critical points there are noticable deviations between RSRG and MC results. Summing up all results one can conclude that the RSRG method can be used successfully for investigations of the surface diffusion in the Ising-like systems. There are excellent coincidence

TABLE 1. Compilation of the critical values for the best *hc* and *sq* RSRG transformations. $p_F, |p_{AF}|, \epsilon_F, \epsilon_{AF}$ are the critical values of the spin pair interaction parameter and the corresponding relative errors.

| Cluster | $k_F$ | $\epsilon_F\%$ | $y_H$ | $y_T$ | $\lvert k_{AF}\rvert$ | $\epsilon_{AF}\%$ | $y_H$ | $y_T$ |
|---|---|---|---|---|---|---|---|---|
| $hc$ $7\times 2$ | 0.683 | 3.79 | 1.91 | 0.72 | 0.647 | 1.67 | 0.71 | 0.36 |
| $hc$ $13\times 2$ | 0.683 | 3.67 | 1.91 | 0.78 | 0.661 | 0.376 | 0.76 | 0.41 |
| $sq$ $13\times 2$ | 0.443 | 0.46 | 1.86 | 0.79 | 0.451 | 2.44 | 0.78 | 0.49 |
| $sq$ $17\times 2$ | 0.440 | 0.13 | 1.87 | 0.84 | 0.444 | 0.78 | 0.84 | 0.50 |

TABLE 2. Compilation of the critical values for the best RSRG transformations for *pt* lattice; $p_F$, $t_{BW}$, $\epsilon_F$, $\epsilon_{BW}$ are the critical values of the spin pair and three-particle interaction parameters and the relative errors, respectively; $|p_{AF}^{min}|$ is the minimal value of the spin pair interaction parameter, when the AF ordering is still possible for $t=0$; $s_0$ is the lattice entropy at $h=T=0$.

| Cluster | $p_F$ | $\epsilon_F\%$ | $y_H$ | $y_T$ | $t_{BW}$ | $\epsilon_{BW}\%$ | $\lvert p_{AF}^{min}\rvert$ | $s_0$ |
|---|---|---|---|---|---|---|---|---|
| $12\times 3S$ | 0.211 | 23.4 | 1.46 | 0.57 | $\pm 0.504$ | 14.4 | 0.786 | 0.324 |
| $13\times 3$ | 0.257 | 6.27 | 1.78 | 0.91 | † | | ‡ | 0.333 |
| $13\times 3S$ | 0.205 | 25.3 | 1.42 | 0.51 | $\pm 0.498$ | 13.0 | 0.785 | 0.328 |

† Baxter-Wu transition is absent, ‡ AF critical ordering is absent.

between data obtained be RSRG and MC methods in whole coverage region for different temperatures. Only in the close vicinities of the critical points for the divergent quantities proportional to the second derivatives of the free energy over its variables, such as isothermal susceptibility, the RSRG approach does not give very accurate results. The accuracy of the RSRG method depends on the number of sites in the block and on the symmetry of the blocks. In general, the most accurate results have been obtained for the most symmetrical and largest blocks. The critical parameters obtained for the best RSRG transformations coincide rather good with the known exact values. We have calculated adsorption isotherms at different temperatures, the coverage dependences of the pair correlation function for the nearest neighbor sites, the coverage dependences of the isothermal susceptibility and the chemical diffusion coefficient at different temperatures. All these quantities have been compared to corresponding MC simulation results. The agreement between the RSRG and MC data is very good and, therefore, we have concluded that the RSRG method is a rather suitable method, which can be used to characterize the thermodynamic and in some cases the kinetic properties of strongly interacting adsorbates.

## Acknowledgments

This work has been supported by the project LN00A015 of the Ministry of Education of Czech Republic and CONICET (Argentina), the National Agency for Promotion of Science and Technology (APCyT, Argentina), Prog.No C-03-03232.

## References

1. Tarasenko, A.A. and Chumak, A.A. (1982) Effect of the Repulsion of Adsorbed Atoms on their Diffusion over the Surface of a Crystal, *Soviet Physics Solid State*, **24**, 1683–1685.
2. Tarasenko, A.A. and Chumak A.A. (1980) Effect of the Repulsion of Adsorbed Atoms on their Diffusion over the Surface of a Crystal and Fluctuations in the Emission Current,*Soviet Physics Solid State*, **22**, 1716-1719.
3. Tarasenko, A.A. and Chumak, A.A. (1989) Diffusion in Two-Dimensional Strongly Interacting Systems and Renormalization Group, *Poverkhnost'. Fiyika, khimija, mekhanika*, **11**, 98–102, (in Russian).
4. Tarasenko, A.A., Jastrabik, L. and Uebing, C. (1998) Adatom Diffusion on a Honeycomb Lattice: Real-Space Renormalization Group Approach, *Phys. Rev.*, **B 57**, 10166–10174.
5. Tarasenko, A.A., Jastrabik, L., Nieto, F. and Uebing, C. (1999) Adatom Diffusion on a Square Lattice: Comparison of Real-Space Renormalization Group and Monte Carlo Approaches, *Physical Review*, **B59**, 8252–8261.
6. Myshlyavtsev, A.V., Stepanov, A.A., Uebing, C. and Zhdanov V.P. (1995) Surface Diffusion and Continuous Phase Transitions in Adsorbed Overlayers, *Physical Review*, **B52**, 5977–5984.
7. Kher, K. and Binder, K. (1987) Simulation of Diffusion in Lattice Gas and Related Kinetic Phenomena, in *"Applications of the Monte Carlo Method in Statistical Physics*, ed. by K. Binder, Vol. **36**, *Topics in Current Physics*, Springer-Verlag, Berlin.
8. Ala-Nissila, T. Han, V.K. and Ying, S.C. (1992) Diffusion Anomaly Near Structural Phase Transitions, *Physical Review Letters*, **68**, 1866–1868.
9. Chumak, A.A. and Tarasenko, A.A. (1980) Diffusion and Density Fluctuations of Atoms Adsorbed on Solid Surfaces, *Surface Science*, **91**, 694–706.
10. Reed, D.A. and Ehrlich, G. (1981) Surface Diffusion, Atomic Jump Rates and Thermodynamics, *Surface Science*, **102**, 588–609.
11. Reed, D.A. and Ehrlich, G. (1981) Surface Diffusivity and the Time Correlation of Concentration Fluctuations, *Surface Science*, **105**, 603–628.
12. Zhdanov, V.P. (1991) *Elementary Physicochemical Processes on Solid Surfaces*, Plenum, New York.
13. Niemeyer, Th. and Van Leeuwen, J.M.J. (1974) Wilson Theory for 2-Dimensional Ising Spin Systems, *Physica*, **71**, 17–40.
14. Nauenberg, M. and Nienhuis, B. (1974) Renormalization-Group Approach to the Solution of the General Ising Models, *Physical Review Letters*, **33**, 1598–1601.
15. Niemeyer, Th. and van Leeuwen, J.M.J. (1976) Renormalization Theory for Ising-like Spin Systems in *Phase Transitions and Critical Phenomena*, ed. by Domb, C. and Green, M.S. Vol. VI, chap. 7, Academic Press, New York.

# SURFACE DIFFUSION NEAR PHASE TRANSITIONS

F. NIETO
*Departamento de Física and Centro Latinoamericano de Estudios Ilya Prigogine, Universidad Nacional de San Luis, CONICET, Chacabuco 917, 5700, San Luis, Argentina*

A. TARASENKO
*Institute of Physics, National Academy of Sciences of Ukraine, prospekt Nauki 46, 252028 Kyiv 28, Ukraine*

AND

C. UEBING
*Department of Physics and Astronomy, Rutgers University 136 Frelinghuysen Rd., Piscataway 08854-8019, U.S.A.*

## 1. Introduction

In this contribution, recent progresses in the understanding of surface diffusion processes within adsorbed monolayers will be reviewed. We will focus on the influence of phase transitions within the adsorbed monolayers. In most of the studies to be discussed below, lattice gas models have been used to a large extent. Unfortunately, in many cases even lattice gas models are too complicated for exact analytical treatments of the diffusion problem, especially if ad-ad interactions are taken into consideration, and only in some exceptional cases general equations for the description of surface diffusion can be derived. However, Monte Carlo simulations can easily be utilized to analyze surface diffusion processes. In recent years, the widespread availability of extremely powerful supermassive parallel computers has allowed many interesting studies on this topic. It is likely that the more recent work reviewed here will stimulate further work in this and related fields. The main emphasis of this article is on transport phenomena at or near equilibrium, i.e., on diffusion of particles within (nearly) equilibrated adsorbed monolayers on homogeneous surfaces.



*M.C. Tringides and Z. Chvoj (eds.),*
*Collective Diffusion on Surfaces: Correlation Effects and Adatom Interactions,* 83–95.

## 2. Simulations of Surface Diffusion

For the computer simulation of surface diffusion, two powerful methods are frequently used, the molecular dynamics (MD) scheme (see e.g., [1–21]) and the Monte Carlo (MC) scheme. Since most of the studies to be discussed in Sec. 3 are based on lattice gas models investigated via MC simulations, we will discuss this method in some detail here.

Within the lattice gas scheme the basic steps of surface diffusion are jumps of adatoms from filled initial sites $i$ to adjacent vacant sites $f$. The activation energy for such jumps can be calculated as the energy difference between saddle point $\epsilon^*_{i\to f}$ and single site energy of the initial site $\epsilon_i$, [22]

$$\Delta E = \epsilon^*_{i\to f} - \epsilon_i. \tag{1}$$

The associated jump probability $P_J$ is given by

$$P_J = \frac{1}{\kappa} \exp\left(-\frac{\Delta E}{\cdot k_{\mathrm{B}} T}\right), \tag{2}$$

with $\kappa$ as a normalization factor. This choice ensures full microscopic reversibility and fulfils the condition of detailed balance [23]. However, the activation energies $\Delta E$ and the jump probabilities $P_J$ are partly arbitrary, since the detailed balance condition does not specify these quantities uniquely [24]. In order to optimize the computational time of a Monte Carlo algorithm, a suitable normalization of jump probabilities is indispensable. A natural choice for $\kappa$ would be

$$\kappa = \kappa_{\max} = \exp\left(-\frac{\Delta E_{\min}}{k_{\mathrm{B}} T}\right). \tag{3}$$

Here $\Delta E_{\min}$ represents the jump probability for the most favorable physically realizable jump [22]. This choice of $\kappa$ generates the highest possible jump probabilities, i.e., minimizes the number of unsuccessful attempts, while avoiding jump events with $P_j > 1$. In some cases this choice of $\kappa$ leads to an impractically large number of Monte Carlo steps required for the equilibration of the lattice gas and for the determination of the desired surface diffusion coefficients. Especially at very low temperatures and in ordered regions of the relevant phase diagram, hop events occur very infrequently. Therefore, smaller normalization factors $\kappa < \kappa_{\max}$ can be chosen. However, it should be verified very carefully that in such cases the fraction of jumps with jump probabilities $P_j > 1$ is still negligibly small ($< 0.1\%$) and does not exhibit any notable influence on the diffusion measurements and on thermodynamic properties of the system as well.

The procedure for simulating jumps in the canonical ensemble has been described in some detail in Refs. [22, 25] and, therefore, we will present only the general schema of the computations. First, for a given lattice gas configuration, an initial site $i$ is randomly picked. If filled, an adjacent final site $j$ is randomly selected. If the destination is vacant, a jump may occur with probability $P_J$ (Eq. 2), otherwise, no jump occurs. Thermodynamic equilibrium is established before starting a diffusion run at the desired fixed coverage $\theta$. Approach to equilibrium is monitored by following the configurational energy and, in case of ordering, by measuring corresponding order parameters of the system. Equilibrium is assumed to be established when these quantities fluctuate about their average values. In a recent publication, we have shown that in some systems even minute deviations from thermodynamic equilibrium may substantially influence the surface diffusion coefficients [26].

In the MC method, the tracer diffusion coefficient $D_t$ can be easily determined from measurements of the mean square displacements of $N$ tagged adatoms according to

$$D_t = \lim_{t\to\infty}\left[\frac{1}{2dt}\left\langle |\vec{r}_i(t) - \vec{r}_i(0)|^2\right\rangle\right], \tag{4}$$

where $d$ is the Euclidean dimension, (in the case of surface diffusion $d = 2$); the vector $\vec{r}(t)$ determines the position of a tagged particle at time t, and $\left\langle (\vec{r}(t) - \vec{r}(0))^2\right\rangle$ is its mean squares displacement.

The chemical diffusion coefficient $D$ can be determined via two different approaches, the fluctuation method as well as the Kubo-Green method. In essence, the fluctuation method measures the particle number autocorrelation function $f_n(t)/f_n(0)$ for a small probed region embedded in the whole two-dimensional lattice. It has been shown in detail how the decay of $f_n(t)/f_n(0)$ can be used to obtain the chemical surface diffusion coefficient, which is called $D_F$ [27, 28]. Thus, this method is a computer simulation of the field emission fluctuation method [27] used experimentally for determining adsorbate diffusion coefficients. For the autocorrelation function one can write

$$\frac{f_n(t)}{f_n(0)} = \frac{\langle \delta N(t)\delta N(0)\rangle}{\left\langle (\delta N)^2\right\rangle}, \tag{5}$$

where $N$ is the number of adatoms in the probed area. In Ref. [28] an analytical expression for the autocorrelation function of the particle density for a square probehole was given as

$$\frac{f_n(t)}{f_n(0)} = \left[\mathrm{erf}\left(\frac{\tau_\circ}{t}\right)^{1/2} + \left(\frac{t}{\pi\tau_\circ}\right)^{1/2}\left(\exp\left(-\frac{\tau_\circ}{t}\right) - 1\right)\right]^2. \tag{6}$$

In the very long time limit, i.e., when the mean squares displacement of the diffusion particles $\Delta r_i(t)$ is long with respect to the lattice spacing,

$\Delta r_i(t) \gg a_\circ$, the correlation function is approximately given by [28]

$$f_n(t) \approx \frac{k_B T K_T \theta^2 a_\circ^2}{4\pi D t}. \tag{7}$$

Here $K_T$ denotes the isothermal compressibility of the adlayer. $\theta = \langle N \rangle / A$ is the mean density of adatoms, i.e., the mean number of adatoms per area $A$ [28]. Eq. 7 implies that the correlation function decays as $t^{-1}$ at long times. However, this equation is valid only in the hydrodynamic regime, i.e., on a time and distance scale being large compared to a mean-free time and mean-free path ($\approx$ lattice spacing) of the lattice gas model [28].

The second method of determining the chemical diffusion coefficient $D$ is based on the Kubo-Green equation, [29] here written as

$$D_{KG} = \left( \frac{\partial \mu / k_B T}{\partial \ln \theta} \right)_T \cdot D_j = \left[ \frac{\left\langle (\delta N)^2 \right\rangle}{\langle N \rangle} \right]^{-1} \cdot D_j, \tag{8}$$

where $\mu$ is the chemical potential, $\theta$ the coverage and $D_j$ a "jump diffusion coefficient," which is given by

$$D_j = \lim_{t \to \infty} \frac{1}{4t} \int dt \left\langle \sum_{i=1}^{N} v_i(0) \sum_{j=1}^{N} v_j(t) \right\rangle. \tag{9}$$

$\langle (\delta N)^2 \rangle$ is the mean square number fluctuation in an area $A$ containing $\langle N \rangle$ particles on the average. $[\partial (\mu / k_B T) / \partial \ln \theta]_T$ is the well-known thermodynamic factor.

## 3. Influence of Phase Transitions on Surface Diffusion

In this section we focus on recent investigations of surface diffusion processes in the vicinity of phase transitions and critical points. In essence, we will cover the last decade. Earlier work, however, has been reviewed in [30] and will not be discussed here. This topic is of general interest, since at a phase transition the properties of a two-dimensional system changes from one phase to another abruptly. In the case of a first-order phase transition, the density (or the magnetization) changes abruptly, while in the case of a continuous or second-order phase transition, the symmetry of the phases changes. Most of the studies to be discussed below deal with surface diffusion in the vicinity of second-order phase transitions and critical points (Sec. 3.1). Unfortunately, only a few studies have addressed the interesting issue of how a first-order phase transition influences surface diffusion. Some of those studies will be discussed in Sec. 3.2.

### 3.1. INFLUENCE OF SECOND-ORDER PHASE TRANSITIONS

In general, the determination of the chemical surface diffusion coefficient requires the solution of a kinetic equation for the system of many strongly interacting particles. However, if interactions are restricted to affect only the ground state and not the activated state of the diffusing adparticles, the problem can be reduced to the calculation of purely thermodynamic quantities [29,31–34]. In this case, the problem of determining the adparticle diffusion coefficient reduces to the calculation of the free energy of the system and the critical behavior of the diffusion coefficient is fully determined by the divergence of the mean squares adparticle density fluctuation. In recent years this problem was attacked using almost all theoretical methods applicable to critical phenomena: mean-field approximation [35], Bethe-Peierls approximation [35], real-space renormalization group [36], transfer-matrix technique, and Monte-Carlo method [37]. Scaling arguments gives an ordinary logarithmic critical slowdown of the diffusion coefficient [38]

$$D \propto 1/\left|\ln|\theta - \theta_c|\right| . \tag{10}$$

The same critical slowdown is found in Ref. [39], where an exact expression for the diffusion coefficient on reconstructive surfaces has been obtained. It is probably interesting to note that this expression (Eq. 5 of Ref. [39]) is the only exact expression for the critical behavior of the diffusion coefficient of interacting particles in two dimensions (exact expressions have been also obtained for 1D Ising chains [33, 40], where no critical slowdown of the diffusion coefficient exists). A general expression for the diffusion coefficient was analyzed in [37] for the hard hexagon model, yielding a power-law critical slowdown of $D$. Classical mean field type approximations predict a stepwise decrease of the diffusion coefficient n the ordered phase (see, for example, Refs. [35]) and it is obviously clear that mean field approximations are too crude even for a qualitative analysis.

#### 3.1.1. *Chemical diffusion on square and rectangular lattices*

At equilibrium (i.e., in the absence of macroscopic particle density gradients), $D_t$ and $D_c$ often strongly depend both on temperature and coverage due to the potential barrier for diffusion and lateral adsorbate-adsorbate interactions [22,30,41]. A few years ago, several studies were aimed to enhance our understanding of collective adsorbate diffusion in the vicinity of second order or continuous phase transitions on square lattices. The emphasis of these studies was on the pecularities in the coverage dependence of $D_c$ near the critical points [37, 38]. Results of Monte Carlo (MC) simulations for a square lattice with repulsive interactions between nearest neighbor particles indicate that the phase transition induced effects on $D$ are localized in

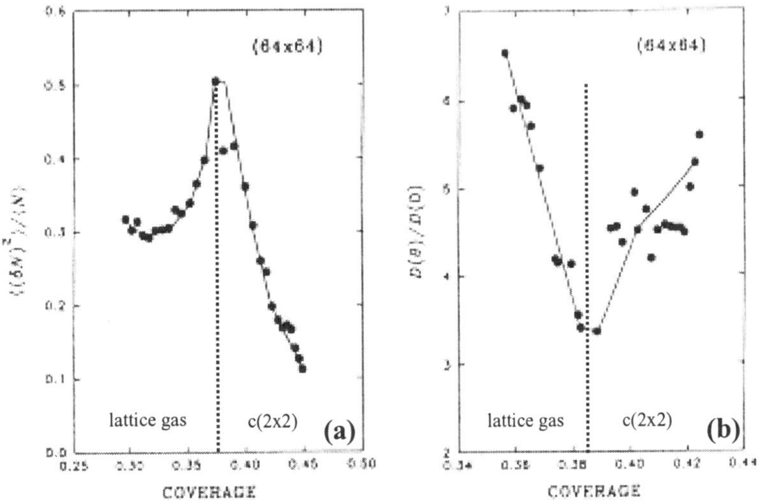


*Figure 1.* (a) Mean square number fluctuation $\langle(\delta N)^2\rangle / \langle N \rangle$ as a function of coverage for square lattices of 64 × 64 sites. Filled circles denote MC results. (b) Normalized chemical diffusion coefficient $D(\theta)/D(0)$ as a function of coverage for $\varphi_{NN}/k_{\mathrm{B}}T = 5.28$, after [37].

rather narrow regions near critical points [37]. The coverage dependence of the normalized mean square fluctuations (Fig. 1a), obtained at a temperature $\varphi_{NN}/kT = 5.28$ well below the critical temperature $\varphi_{NN}/kT_c = 1.76$, exhibits a narrow maximum centered around $\theta_c = 0.385$. In Ref. [37] it was found that this maximum increases with increasing lattice size $L$. This finding qualitatively agrees with results obtained via transfer-matrix techniques [37]. The coverage dependence of the chemical diffusion coefficient, obtained via the Kubo-Green method, clearly exhibits a minimum at the second-order phase transition (Fig. 1b) [37].

A comparison between normalized chemical diffusion coefficients of repulsively interacting adatoms on a square lattice obtained via RSRG treatment and MC simulations is shown in Fig. 2. The fit between RSRG and MC data is surprisingly good for the entire range of temperatures and coverages. Even for low temperatures, $T < T_c$, the deviations between data obtained by these quite different methods are very small. Only in the vicinity of the critical point at half coverage there are visible discrepancies.

Employing the scaling arguments, it has been shown that in the vicinity of continuous surface phase transitions, the temperature-dependence of $D_c$ has power-law or inverse logarithmic singularities, $D_c \propto |\Delta T|^{\alpha/(1-\alpha)}$

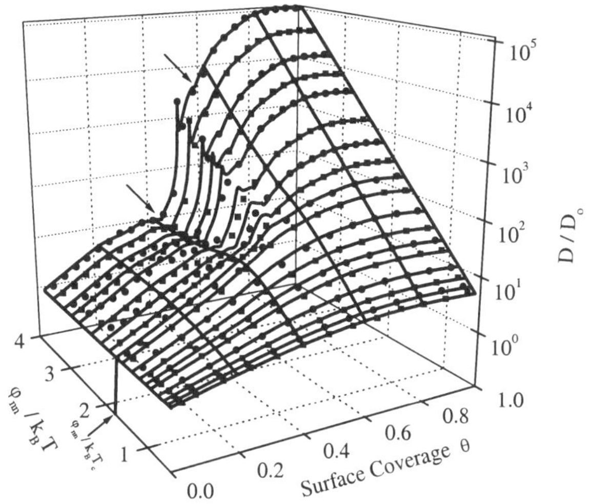


*Figure 2.* 3D plot of the normalized chemical diffusion coefficient $\ln D/D_0$ vs. $\theta$ and $|\varphi_{nn}|/k_BT$, after [42]. Note the logarithmic axis. Solid lines represent RSRG results, symbols denote MC results. Arrows indicate the minima of $D$ at the second-order phase transitions below $T_c$.

or $\propto 1/|\ln|\Delta T||$ ($\alpha > 0$ is the specific heat exponent). Very recently such anomalies have also been seen in the temperature-dependence of $D$ (but not for $D_t$) for oxygen adsorbed on W(110) (Fig. 3) [43]. Monte Carlo simulations indicate that these singularities (i) occur in a very narrow temperature interval near the continuous phase transition between disordered lattice gas and ordered structures, (ii) can be reproduced only if the lattice size is large ($L > 500$) and (iii) are unequivocally attributable to the well known anomalies of the thermodynamic factor in the vicinity of continuous phase transitions [43]. Outside the critical region, the temperature-dependence of $D_c$ is regular and the deviations from the ideal Arrhenius behavior are relatively weak.

The weak anomaly in the temperature-dependence of $D$ just near $T_c$ seen for large lattice sizes only, can be explained (at least in part) by the temperature-dependence of the normalized mean squares fluctuations $\left\langle (\delta N)^2 \right\rangle / \langle N \rangle$ (Fig. 4). The behavior of this quantity is smooth for smaller lattices ($L = 64$ and 128) but shows a weak anomaly, i.e., a tiny maximum, for large lattices (see the insert in Fig. 4 for $L = 512$).

The temperature-dependence of the coefficients of chemical and tracer diffusion of oxygen atoms adsorbed on the W(110) surface has recently also

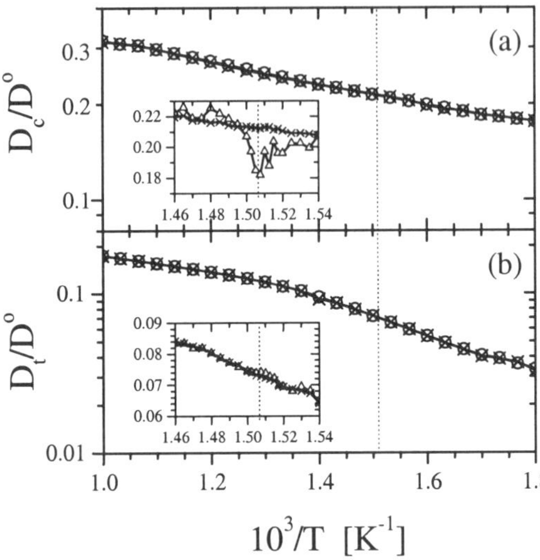


*Figure 3.* Arrhenius plots for the normalized chemical (a) and tracer diffusion coefficients (b), $D_c$ and $D_t$ (or $D^*$, as already mentioned), for oxygen diffusion on W(110), after [43]. The main panels show the results for the $64 \times 64$ (○) and $128 \times 128$ (×) lattices. The inserts exhibit the data for $128 \times 128$ (×) and $512 \times 512$ (Δ) lattices. The dotted lines indicate $T_c$. The simulations are performed for $\theta = 0.45$ by using $\varphi_{\mathrm{N}} = -8.8$, $\varphi_{2\mathrm{N}} = \varphi_{3\mathrm{N}} = 7.1$ and $\varphi_{4\mathrm{N}} = -2.3$ kJ/mol as adsorbate-adsorbate interactions in the ground states. The lateral interaction in the activated state is neglected ($\varphi_N^* = 0$).

been simulated at $\theta = 0.45$ by Vattulainen et al. [44] These authors have obtained a wide maximum of the corresponding activation energies $E_a$ centered at $T = T_c$ both for $D_c$ and $D_t$ and attributed it to peculiarities of the average jump rate $\Gamma$ in the vicinity of the order-disorder phase transition. The results presented above are qualitatively different compared to the findings by Vattulainen et al., as discussed in detail in Ref. [45]. A more detailed discussion of this controversy is presented in Ref. [43]

### 3.2. INFLUENCE OF FIRST-ORDER PHASE TRANSITIONS

Only a few attempts have been undertaken to investigate the influence of phase transitions of first-order on surface diffusion. In Ref. [46] a lattice gas on a square lattice with nearest-neighbor attractive ($\varphi_{\mathrm{NN}} > 0$) and next nearest-neighbor repulsive interactions ($\varphi_{\mathrm{NNN}} < 0$) has been investigated in detail. This model shows first and second-order phase transitions at low temperatures, depending on the ratio $R \equiv \varphi_{\mathrm{NNN}}/\varphi_{\mathrm{NN}} \leq 0$.

Figure 5 shows typical adsorption isotherms $\theta = f(\mu)$ for equilibrated

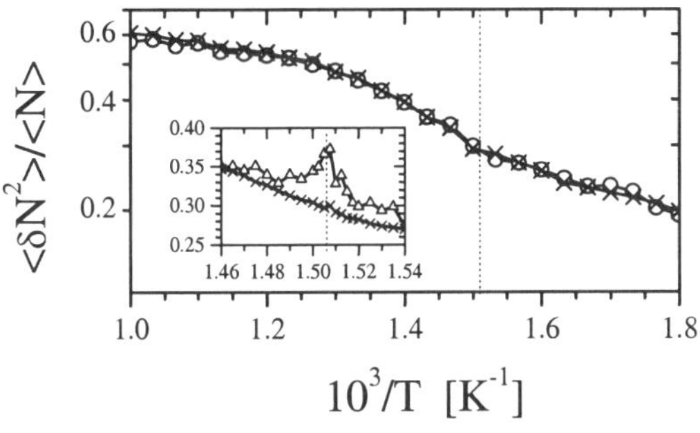


*Figure 4.* Arrhenius plot for the normalized mean squares fluctuations $\left\langle (\delta N)^2 \right\rangle / \langle N \rangle$ for oxygen diffusion on W(110), after [43]. The main panels show the results for the $64 \times 64$ (o) and $128 \times 128$ ($\times$) lattices. The inserts exhibit the data for $128 \times 128$ ($\times$) and $512 \times 512$ ($\Delta$) lattices. The dotted lines indicates $T_c$. The simulations are performed at $\theta = 0.45$ for $\varphi_{\mathrm{N}} = -8.8$, $\varphi_{2\mathrm{N}} = \varphi_{3\mathrm{N}} = 7.1$, and $\varphi_{4\mathrm{N}} = -2.3$ kJ/mol.

lattices gases with nearest-neighbor attraction and next nearest-neighbor repulsion. For $R = -2$ the isotherms show a wide plateau around $\theta = 0.5$ which corresponds to the p$(2 \times 1)$ lattice gas phase [26], which is separated from disordered lattice gas by second-order phase transitions. Upon increasing the temperature, the plateau disappears above the critical temperature $T_c$. It is $\varphi_{\mathrm{NN}}/k_{\mathrm{B}}T_c = 0.90$ [26]. With decreasing next nearest-neighbor repulsions $\varphi_{\mathrm{NNN}}$ the width of the plateau shrinks and finally disappears for the case $R = -0.5$. Second-order phase transitions are still present for $R = -1$, but disappear for $R = -0.5$. Obviously the p$(2 \times 1)$ ordered lattice gas phase is stabilized by the next nearest-neighbor repulsive interactions. For $R = 0$ and $\varphi_{\mathrm{NN}}/k_{\mathrm{B}}T = 2.40$ and also for $R = -0.5$ and $\varphi_{\mathrm{NN}}/k_{\mathrm{B}}T = 4.01$, almost vertical jumps of the surface coverage $\theta$ are observed, which are attributed to first-order phase transitions between low density lattice gas (LG) and high density lattice fluid (LF).

Figure 6a shows the normalized chemical diffusion coefficient obtained via Kubo-Green ($D_{KG}$) and fluctuation method ($D_F$) vs $\varphi_{\mathrm{NN}}/k_{\mathrm{B}}T$ at $\theta = 0.5$. The $T$ range goes from that corresponding to disordered lattice gas deep into the p$(2 \times 1)$ ordered region for $R = -2$, $R = -1$. For $R = 0$ the lowest temperature studied lies within the miscibility gap of the phase diagram. The data were obtained in the canonical ensemble after full equilibration. Here we do not intend to discuss the general temperature dependence of the diffusion coefficient, which behaves as intuitively expected (see for instance Ref. [46]). More interesting than this is the comparison of Kubo-Green and fluctuation results. For $R = -2$ and $R = -1$ $D_{KG}$ and $D_F$ agree quantita-

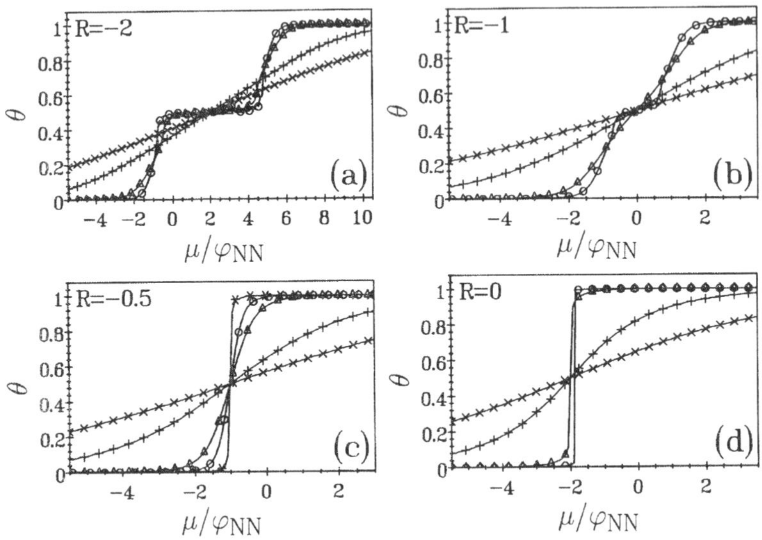


*Figure 5.* Adsorption isotherm $\theta(\mu/\varphi_{\text{NN}})$ for the lattice gas model with various values of $R \equiv \varphi_{\text{NNN}}/\varphi_{\text{NN}}$ as indicated: ($*$): $\varphi_{\text{NN}}/k_{\text{B}}T = 4.01$; ($\circ$): $\varphi_{\text{NN}}/k_{\text{B}}T = 2.40$; ($\triangle$): $\varphi_{\text{NN}}/k_{\text{B}}T = 1.60$; ($+$): $\varphi_{\text{NN}}/k_{\text{B}}T = 0.48$; ($\times$): $\varphi_{\text{NN}}/k_{\text{B}}T = 0.24$. For the calculations a $64 \times 64$ lattice was used. Equilibration was carried out in the grand canonical ensemble, after [46].

tively over the entire $T$ range. This is true for $R = -0.5$ and $R = 0$ only at relatively high temperatures. At lower temperatures there are significant deviations between both quantities, $D_F \ll D_{KG}$. This striking finding is completely unexpected, since it has been shown in [26] that both methods give identical results for lattice gases with various sets of interaction energies. The reasons for these discrepancies are not fully understood. However, it is very likely that these discrepancies are related to the first-order phase transition at low temperatures at $R = -0.5$ and $R = 0$ well above $T_c$. Simulations for $256 \times 256$ lattices indicate that the observed deviations are not due to finite size effects.

The deviations between $D_F$ and $D_{KG}$ persist at temperatures well above the first-order phase transitions between LG and LF at $R = -0.5$ and $R = 0$. Characteristic snapshots such as those shown in Ref. [46] indicate that even at such high temperatures the lattice gas starts to decompose into tiny domains of low and high density, respectively. It is intuitively obvious that even in a (thermodynamically) disordered region of the phase diagram, i.e., corresponding to disordered lattice gas, homogeneity at a given instant

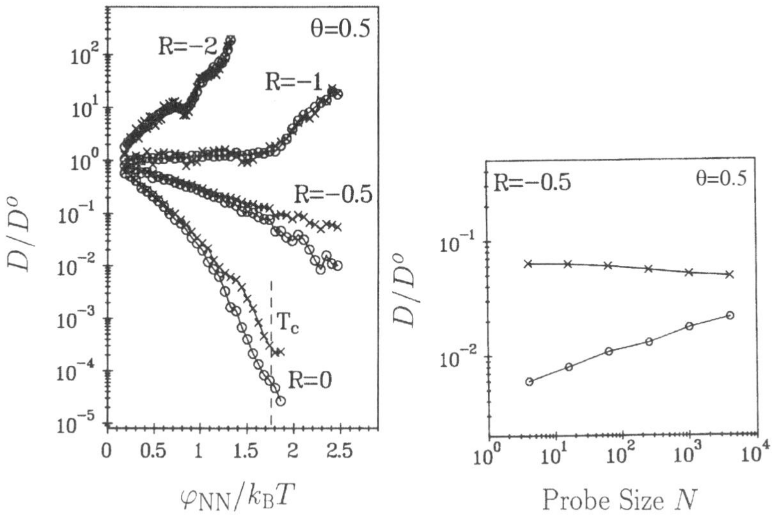


*Figure 6.* (a) Temperature dependence of $D_F$ (∘) and $D_{KG}$ (×) at $\theta = 0.5$ for various values of $R$ as indicated, after [46]. For the calculations 64 × 64 lattices were used. For $R = 0$ the critical temperature is given by $\varphi_{\rm NN}/k_{\rm B}T_c = 1.76$ (dashed line). (b) $D_F$ (∘) and $D_{KG}$ (×) vs size of the probed region at $\theta = 0.5$ and $\varphi_{\rm NN}/k_{\rm B}T = 2.40$ for the lattice gas model with $R = -0.5$, after [46]. For the calculations 256 × 256 lattices were used.

exists only on a length scale sufficiently larger than remanent correlation lengths. Therefore, some of the diffusion runs were repeated for lattices with 256 × 256 sites and probe sizes of up to 64 × 64 sites [46]. The results of these calculations are shown in Fig. 6b. It is obvious that both $D_F$ and $D_{KG}$ depend slightly on the probe size in such a way that the deviations between both quantities decrease with increasing probe size. However, even for probes of 64 × 64 sites the deviations are still measurable. Therefore, it can be concluded that the results of diffusion measurements may depend on the length scale of the measuring technique. The effect of the length scale is well pronounced in systems with substantial remanent "graininess," i.e. in the vicinity of a first-order phase transition [46].

## 4. Conclusions

The simulation of surface diffusion via Monte Carlo and molecular dynamics methods is a rapidly progressing field of research. It is evident that the Monte Carlo method has significantly contributed to the understanding of surface diffusion, especially in systems with phase transitions. Additional studies are highly desirable, for instance, to improve our understanding of surface diffusion in systems with first-order phase transitions.

## Acknowledgments

This work has been supported by the project No LN00A015 of the Czech Academy of Sciences, CONICET (Argentina), the National Agency for Promotion of Science and Technology (APCyT, Argentina), Prog. No C-03-03232, and by the DFG (Ue 34/4-1).

## References

1. V.P. Zhdanov, *Surf. Sci.* **214**, 289 (1989).
2. S.J. Lombardo and A.T. Bell, *Surf. Sci. Rep.* **13**, 1 (1991).
3. G. Wahnstrom and V.P. Zhdanov, *Surf. Sci.* **247**, 24 (1991).
4. K.D. Dobbs and D.J. Doren, *J. Chem. Phys.* **97**, 3722 (1992).
5. D. Cohen and Y. Zeiri, *J. Chem. Phys.* **97**, 1531 (1992).
6. L. Utrera and R. Ramirez, *J. Chem. Phys.* **96**, 7838 (1992).
7. D.E. Sanders and A.E. DePristo, *Surf. Sci.* **260**, 116 (1992).
8. R.M. Lynden-Bell, *Surf. Sci.* **259**, 129 (1991).
9. T.R. Mattsson, U. Engberg, and G. Wahnstrom, *Phys. Rev. Let.* **71**, 2615 (1993).
10. S. Pai and D. Dore, *Surf. Sci.* **291**, 185 (1993).
11. P. Blandin and C. Massobrio, *Surf. Sci.* **279**, L219 (1992).
12. J. Ellis and J.P. Toennies, *Surf. Sci.* **317**, 99 (1994).
13. F. de Jong and A.P.J. Jansen, *Surf. Sci.* **317**, 1 (1994).
14. P.A. Gravil and S. Holloway, *Surf. Sci.* **310**, 267 (1994).
15. Lichang Wang, Qingfeng Ge, and G.D. Billing, *Surf. Sci.* **304**, L413 (1994).
16. K.D. Dobbs and D.J. Doren, *J. Chem. Phys.* **99**, 10041 (1993).
17. I.I. Suni and E.G. Seebauer, *Surf. Sci.* **301**, L235 (1994).
18. G.A. Evangelakis and N.I. Papanicolaou, *Surf. Sci.* **347**, 376 (1996).
19. G. Boisvert and L.J. Lewis, *Phys. Rev. B* **54**, 2880 (1996).
20. A.F. Voter, *J. Chem. Phys.* **106**, 4665 (1997).
21. J.L. Riccardo and W.A. Steele, *J. Chem. Phys.* **105**, 9674 (1996).
22. C. Uebing and R. Gomer, J. Chem. Phys. **95**, 7626, 7636, 7641, 7648 (1991).
23. M. Tringides and R. Gomer, *Surf. Sci.* **166**, 419 (1986).
24. K. Kehr and K. Binder, in *Applications of the Monte Carlo Method in Statistical Physics*, Vol. 36 of *Topics in Current Physics*, 2 ed., edited by K. Binder (Springer-Verlag, Berlin, 1987), p. 181.
25. M. Tringides and R. Gomer, *Surf. Sci.* **145**, 121 (1984).
26. C. Uebing and R. Gomer, *J. Chem. Phys.* **100**, 7759 (1994).
27. R. Gomer, *Surf. Sci.* **38**, 373 (1973).
28. G. Mazenko, J.R. Banavar, and R. Gomer, *Surf. Sci.* **107**, 459 (1981).
29. D.A. Reed and G. Ehrlich, *Surf. Sci.* **105**, 603 (1981).
30. R. Gomer, *Rep. Prog. Phys.* **53**, 917 (1990).
31. M. Bowker and D.A. King, *Surf. Sci.* **71**, 583 (1978).
32. D.A. Reed and G. Ehrlich, *Surf. Sci.* **102**, 588 (1981).
33. A.A. Chumak and A.A. Tarasenko, *Surf. Sci.* **91**, 694 (1980).
34. A.A. Tarasenko, P.M. Tomchuk, and A.A. Chumak, *Fluctuations in the bulk and on the surface of solids* (Naukova dumka, Kijv, 1992), (in Russian).
35. A.A.Tarasenko and A.A.Chumak, *Fiz. Tverd. Tela (Leningrad)* **24**, 2972 (1982); **22**, 2939 (1980), [Sov. Phys. Solid State **24**, 1683 (1982); **22**, 1716 (1980)].
36. A.A. Tarasenko and A.A. Chumak, *Poverkhnost' Fiyika, Khimija, Mekhanika* **12**,

37 (1991); **11**, 98 (1989), (in Russian).
37. A.V. Myshlyavtsev, A.P. Stepanov, C. Uebing, and V.P. Zhdanov, *Phys. Rev. B* **52**, 5977 (1995).
38. V.P. Zhdanov, *Phys. Lett. A* **161**, 556 (1992).
39. A.A. Chumak and A.A. Tarasenko, *Surf. Sci.* **364**, 424 (1996).
40. W. Zwerger, *Z. Phys. B* **42**, 333 (1981).
41. V.P. Zhdanov, *Elementary Physicochemical Processes on Solid Surfaces* (Plenum, New York, 1991).
42. A. Tarasenko, L. Jastrabík, F. Nieto, and C. Uebing, *Phys. Rev. B* **59**, 8252 (1999).
43. V.P. Zhdanov and C. Uebing, *J. Chem. Phys.* **109**, 3197 (1998).
44. I. Vattulainen, J. Merikoski, T. Ala-Nissila, and S.C. Ying, *Phys. Rev. Lett.* **79**, 257 (1997).
45. C. Uebing and V.P. Zhdanov, *Phys. Rev. Let.* **80**, 5455 (1998).
46. C. Uebing and R. Gomer, *Surf. Sci.* **331-333**, 930 (1995).

# SURFACE HETERODIFFUSION IN ADSORBED AND COADSORBED OVERLAYERS OF Li, Sr, AND Cu ON THE W AND Mo (112) SURFACES

A.T. LOBURETS and N.B. SENENKO
*Yu.V. Kondratyuk State Technical University, UA-63601, Poltava, Ukraine*
Yu.S. VEDULA and A.G. NAUMOVETS
*Institute of Physics, Natl. Acad. Sci. of Ukraine, UA-03028, Kiev, Ukraine*

**Abstract.** Results of systematic comparative investigations of surface diffusion for a number of adsorbates– Li, Sr, and Cu– on the W and Mo (112) surfaces are presented and discussed. The diffusion parameters have been extracted from experimental data on time evolution of coverage profiles shaped initially as coverage steps. The results reveal, on the one hand, a considerable sensitivity of the diffusion kinetics to specifics of interactions in the overlayer, and, on the other hand, a number of common regularities reflecting an interplay between the diffusion and phase transitions in the spreading layer. Some data are also presented on the diffusion in coadsorbed overlayers.

## 1. Introduction

Surface diffusion (SD) occurring in a concentration (coverage) gradient – the chemical diffusion, or heterodiffusion – is a wide-spread phenomenon that attracts an abiding basic and practical interest [1,2]. Physically, this is a non-equilibrium process in a system of particles which in most real situations are bonded not only to a substrate, but also to each other by interactions whose magnitudes can vary within broad limits [3–5]. For this reason the diffusion is generally accompanied by phase transitions in the spreading overlayer, and the diffusion zone (the region of varying concentration) represents a dynamical phase portrait of the overlayer [6]. This makes such a surface diffusion an intriguing process with rich physics. In nature and applications, surface heterodiffusion determines the kinetics of many important processes and is, in particular, a serious factor that can affect the service life of various small-sized devices.

In spite of a considerable effort mounted recently to investigate surface heterodiffusion, the experimental basis available for its physical understanding remains yet rather scarce. The experiment in this field is laborious and imposes heavy demands on cleanliness of vacuum conditions, surface perfection, spatial and coverage resolution. The existence of some quantity of uncontrollable defects on macrocrystal surfaces is practically unavoidable under present-day technologies of surface preparation. All this substantially complicates the comparison of results obtained in different experiments and by different techniques.



*M.C. Tringides and Z. Chvoj(eds.),*
*Collective Diffusion on Surfaces: Correlation Effects and Adatom Interactions,* 97–106.

The main aim of this work was to compare, on the one hand, the diffusivity of the same adsorbates on different substrates having almost identical atomic structures and, on the other hand, the diffusivity of different adsorbates on the same substrate. The W(112) and Mo(112) surfaces were chosen as substrates. Both the metals have a b.c.c. structure and their lattice constants differ by a mere 0.6%. The (112) b.c.c. surfaces are known to possess a highly anisotropic ("channeled") corrugation. As a result, most adsorbates diffuse at not too high temperatures mainly along the channels built of closely packed rows of atoms [see Ref. 7 and references therein]. The diffusates under study were Li, Sr, and Cu representing metals of three different groups of the periodic table. The experimental data for some systems used here for comparison were obtained in our earlier works, while the data for other systems are presented here for the first time. We have also investigated surface diffusion in some mixed (coadsorbed) overlayers and discuss the results in this paper.

## 2. Experimental

To record the diffusion evolution of the coverage profiles, we have used a technique which can be called scanning contact potential microscopy (SCPM) [1,8]. Briefly, it consists of measuring the work function distribution over the surface under study by scanning it with a narrow (1 to 15 $\mu$m in diameter) electron beam. The beam serves to locally determine the contact potential difference between the electron gun cathode and the sample (anode) with the aid of Anderson's method, which is based on recording the position of the retarding-field section of the current-voltage curve. In a separate experiment, one carefully calibrates the dependence of the work function on the adsorbate coverage, which then allows conversion of the work function distribution over the surface into the coverage distribution termed also *coverage profile*. The initial coverage profile is usually set up as a step with the edge directed across the atomic channels, and its evolution with annealing time is evaluated on the basis of the Boltzmann–Matano equation to elicit the adsorbate diffusivity as a function of coverage. Throughout this paper, the coverage $\theta$ is given as the ratio of the adatom concentration to the concentration of the substrate surface atoms. The coverage at $\theta$=1 is defined as one monolayer. The structure of the overlayers was investigated in separate works using LEED. All the experimental results presented below relate to adsorbate diffusion along the atomic channels, i.e., in the [111] direction. The rest of the details relating to experiments and evaluation procedure can be found in our previous works [1,8,9].

## 3. Results and discussion

### 3.1. THE SAME ADSORBATE ON DIFFERENT SUBSTRATES: Li, Sr and Cu ON W AND Mo (112)

In this section, we compare the diffusion kinetics of the same adsorbate (Li, Sr, Cu) on two substrates – W and Mo(112), which differ by their chemical nature, but have almost identical atomic structures.

3.1.1. *Li on W and Mo(112)*
Figure 1 correlates the coverage dependences of Li diffusivity on W and Mo(112). The data for Li on Mo(112) are taken from our earlier work [10]. So far, the Li diffusivity on W(112) has been investigated by us at $\theta \leq 0.9$ only.

Let us first consider the results relating to coverages $\theta \leq 0.5$. This $\theta$ region is characteristic of the formation of Li chains oriented across the atomic channels on W and Mo(112) [3,11,12]. The diffusing species (mass carriers) are believed to be just the chains or, at least, some chain sections. Thus, the diffusion process has an essentially collective character, the random walks of the chains being effected through their meandering [13]. For Li on Mo(112), the diffusivity is nearly independent of coverage at $0 < \theta < 0.45$. For Li on W(112), there is a shallow diffusivity minimum in the range $0.2 < \theta < 0.4$, and a slightly enhanced diffusivity is observed at $\theta < 0.15$. A possible reason for this may be a higher mobility of individual Li adatoms or rather short chain sections (Li oligomers) compared to that of long Li chains formed at higher coverages.

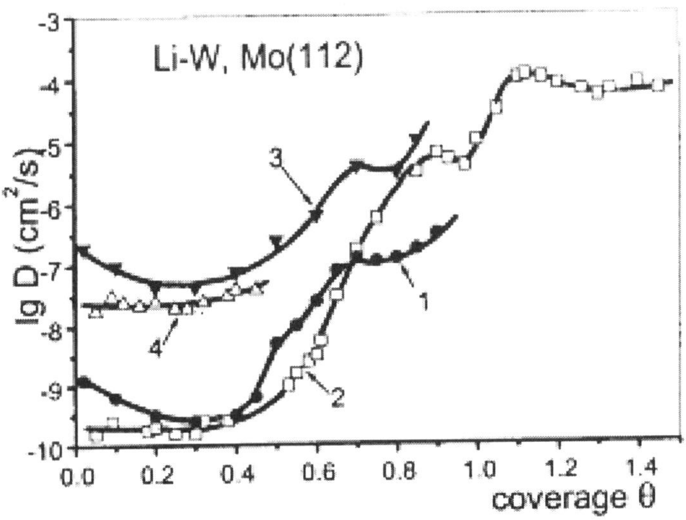


*Figure 1.* Coverage dependences of Li diffusivity on the (112) surfaces of W and Mo. 1,2– Li on W and Mo, respectively, at T=300 K; 3,4– the same at T=375 K.

At $0.5 < \theta < 0.9$, a pronounced growth of Li diffusivity by a few orders of magnitude occurs on both Mo and W(112). This effect can be ascribed to the progressing compression of Li overlayers along the channels, which are observed at such coverages, and to the beginning of commensurate-incommensurate transition [3,11,12]. This should give rise to the appearance of incommensurate domain walls between the commensurate domains. The walls behave as topological solitons, predicted to have an enhanced mobility ensuring a high diffusion rate [13].

On the whole, one can state from Fig. 1 that Li diffusivity on W and Mo(112) is rather close (at the same temperatures) in the coverage interval $0.1 < \theta < 0.4$. A specificity for the chemical nature of the substrate manifests itself more markedly at $\theta < 0.15$ and $\theta > 0.45$. At $\theta > 0.45$, the coverage dependences of Li diffusivity on W and Mo(112) are qualitatively similar, but the absolute values of the diffusivity differ at some coverages by about an order of magnitude. To provide an explanation for this specificity, it would be necessary to carry out calculations of the diffusion parameters taking into account peculiarities of the potential corrugation and lateral interactions on either substrate. Both the corrugation and lateral interactions should be dependent on the surface electronic structure.

3.1.2. *Sr on W and Mo(112)*

The diffusivities of Sr on W and Mo(112) at T=600 K are compared in Fig. 2. The overall result is that Sr on Mo(112) diffuses somewhat faster (by a factor of $\approx 1.5$ to 10, depending on coverage) than on W(112). Taking into account that the accuracy of temperature determination is estimated at $\pm 5$ K, the *D* values obtained on both

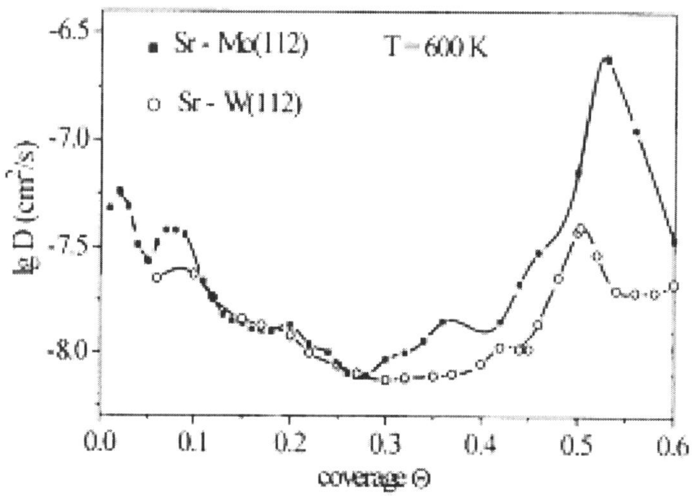


*Figure 2.* Coverage dependences of Sr diffusivity on the (112) surfaces of W and Mo at T=600 K.

substrates at coverages $\theta \leq 0.3$ can be considered as rather close to each other. The coverage dependences of *D* on W and Mo are also very similar: (i) both the curves show an increase in Sr diffusivity at $\theta \leq 0.1$, which can be attributed to a higher mobility of single Sr adatoms and shorter sections of Sr chains (clusters) [6,9]; (ii) there is a region of low diffusivity at medium coverages ($0.15 < \theta < 0.4$) interpreted as a consequence of the first-order phase transition in Sr overlayer [14] causing the necessity of detachment of diffusing species from islands of the denser phase (in this case the overlayer is heterogeneous and the macroscopic diffusivity represents actually an averaged parameter

[15]); (iii) a pronounced $D$ maximum is seen at $\theta \geq 0.5$, which is ascribed to the operation of the soliton diffusion mechanism.

A common feature for Sr diffusion on W and Mo(112) is also a strong reduction of the difference between the Sr adsorption energy and the diffusion activation energy at $\theta \rightarrow 1$. This results in desorption of Sr atoms after passing a rather short diffusion path on the surface (see Sec. 3.2).

### 3.1.3. *Cu on W and Mo(112)*

Both Li and Sr are strongly electropositive elements with respect to W and Mo, which results in pronounced effects of dipole-dipole repulsive interaction in their overlayers. By contrast, the polarity of the adsorption bond in Cu overlayers on W and Mo is quite weak, and the dominating lateral interaction in them is attractive. Due to this, the Cu monolayers are generally filled by way of successive first-order phase transitions. Cu structures have been investigated by LEED on W(112) only [16], but the strong resemblance of the work function and diffusion data for Cu on W and Mo(112) suggests that the structures should be similar on both the substrates [17].

Cu diffusivities on W and Mo(112) determined at T=600 K are plotted in Fig. 3 as functions of Cu coverage. The data have been obtained up to three monolayers. It must be remembered that according to the $\theta$ definition accepted above the first Cu monolayer just fills the atomic channels on W and Mo(112). A complete screening of the substrate by Cu adatoms is not achieved until a double layer ($\theta$=2) is formed. As follows from Fig. 3, the variation of Cu diffusivity in the interval $0 < \theta < 3$ is highly nonmonotonic and spans about three orders of magnitude. There are sharp peaks of $D$ tied to coverages close to $\theta \rightarrow 0$, $\theta \approx 1$ and $\theta \approx 2$. Most probably the $D$ maxima are due to enhanced mobility in the two-dimensional gas phases, consisting of individual Cu adatoms and small clusters, formed at the initial stage of filling each new monolayer. The changes in the adsorption energy, which can be rather sharp in the transition from a monolayer to the following one, can also contribute to an enhanced value of the thermodynamic factor in the diffusivity [6]. At least in the second Cu monolayer on W(112), the final stage of its packing was evidenced to proceed via formation of an incommensurate structure [16]. This structure appears because the misfit between the atomic radii of Cu and W amounts to $\approx 7\%$. One cannot rule out that, at least for the $D$ peak at $\theta \approx 2$, a contribution to the high diffusion rate may also stem from the soliton diffusion mechanism in this incommensurate phase. It is interesting to note that each successive diffusivity peak increases in height by one to two orders of magnitude with respect to the preceding one. This seems to be caused by the change in the chemical composition of the atomic channels in which copper adatoms are transported.

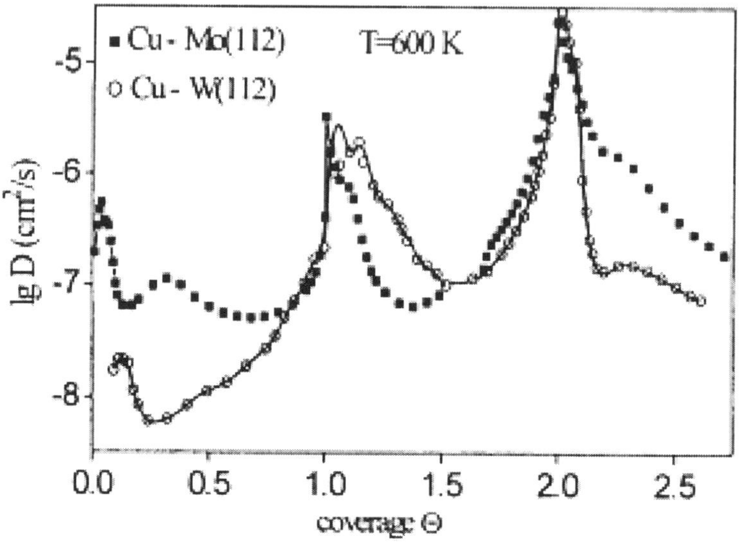


*Figure 3.* Coverage dependences of Cu diffusivity on (112) surfaces of Mo and W at T=600 K.

Within the limits of the first monolayer, the channels are comprised of W (or Mo) atoms only. At $1<\theta<2$, the bottom of the channels still consists of W or Mo atoms, but their sides are built of Cu atoms. Finally, at $\theta>2$ the channels consist of Cu atoms only.

The data depicted in Fig. 3 show that regularities of Cu diffusion on W and Mo(112) are generally similar, although there are noticeable and rather intricate quantitative distinctions in the diffusivities. In particular, the diffusivity on W(112) at $\theta=0.2\div0.5$ is by an order of magnitude *lower* than on Mo(112). In contrast, the diffusivity on W(112) at $\theta=1.2\div1.3$ is *higher* by an order of magnitude. Obviously, the reasons for these distinctions must lie in specificity of adatom interactions on W and Mo, which entail differences in the surface potential corrugation and in the structure of 2D phases (including properties of clusters formed in 2D gas phases). These points need further, more detailed investigations.

## 3.2. DIFFERENT ADSORBATES ON THE SAME SUBSTRATE

The results presented in Figs. 1–3 also allow one to compare the diffusion kinetics of Li, Sr, and Cu on the same substrates. For reasons of space we cannot give here the diffusion activation energies and pre-exponential factors for different systems. So let us just compare the diffusion temperatures corresponding to approximately equal diffusivity values and, on the other hand, the diffusivities of different adsorbates at the same temperature. One can see that the diffusivities, generally, decrease in the order $D_{Li}>D_{Cu}>D_{Sr}$. This change occurs in the opposite direction with respect to the sequence of the adsorption energies of Li, Cu and Sr (taken on W(112) at $\theta\to0$): $q_{Li}$=2.8 eV [18],

$q_{Cu}$=3.3 eV (our data), and $q_{Sr}$=3.6 eV [14]. At appropriate temperatures, one can easily observe the diffusion of Li and Cu within the second and third monolayers over macroscopic distances (up to millimeters in our experiments). This is, however, practically unattainable for Sr within realistic times, since the gap between its adsorption energy, $q$, and its diffusion activation energy, $E_d$, strongly narrows at $\theta \geq 0.7$. The same behavior was found for other alkaline-earth and rare-earth adsorbates (Ba, La, Dy) [7]. It should be reminded that the root-mean-squares diffusion path of a particle until desorption depends on $(q-E_d)$ in the following way: $(<x^2>)^{1/2} \propto \exp[(q-E_d)/2kT]$. Thus, a common feature is that alkaline-earth and rare-earth elements have a low spreading ability at $\theta \geq 1$.

## 3.3. DIFFUSION IN COADSORBED SYSTEMS

Diffusion in coadsorbed systems is of considerable interest in many aspects, both basic and practical. In particular, catalysis is a field where this case of diffusion is of primary concern. However, the currently available evidence on interdependence between diffusivities of coadsorbed species is quite scarce. In this work, we have studied on the W(112) surface: (i) the effect of preadsorbed Sr on Li diffusion; (ii) the effect of preadsorbed Cu on Sr diffusion. The experiments were carried out in such a way that first the *whole* substrate surface had been precovered with a *uniform* layer of Sr (or Cu) and then *a step-like coverage profile* of Li (or Sr) was created on such a surface. The evolution of these profiles gave information on the diffusion kinetics of Li and Sr in the presence of coadsorbates (Sr and Cu, respectively).

Experimental data about Li diffusion on Sr/W(112) are depicted in Fig. 4. A conspicuous effect is the strong reduction of Li diffusivity by preadsorbed Sr manifesting itself even when Sr coverage is as small as 0.02 or 0.045 ML. For example, at $\theta_{Li}$=0.2÷0.4, the decrease of $D_{Li}$ comprises about one order of magnitude at $\theta_{Sr}$=0.02 and two orders of magnitude at $\theta_{Sr}$=0.045. In the interpretation of these results, one should first of all consider that Sr adatoms have a much lower mobility than Li (see Sec. 3.2). Therefore Sr adatoms should generally act as stoppers (inhibitors) in Li diffusion. The second important factor is caused by the channeled structure of b.c.c. (112) surfaces. The result is that preadsorbed Sr atoms very effectively block the channels, which are the transport paths for diffusing Li adatoms. The bypassing of such stoppers within the channels (furrows) demands some additional activation. The third factor that may enhance the inhibiting effect of Sr is the essentially collective character of Li diffusion in a broad interval of coverages. Since the diffusing species are not just individual adatoms but adatom clusters, chains or even more complex formations like domain walls (solitons), one stopper can pin a collective mass carrier which includes a considerable number of adatoms [19].

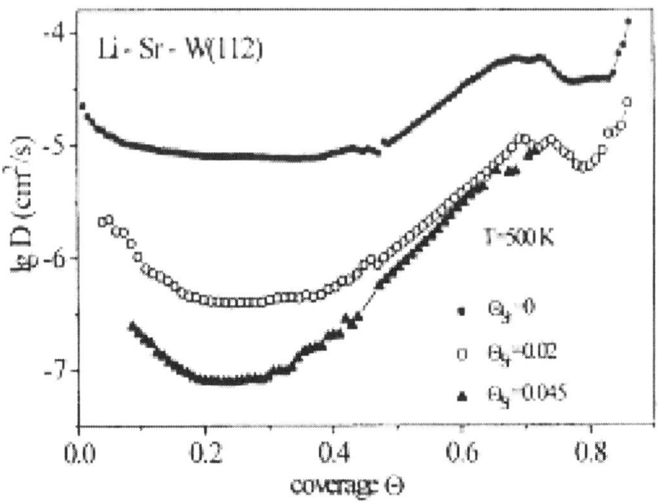


*Figure 4.* Effect of preadsorbed Sr on coverage dependence of Li diffusivity on the W(112) surface. The upper, middle and bottom curve correspond to Sr coverages equal to 0, 0.02, and 0.045, respectively.

Consider now the data for Sr diffusion on Cu/W(112) (Fig. 5). In spite of the fact that Cu diffusivity at $\theta_{Cu}\approx 1$ is considerably *higher* than Sr diffusivity in the broad coverage range $0<\theta_{Sr}<0.4$ (cf. Figs. 2 and 3), the preadsorbed Cu monolayer *decreases* the diffusivity of Sr on W(112). The particular case illustrated in Fig. 5 relates just to $\theta_{Cu}\approx 1$. As pointed out in Sec. 3.1.3, in this situation the bottom of the surface atomic channels consists of W atoms while their sides are built of Cu atoms. The coverage dependence of the diffusivity of Sr on Cu/W(112) is qualitatively similar to that for Sr on W(112), although the range of $D_{Sr}$ variation is somewhat narrower in the former case. Further more detailed investigations of the structure of coadsorbed layers and adatom interactions in them are necessary to convincingly interpret these results. In particular, one cannot rule out the possibility that such diffusion mechanisms as e.g. the atom exchange mechanism can come into play or that the diffusing species (clusters and chains) can actually consist of adatoms of different chemical nature.

## 4. Conclusions

We have performed a series of systematic comparative investigations of surface diffusion for a number of adsorbates – Li, Sr, and Cu – on the W and Mo (112) surfaces.

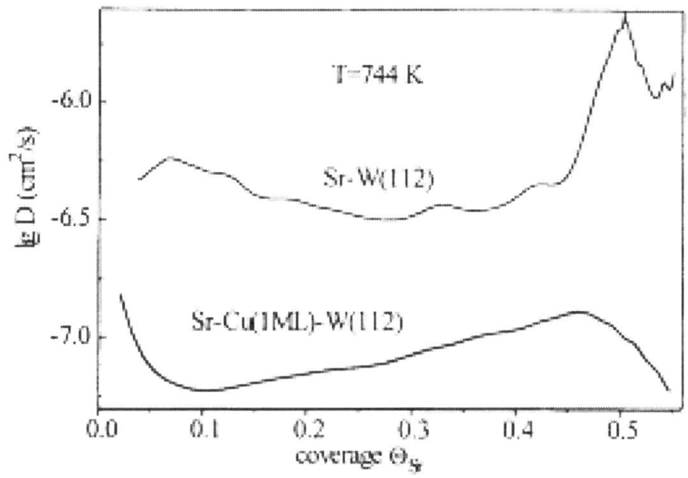


*Figure 5.* Effect of preadsorption of one copper monolayer on coverage dependence of Sr diffusivity on the W(112) surface. The upper curve shows data for Sr on clean W(112). T=744 K.

The comparison of the results shows a high sensitivity of the diffusion kinetics to the chemical nature of the system adsorbate+substrate. At the same time, there are also remarkable common regularities stemming from the correlation between the diffusion rate and phase transitions in the spreading layer (the same regularities were also found for diffusion of rare-earth elements [7]). The collective effects at low coverages originate from formation of adatom clusters which are known to show a striking diversity of diffusion parameters and mechanisms [20–25]. The spatial extension of 2D phases emerging in the diffusion zone and concentration gradient in them are fitted to their "throughput" (diffusivity). Experiments give also increasingly more examples of collective effects in surface diffusion at medium and high coverages. The diffusion in mixed overlayers represents an especially intricate situation whose investigation is actually at its beginning. Some experimental results presented above raise more questions than provide answers and demonstrate once again that the level of our understanding of regularities and mechanisms of surface diffusion is still far from satisfactory. This field, in all its complexity and importance, still remains a challenge in surface science.

## Acknowledgments

The financial support of this work by the Volkswagen-Stiftung (project "Oberflaechendefekte, Ordnung und Diffusion in Adsorbatschichten"), by the Ministry of Education and Science of Ukraine (project # 2.4/776 "Zlagoda") and by the INTAS-Ukraine Program (project 95-0186) is gratefully acknowledged.

## References

1. Naumovets, A.G., and Vedula, Yu. S. (1985) Surface diffusion of adsorbates, *Surf. Sci. Rep.* **4**, 365–434.
2. Gomer, R. Diffusion of adsorbates on metal surfaces, Rep. Prog. Phys. 53 (1990) 917.
3. Braun, O.M. and Medvedev, V.K. (1989) Interactions between particles adsorbed on metal surfaces, *Sov. Phys. Uspekhi* **32**, 328–348.
4. Naumovets, A.G. (1994) Two-dimensional phase transitions in alkali-metal adlayers, in D.A. King and D.P. Woodruff (eds.), *The Chemical Physics of Solid Surfaces, vol. 7,* Elsevier, Amsterdam, pp. 163–213.
5. Einstein, T.L. (1996) Interactions between adsorbate particles, in: *Handbook of Surface Science,* ed. by W.N. Unertl (Elsevier, Amsterdam) p. 577.
6. Loburets, A.T., Naumovets, A.G., and Vedula, Yu. S. (1997) Surface diffusion and phase transitions in atomic overlayers, in M.C. Tringides (ed.), *Surface Diffusion: Atomistic and Collective Processes,* Plenum, New York, pp. 509–528.
7. Loburets, A.T., Naumovets, A.G., and Vedula, Yu. S. (1998) Diffusion of dysprosium on the (112) surface of molybdenum, *Surf. Sci.* **399**, 297–304.
8. Loburets, A.T. (1999) An experimental technique for recording adsorbate surface concentration profiles based on local measurements of the contact potential, *Metallofiz. Noveishie Tekhnol.* **21**, 43–46.
9. Loburets, A.T., Naumovets, A.G., Senenko, N.B., and Vedula, Yu. S. (1997) Surface diffusion and phase transitions in Sr overlayers on W(112), *Zs. Phys. Chemie* **202**, 75–85.
10. Naumovets, A.G., Paliy, M.V., Vedula, Yu. S., Loburets, A.T., and Senenko, N.B. (1995) Diffusion of Li and Sr on Mo(112), *Progr. Surf. Sci.* **48**, 59–70.
11. Gupalo, M.S., Medvedev, V.K., Palyukh, B.M., and Smereka, T.P. (1979) Adsorption of lithium on the (112) face of molybdenum crystal, *Sov.* Phys. *Sold State* **21**, 568–573.
12. Fedorus, A., Kolthoff, D., Koval, V., Lyuksyutov, I., Naumovets, A.G., and Pfnuer, H. (2000) Phase transitions in the adsorption system Li/Mo(112).- *Phys. Rev. B.* **62**, 2852–2861.
13. Lyuksyutov, I.F., Naumovets, A.G., and Pokrovsky, V. (1992) *Two-Dimensional Crystals,* Academic Press, Boston.
14. Medvedev, V.K., and Yakivchuk, A.I. (1975) Structure and electron-adsorption properties of strontium on the (112) surface of tungsten, *Ukrainskii Fiz. Zhurn.* **20**, 1900–1908.
15. Ala-Nissila, T. and Ying, S.C. (1992) Theory of classical surface diffusion, *Progr. Surf. Sci.* **39**, 227–323.
16. Stolzenberg, M. (1992) *Zweidimensionalle Phasenu berga enge in Cu- und Te Adsorbatsystemen,* Dissertation, Physikalisches Institut, TU Clausthal, Kapitel 7.
17. Loburets, A.T. (1999) Surface diffusion and phase transitions in copper overlayers on the (112) surfaces of molybdenum and tungsten, *Metallofiz. Noveishie Tekhnol.* **21**, 47–51.
18. Medvedev, V.K., Naumovets, A.G., and Smereka, T.P. (1973) Lithium adsorption on the (112) face of tungsten, *Surf. Sci.* **34**, 368–384.
19. Vedula, Yu.S., Lyuksyutov, I.F., Naumovets, A.G., and Poplavsky, V.V. (1982) Surface defects and collective character of diffusion in the system Li-O-W(011), *JETP Letters* **36**, 88–91.
20. Kellogg, G. (1994) Field ion microscope studies of single-atom surface diffusion and cluster nucleation on metal surfaces, *Surf. Sci. Rep.* **21**, 1–88.
21. Wang, S.C., and Ehrlich, G. (1997) Diffusion of large surface clusters: direct observations on Ir(111), *Phys. Rev. Lett.* **79**, 4234–4237.
22. Wang, S.C., Kuerpick, U., and Ehrlich, G. (1998) Surface diffusion of compact and other clusters: $Ir_x$ on Ir(111), *Phys. Rev. Lett.* **81**, 4923–4926.
23. Chirita, V., Muenger, E.P., Greene, J.E., and Sundgren, J.-E. (1999) Reptation: a mechanism for cluster migration on (111) face-centered-cubic metal surfaces, *Surf. Sci.* **436**, L641–L647.
24. Montalenti, F., and Ferrando, R. (1999) Leapfrog diffusion mechanism for one-dimensional chains on missing-row reconstructed surfaces, *Phys. Rev. Lett.* **82**, 1498–1501.
25. Tsong, T.T. (2000) Atomic, molecular and cluster dynamics on flat and stepped surfaces, *Progr. Surf. Sci.* **64**, 199–209.

# METAL HETERODIFFUSION ON METALLIC SURFACES: CASE OF Pb ON Cu

C. Cohen, G. Prévot and D. Schmaus
*Groupe de Physique des Solides, UMR CNRS 7588,*
*Universités Paris VI and VII, tour 23,*
*2, place Jussieu, 75251, Paris cedex 05, France*

**Abstract.** We have investigated the role of different parameters, concerning both the substrate (surface structure and anisotropy) and the deposit (coverage $\theta$, structure and morphology), on the diffusion of Pb on Cu. Pb and Cu are not miscible in bulk; however, surface alloying is observed, in particular at low coverage. The growth is of the "Stranski-Krastanov" type. Above saturation coverage $\theta_S$, three-dimensional (3D) Pb islands form on top of a Pb wetting single layer. We have studied experimentally the spreading of Pb deposits, using various techniques. The results were interpreted through analytical models and confronted to molecular dynamics simulations. We find that for isolated Pb atoms, diffusion takes place both through jumps of Pb adatoms and exchange with surface Cu atoms. Increasing the Pb coverage towards $\theta_S$, we observe that Pb diffusion depends strongly on $\theta$, evidencing the influence of Pb-Pb interactions. The appearance of ordered structures markedly affects the chemical diffusion coefficient. Above $\theta_S$, the structure and morphology of the wetting layer influences Pb diffusion.

## 1. Introduction

The morphologies observed during thin film deposition are often governed by kinetics, i.e., by the diffusion of the species during deposition. In turn, the morphologies that form may influence the diffusion processes. In this paper, we report on such effects in the case of the heterodeposition of Pb on Cu single crystals. This system exhibits features that made it suitable for our studies:

- Pb and Cu, which both have a fcc structure but very different lattice parameters, are not miscible in bulk up to the Pb melting temperature ($T_{Bulk} = 600$ K). The latter being much lower than that of the Cu substrate, one may study Pb diffusion and growth processes over a wide temperature range, on top of a still well ordered substrate.
- Let us define the coverage $\theta$ as the ratio between the density of deposited Pb atoms and the density of substrate atom sites. At low $\theta$, in the submonolayer range, surface alloying has been evidenced by scanning tunneling microscopy (STM) around or above room temperature (RT) [1,2]. However, when deposits are achieved at lower



*M.C. Tringides and Z. Chvoj (eds.),*
*Collective Diffusion on Surfaces: Correlation Effects and Adatom Interactions,* 107–116.

temperatures, and analysed before any annealing, no alloying is observed. In turn, small two-dimensional Pb clusters form on top of the substrate surface, even at very low coverage [3]. This change of deposit morphology is obviously related to the mechanisms of Pb diffusion.

- The structure of the two-dimensional Pb deposit that forms at submonolayer coverage also depends on $\theta$. Around RT, one first observes a disordered surface alloy [1,2], then various ordered structures [4-7] which "melt" at temperatures somewhat below $T_{Bulk}$ [6]. The system is thus well adapted for studying the influence of order-disorder phase transition on surface heterodiffusion.

- After the completion of a wetting Pb monolayer at coverage $\theta_S$, 3D Pb islands form [4]. One can then study Pb diffusion on top of the Pb wetting layer, a process that governs the "Ostwald ripening" of the islands.

In order to address these various questions, we have undertaken diffusion studies as a function of $\theta$ and of temperature T.

## 2. Experimental

The experiments were mostly performed in an ultra-high vacuum chamber coupled to a 2.5 MV van de Graaff ion accelerator. This chamber, described in [8], comprises a 3 axes goniometer with x-y translation, a four-grid LEED-Auger system, a sputtering ion gun, and evaporators. The sample (a Cu crystal disk) can be heated up to 1100 K and cooled down to 120 K. After in-situ sputtering and annealing, its cleanliness, surface order, and homogeneity were controlled by Auger electron spectroscopy (AES), low energy electron diffraction (LEED) and Rutherford backscattering spectrometry (RBS) in channeling geometry.

In most experiments, a uniform Pb deposit of well defined geometry and of millimeters dimension was achieved at the center of the sample by using an well-adapted diaphragm. The Pb concentration profiles along two perpendicular straight lines crossing at the deposit center were determined, before and after annealing, by RBS with mono-energetic $^4He^+$ incident ions in the MeV energy range. The lateral resolution, around 0.5 mm, is given by the beam spot size and is measured accurately. The stability of the beam position and the precision of the sample translation are better than 0.1 mm. The absolute calibration of the Pb concentration is obtained with a precision of 2%, using a RBS standard [9]. We often performed the RBS measurements in channeling geometry in order to minimize the overall backscattering yield, and thus pile-up problems. With this geometry, we were able to measure Pb amounts of $10^{13}$ at/cm$^2$ (i.e., about $10^{-2}$ monolayer) within two minutes with statistical uncertainties of the order of 3%.

## 3. Low Coverage: Diffusion Mechanisms of Isolated Pb Atoms

At low coverage, $\theta \leq 0.10$, we observed that the spreading velocity of Pb deposits depends little on $\theta$. The shape of the deposits after annealing can be adjusted by a two-dimensional Fick's diffusion equation with diffusion coefficients independent of $\theta$. These coefficients can thus be associated with the diffusion of isolated Pb atoms.

On the isotropic (100) Cu face, the deposit spreading is also isotropic and thus characterized by a unique diffusion coefficient, D. Between 473 and 638 K, D varies exponentially with 1/T and is characterized by a rather high activation energy (0.68 eV) [10]. Assuming an attempt frequency of the order of $10^{13}$ Hz, the preexponential factor corresponds to jumps of atomic size.

We have also studied, between 500 and 800 K, the anisotropy of the Pb spreading on the anisotropic (110) Cu face [11]. The initial deposits were square shaped, with the sides aligned, respectively, along the [-110] and [001] directions. Figure 1 shows typical spreading measured along these two directions. The Pb diffusion is anisotropic, and more rapid along the"open" [-110] direction. However, as seen in the Arrhenius plot of Fig. 2, the diffusion coefficients $D_{[-110]}$ and $D_{[001]}$ obtained by fitting the data can both be represented by an exponential law with the same activation energy, close to 0.6 eV. Thus, in the temperature range explored, the anisotropy is nearly independent of T, but the preexponential factor is about 2.4 times higher along the [-110] direction.

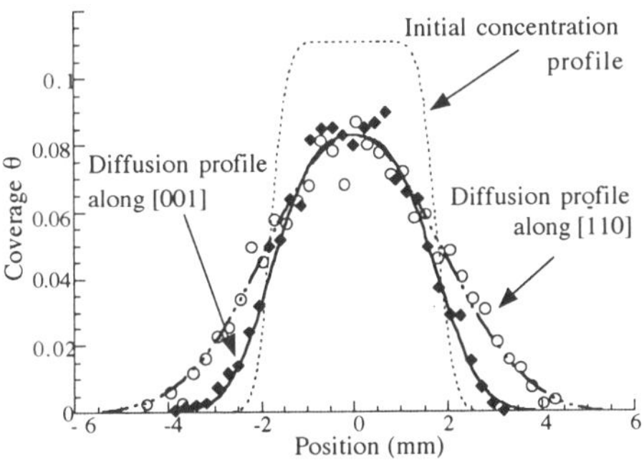


*Figure 1.* Experimental Pb concentration profiles, measured by Rutherford Backscattering Spectrometry, as deposited (square deposit), and after 4.5 h annealing at 620 K. The three measurements and their fits presented here correspond to the convolution of the actual Pb profile by the lateral resolution (experimentally measured) associated with the probing beam spot size.

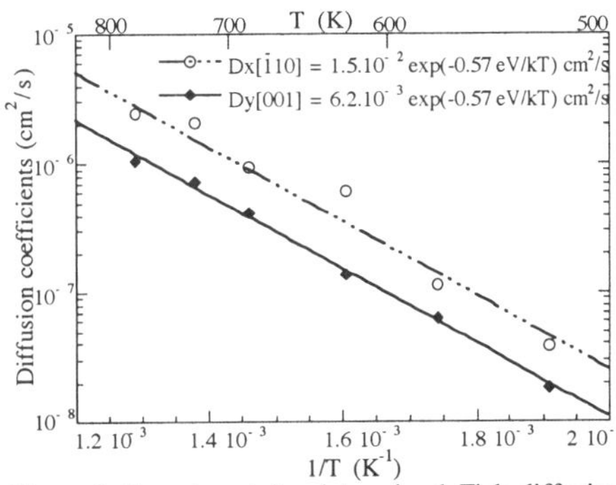


*Figure 2.* Experimentally determined Fick diffusion coefficients for Pb on Cu(110) as a function of temperature.

The only diffusion mechanism leading to diffusion in both directions with the same activation energy is the exchange process. Moreover, such a mechanism, already observed for heterodiffusion on various surfaces [12, 13], is consistent with the fact that, at and above RT, the Pb atoms are inserted in the Cu surface plane [1,2]. We propose that a Pb inserted atom exchanges with a Cu adatom and can then be reinserted through exchange with another Cu atom. In order to account for the fact that the preexponential factor is higher along the [-110] direction, we also assume that the Pb atom can suffer some jumps along this direction (and only along it), before its reinsertion.

The limiting process, with the highest associated energy $E_d$, is de-insertion, and $E_d$ is the activation energy that is obtained from the Arrhenius plots in Fig. 2. $E_d$ is the sum of the activation energy corresponding to an exchange between an inserted Pb atom and a Cu adatom, and of the formation energy of this Cu adatom near the inserted Pb atom. The spreading asymmetry depends on the ratio between the probabilities for a Pb adatom to jump or to be reinserted, and thus on the difference between the activation energies $E_j$ and $E_r$ corresponding to these two events. The fact that, in the temperature range studied, the ratio of the preexponential factors corresponding, respectively, to the [-110] and

[001] directions is only 2.4 and varies little with T, demonstrates that $E_r$ - $E_j$ is not very high at the scale of kT. At lower temperatures, one should expect a much higher anisotropy, increasing rapidly when decreasing T. We have shown [11] that the ratio $\alpha = D_{[-110]} / D_{[001]}$ writes:

$$\alpha = [1 + (\nu_j/\nu_r)\exp((E_r - E_j)/kT)]/2 \tag{1}$$

where $\nu_j$ and $\nu_r$ are, respectively, the attempt frequencies associated to jump and reinsertion events. Assuming $\nu_j = \nu_r$, the best fit of formula (1) to our experimental results leads to a difference $E_r - E_j = 0.07$ eV.

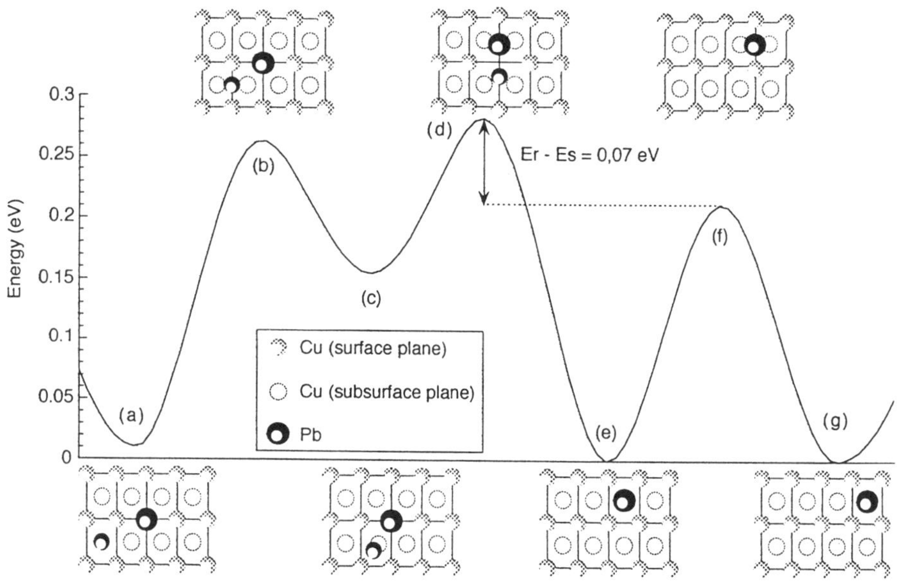


*Figure 3.* Quenched molecular dynamics simulation of the diffusion of a Pb atom on Cu(110). Schematic representation of the energy variations along the diffusion path. (a), (c), (e) and (g) are local minimum energy positions. (b), (d) and (f) are saddle point configurations. Pb de-insertion: (a) → (e); jumps of Pb adatoms: (e) → (g).

In order to check the diffusion mechanism described above, we have undertaken molecular dynamics (MD) simulations [14]. These simulations rely on a phenomenological many-body potential derived in the framework of the second-moment approximation of the tight-binding method, with parameters fitted to the physical properties of the pure metals and to the copper-lead phase diagram. Static calculations at T = 0 K allowed us to determine the energy and relaxed atomic positions of the equilibrium and saddle point configurations of various possible diffusion mechanisms. These static calculations confirm that the "exchange and jump" mechanism, assumed to interpret our experimental results, is by far the dominant process. We show in Fig. 3 a schematic representation of the energy variations along the diffusion path derived from the calculations. The agreement between the calculated activation energies and the values that fit our experimental data is surprisingly good. In particular, the calculations lead to $E_r - E_j = 0.07$ eV and to $E_d = 0.59$ eV.

However, simulations performed in the temperature range 400-700 K show that multiple jumps occur frequently, especially at the lowest temperatures studied, thus leading to Pb diffusion more anisotropic but also more steeply dependent on temperature than could be expected from the static calculations. This is illustrated in Fig. 4 that compares the temperature-dependence of the asymmetry ratio $\alpha$ experimentally measured and fitted by relation (1) to the predictions of the MD simulations. Even if the discrepancies mentioned just above are obvious (they may be due to the limited accuracy of the many-body potential used), the major features of measurements and simulations are very similar. The strong increase of $\alpha$ at low temperatures is certainly responsible for morphologies observed at low $\theta$ for deposits performed at low T (two-dimensional Pb clusters on top of the substrate surface [3]). At low temperatures, a deposited Pb atom has a strong probability to perform many jumps in the adatom position before being inserted. Consequently, it also has a strong probability to meet other Pb atoms and to participate to the formation of Pb adatom clusters, stable at such temperatures, that prevent the surface alloy formation.

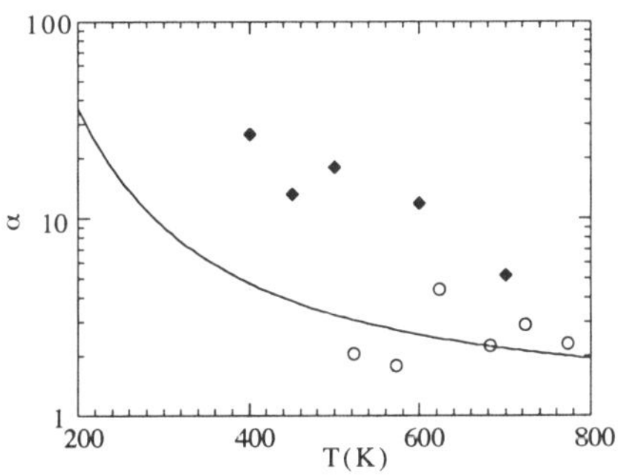


*Figure 4.* Anisotropy ratio $\alpha = D_{[-110]}/D_{[001]}$. Comparison between experimental results (open circles), analytical model (line) [15] with $E_r - E_j = 0.07$ eV and Molecular Dynamics simulations (full diamonds).

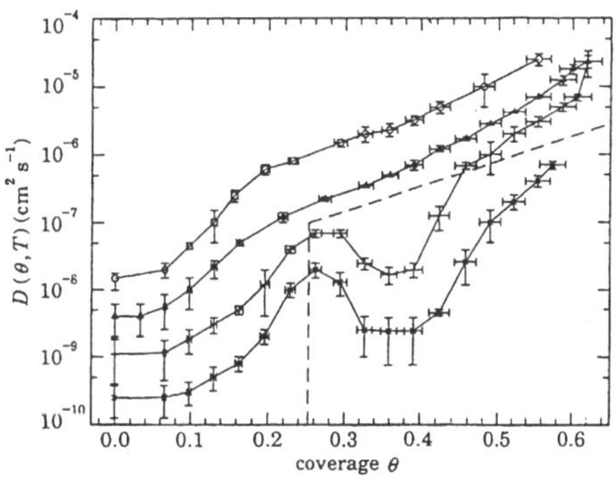


*Figure 5.* Experimentally determined chemical diffusion coefficient D for Pb on (100) Cu as a function of coverage $\theta$ for various temperatures (○ 638 K, △ 573 K, + 523 K, × 473 K). The dashed line delimits schematically the region (high $\theta$, low T) where Pb structures are observed.

## 4. Influence of $\theta$ and Ordered-phase Formation on the Spreading of Pb Submonolayers

For coverage $\theta > 0.10$ we observed that the spreading velocity of Pb deposits on Cu(100) depends markedly on $\theta$, demonstrating that collective effects (related to Pb-Pb interactions) have to be considered. We thus decided to study systematically this dependence at various T, between 473 and 638 K, in a coverage range extending up to $\theta = \theta_S$, ($\theta_S = 0.6$ on Cu(100)), i.e., for purely two-dimensional Pb deposits [10]. We took advantage of the isotropy of the substrate surface. We thus searched for a function $D(\theta,T)$ that could fit all the measured diffusion profiles, whatever the initial concentration of the deposit, and whatever the annealing time t, by numerical integration of the Fick's law in polar geometry:

$$\partial\theta/\partial t = (1/r)\partial/\partial r(rD(\theta,T)\partial\theta/\partial r) \qquad (2)$$

The integration has been performed by the finite difference method. We first fitted the deposits with the lowest concentration, obtaining a reliable function $D(\theta,T)$ in the corresponding coverage range; then the knowledge of this function was extended towards higher and higher $\theta$ values by fitting deposits with higher and higher initial coverage.

The diffusion coefficients $D(\theta,T)$ leading to the best overall fits are displayed in Fig. 5 as a function of $\theta$, for the various temperatures T at which the experiments were performed. At all temperatures, one observes a steep overall increase (three orders of magnitude) of D when $\theta$ increases up to $\theta_S$, evidencing the existence of strong Pb-Pb repulsive interactions at high coverage [15]. However, for T = 473 K and T = 523 K, i.e., for the two annealing temperatures at which some ordered structures are still present, there is a marked lowering of D in the domain of $\theta$ where these structures exist.

The phase diagram of Pb on Cu is complex, and the lowering observed is thus hard to interpret. We have only attempted to check if such behavior could be predicted in the simpler situation of a square lattice gas model with double occupancy of sites forbidden and nearest-neighbors repulsive interactions (Ising antiferromagnetic model). This system produces an ordered c(2×2) structure, around $\theta = 0.5$, below a critical temperature $T_C$. We have determined the coverage dependence of the chemical diffusion coefficient $D(\theta,T)$, respectively, above and below $T_C$. This was achieved exactly through Monte-Carlo simulations, or analytically, in the frame of the Bethe-Peierls approximation. Above $T_C$, $D(\theta,T)$ increases steadily with $\theta$. Below $T_C$, a lowering of $D(\theta,T)$ is indeed observed in the coverage domain where the ordered structures form. We thus reproduce the behavior observed experimentally. The chemical diffusion can be written:

$$D(\theta,T) = \partial(\mu/kT)/\partial\ln(\theta)\ f_{corr}(\theta)\ l^2\ \Gamma(\theta) \qquad (3)$$

where $f_{corr}$ is a correlation factor, l a jump length, and $\Gamma(\theta)$ an effective jump frequency. The so-called "thermodynamical factor" $\partial(\mu/kT)/\partial\ln(\theta)$ which expresses the fact that particle currents respond to a gradient of chemical potential $\mu$ rather than to a concentration gradient is maximum when ordering takes place. However, the Monte-Carlo simulations show this maximum is more than compensated by a minimum in $\Gamma(\theta)$ and in the correlation factor $f_{corr}$ associated to the jumps. These minima are related to the marked improvement of short-range order that take place together with long range ordering.

## 5. Diffusion Above $\theta_S$, in Presence of Three-dimensional Pb Islands

These experiments, performed on the Cu(100) face are detailed in Ref. [16]. We show in Fig. 6 the spreading, along one diameter, of a Pb deposit, initially disk shaped with uniform coverage $\theta_{in} > \theta_S$. The annealing temperature (593 K) is higher than the melting temperature $T_M$ = 520 K of the wetting Pb layer, and, under these conditions, we observed an isotropic spreading. The profiles measured for various annealing times t show that a Pb layer with uniform coverage $\theta_S$ develops at the expense of 3D islands. The width of the central zone, with $\theta > \theta_S$, decreases (while keeping a constant coverage $\theta_{in}$), until the time $t_c$ at which no more Pb islands are present. After $t_c$, the spreading is much slower. These observations demonstrate there is an abrupt increase of the spreading velocity with $\theta$ around $\theta_S$.

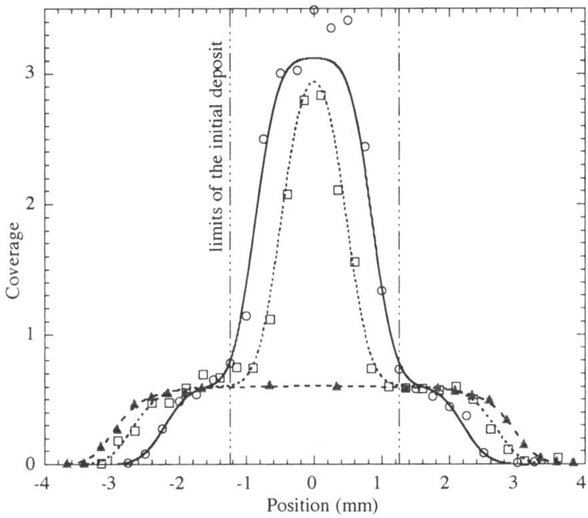


*Figure 6.* Pb coverage profiles measured after various annealing times t at 593 K on a deposit with initial uniform coverage $\theta = 3.2$. circles: t = 10 min; squares: 26 min; triangles: 42 min. All the lines are theoretical profiles corresponding to a value of the product $D_{ML}\varepsilon$ equal to $8.10^{-6}$ $cm^{-2}s^{-1}$ (see text). The adjustment to this value has been obtained by fitting only the broadest profile corresponding to the time t = 42 min at which the central zone with $\theta > \theta_S$ disappears.

We assumed this increase is associated to the diffusion of Pb adatoms on top of a continuous wetting layer, with a constant diffusion coefficient $D_{ML} >> D(\theta < \theta_S)$. We also assumed that, in the region where they exist, the Pb islands are in equilibrium with a constant adatom concentration $\varepsilon$. We then fitted the spreading profiles by integrating numerically Eq. 2. (the fits are represented in lines in Fig. 6). There are two quantities to adjust: $D_{ML}$ and $\varepsilon$. In fact, the adjustment depends only on the product $D_{ML}\varepsilon$. Figure 6 shows that concentration profiles corresponding to different annealing times at the same temperature are well fitted by the same value of $D_{ML}\varepsilon$.

The three upper points of the Arrhenius plot displayed in Fig. 7 represent the values of $D_{ML}\varepsilon$ that fit the spreading profiles above $T_M$. The slope of the plot in this region corresponds to an energy, $E_T = 0.5$ eV, significantly lower than the activation energy, $E_{d0} = 0.68$ eV, found for the diffusion of isolated Pb atoms on Cu(100). $E_T$ is the sum of two energies associated with $D_{ML}$ and $\varepsilon$, respectively: the activation energy $E_d$ for Pb adatom diffusion on the wetting layer and the formation energy of these adatoms from the Pb islands. Thus, $E_d$ is necessarily much smaller than $E_{d0}$. The preexponential factor associated to $D_{ML}\varepsilon$ is consistent with jump lengths of atomic size.

Below $T_M$ the wetting layer is ordered with a c(5√2×√2)R45° two-domain structure. A schematic representation of one domain is displayed in Fig. 8. In our experiments, the Cu(100) crystal was slightly miscut, and, as a consequence, the LEED diagram revealed that one of the two domain orientations was markedly favored. In the mean time, we observed that the spreading below $T_M$ was very anisotropic. Such behavior can be understood when examining the structure represented in Fig. 8: there is an easy path for diffusion along the open [001] direction and periodic obstacles along the [010] orientation. The favored domain orientation for the Pb deposits on our Cu crystal was the one represented in Fig. 8 and, in our experiments, the spreading velocity was indeed higher along the [001] direction. In order to fit the spreading profiles, we integrated numerically a two-dimensional diffusion equation:

$$\partial\theta/\partial t = \partial/\partial x(D_x(\theta)\partial\theta/\partial x) + \partial/\partial y(D_y(\theta)\partial\theta/\partial y) \quad (4)$$

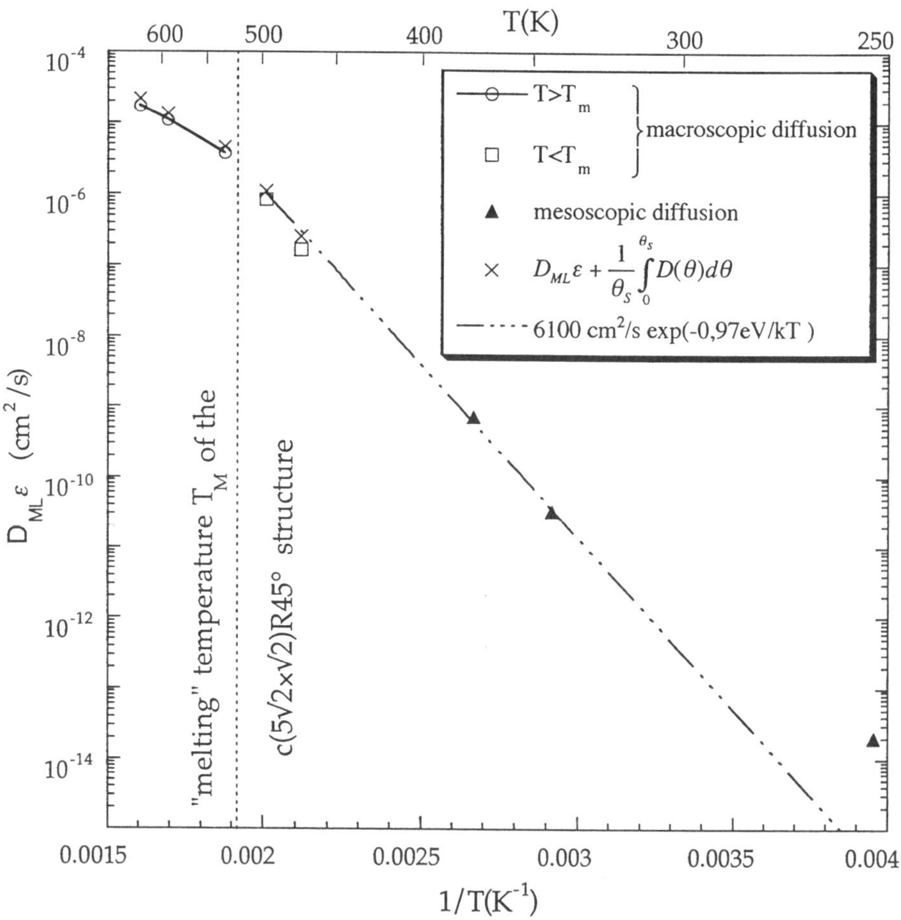


*Figure 7.* Arrhenius representation of the temperature-dependence of the product $D_{ML}\varepsilon$ for all the experiments described in section 5. ○ RBS measurements above $T_M$; □ RBS measurements below $T_M$; ▲ Auger measurements. The triangle at the lowest temperature corresponds to wetting layer reformation from "unannealed" islands. The two other triangles correspond to wetting layer reformation from islands obtained after annealing at 530 K.

We assumed that, for $\theta > \theta_S$, diffusion was fully blocked along the x ([010]) direction, and, that along the perpendicular [001] direction (y), the diffusion coefficient had a constant value $D_{ML}$. The open squares in Fig. 7 correspond to $D_{ML}\varepsilon$ values fitting measurements performed at 500 and 475 K, i.e., below $T_M$. The slope of the line joining these two points is twice higher than the slope of the line joining the points above $T_M$ and corresponds to $E_T = 1$ eV. The two straight lines intersect at $T_M$. Thus, around $T_M$, there is a change of $E_T$ compensated by a huge change of the preexponential factor: the mean jump length fitting the data below $T_M$ is a few thousand Å! Our interpretation is that when the wetting layer is ordered, with a two-domain structure, the crossing of domain boundaries is the limiting process for Pb adatom diffusion. This explains both the increases of $E_T$ and of the jump length that becomes of the order of the size of a domain.

In order to verify these assumptions, we have performed experiments at lower temperature. There, the spreading during realistic experimental times was of mesoscopic dimension, of the order of the distance between Pb islands. We then adopted the following procedure: once a deposit was prepared, we removed the wetting layer between the islands by ion sputtering at 150 K and controlled by AES that, at this temperature, the remaining Pb atoms could not leave the islands. The sample was then heated at fixed temperature and the reformation of the wetting layer was followed by LEED and AES. The LEED observations demonstrated that, since the beginning of the reformation, the dense $c(5\sqrt{2}\times\sqrt{2})R45°$ structure is present. The analysis of the AES results provided the kinetics of the reformation.

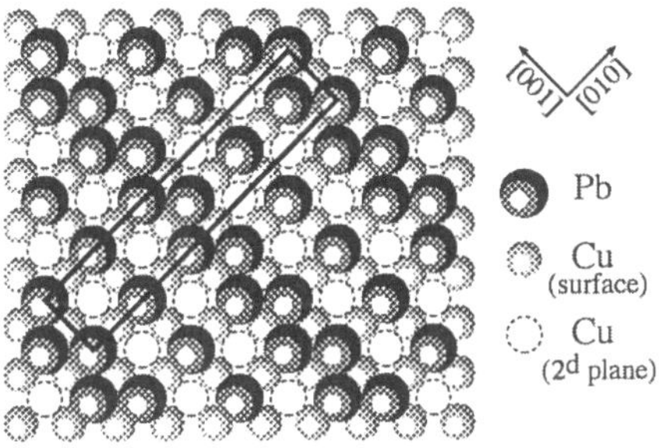


*Figure 8.* Scheme of one domain of the $c(5\sqrt{2}\times\sqrt{2})R45°$ structure formed by Pb atoms on a Cu(100) surface.

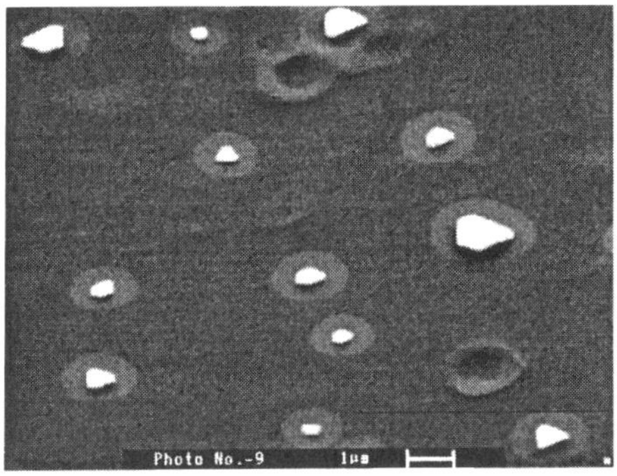


*Figure 9.* Observation by SEM of the reformation at 345 K of the wetting Pb monolayer around the Pb 3D islands. Previous to this observation, the wetting layer had been removed by sputtering at 120 K.

We show in Fig. 9 the observation by scanning electron microscopy (SEM) of a deposit of coverage $\theta = 7\theta_S$. This deposit was formed at RT, then annealed at 530 K in order to lower the island density (by Ostwald ripening), sputtered at 150 K and annealed a few minutes at 345 K to initiate the reformation of the wetting layer. On Fig. 9, one clearly observes this reformation around the islands. In order to fit the reformation kinetics we have numerically integrated Eq. 3. The initial conditions were provided by the islands distribution, as observed by SEM or STM. The fits of the reformation kinetics at 345 and 380 K, on deposits similar to those shown in Fig. 9 (i.e., with a mean distance between islands of a few μ) are very good. The corresponding values of $D_{ML}\varepsilon$ are reported in Fig. 7. They are perfectly aligned with the two values already obtained below $T_M$. This confirms the associated (high) values of $E_T$ and of jump length and thus the limitation of Pb adatom diffusion by domain boundaries.

Another confirmation is provided by studies achieved on a deposit formed at RT and sputtered at low T without any preannealing. The mean distance between islands is then only a few thousand Å, probably comparable to a domain size. The reformation kinetics exhibits an abrupt slowing down when the wetting layer occupies about 40% of the surface that cannot be reproduced by integration of Eq. 4. When fitting only the beginning of the reformation, we obtain a value $D_{ML}\varepsilon$ which, as seen in Fig. 7, lies well above the straight line joining all the points corresponding to $T>T_M$. Our interpretation is this high value corresponds to diffusion in a single domain, i.e., in the absence of domain boundaries, and that the formation of the latter, which takes place when the wet-ting layer develops, is responsible for the very strong slowing down of the spreading.

## Acknowledgments

The authors wish to thank L. Beaunier, W. Chab, Y. Girard, P. Leroux-Hugon, A. L'Hoir, and V. Pontikis for their participation to part of the work. The technical assistance of J.M. Guigner and of J. Moulin was highly appreciated. This work was supported by the Centre National de la Recherche Scientifique under GDR 86.

## References

1. Nagl, C., Platzgummer, E., Haller, O., Schmid, M. and Varga, P. (1995) Surface alloying and superstructures of Pb on Cu(100), *Surf. Sci.* **331-333**, 831-837.
2. Robert, S., Gauthier, S., Bocquet, F., Rousset, S., Duvault, J.L. and Klein, J. (1996) An STM study of the adsorption of Pb on Cu(100): formation of an ordered surface alloy, *Surf. Sci.* **350**, 136-144.
3. Bocquet, F. and Gauthier, S. (1998) A study of the low-temperature submonolayer deposition of Pb on Cu(100), *Surf. Sci.* **416**, 1-8.
4. Henrion, J. and Rhead, G.E. (1972) LEED studies of the first stages of deposition and melting of Lead on low index faces of Copper, *Surf. Sci.* **29**, 20-36.
5. Hösler, W. and Moritz W. (1986) LEED analysis of a dense lead monolayer on copper (100), *Surf. Sci.* **175**, 63-77.
6. Sanchez, A. and Ferrer, S. (1989) Structure and melting of lead overlayers on Cu(100) studied with thermal-energy atom scattering, *Phys. Rev. B* **39**, 5778-5786.
7. de Beauvais, C., Rouxel, D., Michailov, M. and Mutaftschiev, B. (1995) Two-dimensional overgrowth in the low submonolayer range: the case of c(2×2)-Pb/Cu(110), *Surf. Sci.* **324**, 1-7.
8. Abel, F., Cohen, C., Davies, J.A., Moulin, J. and Schmaus, D. (1990) Thin film growth of Sn on Cu(100), *Appl. Surf. Sci.* **44**, 17-27.
9. Cohen, C., Davies, J.A., Drigo, A.V. and Jackman, T.E. (1983) Intercomparison of absolute standards for RBS studies, *Nucl. Instr. Meth.* **218**,147-148.
10. Cohen, C., Girard, Y., Leroux-Hugon, P., L'Hoir, A., Moulin, J. and Schmaus, D.(1993) Surface diffusion of Pb on (100) Cu: coverage dependence and influence of ordered-phase formation, *Europhys. Lett.* **24**, 767-772.
11. Prévot, G., Cohen, C., Moulin, J. and Schmaus, D. (1999) Surface diffusion of Pb on Cu(110) at low coverage: competition between exchange and jump, *Surf. Sci.* **421**, 364-376.
12. Wrigley, J.D. and Ehrlich, G. (1980) Surface diffusion by an atomic exchange mechanism, *Phys. Rev. Lett.* **44**, 661-663.
13. Kellogg, G.L. (1991) Direct observations of adatom-surface atom replacement: Pt on Ni(110), *Phys. Rev. Lett* **67**, 216-219.
14. Prévot, G., Cohen, C., Schmaus, D. and Pontikis, V. (2000) Non-isotropic surface diffusion of lead on Cu(110): a molecular dynamics study, *Surf. Sci.* **459**, 57-68.
15. Uebing, G. and Gomer, R. (1991) A Monte-Carlo .study of surface diffusion coefficients in the presence of adsorbate-adsorbate interactions. I. Repulsive interactions, *J. Chem. Phys.* **95**, 7626-7652.
16. Prévot, G., Cohen, C., Guigner, J.M. and Schmaus, D. (2000) Macroscopic and mesoscopic surface diffusion from a deposit formed by a Stranski-Krastanov type of growth: Pb on Cu(100) at above one layer of coverage, *Phys. Rev. B* **61**, 10393-10403.

# STEP EFFECTS ON SURFACE DIFFUSION:

*A Comprehensive Study of CO on Pt(111)*

XUDONG XIAO
*The Hong Kong University of Science and Technology*
*Hong Kong, China*

**Abstract.** The experimental results of CO diffusion perpendicular and parallel to steps on vicinal Pt(111) surfaces were presented. While diffusion perpendicular to step subjects to an impediment due to the additional trapping at the step sites, diffusion parallel to steps shows enhancement over terrace in various ways. At high temperatures, all three types of stepped surfaces show the contribution to mass transport by the fast step edge diffusion. At low temperatures, a third channel, which exists only for A- and AB-type steps, also contributes to a diffusion enhancement.

## 1. Introduction

Surface diffusion is central to the understanding of technologically important processes such as crystal growth, catalysis, and corrosion. It has also become increasingly clear that even at low densities, steps can significantly influence the macroscopic diffusion (chemical diffusion) rate with strong enhancement along the step direction but with impediment perpendicular to the steps [1-7]. Because of its intrinsic interests as well as potential technological importance, there have been many theoretical interests in modelling these processes [8,9]. At the step, the surface potential is modelled by the features that a) there is an extra binding energy, $\Delta E_B$, at the step site (trapping site) and b) the diffusion activation energy along the step edge, $E_2$ is lower than the corresponding activation energy $E_t$ for diffusion on the terrace. In the direction perpendicular to the step, there exists an additional barrier $\Delta E_s$ (Schwoebel barrier) on one side of the step. This Schwoebel barrier is at the upper edge of the step for metal atoms on metal surfaces, but at the lower step edges for gas molecules on metal surfaces. For the latter, the trapping site is at the upper edge of the step.

For tracer diffusion measured in the FIM and STM experiments [10-13], the step-crossing is controlled by the Schwoebel barrier since the diffusion process terminates once the diffusing atom is incorporated into a step site after step-crossing. In contrast, the chemical diffusion perpendicular to steps can be slowed down by both the Schwoebel barrier and the additional trapping well [9], since the techniques that measure chemical diffusion cannot follow single atomic events. In order to achieve a



*M.C. Tringides and Z. Chvoj (eds.),*
*Collective Diffusion on Surfaces: Correlation Effects and Adatom Interactions,* 117–127.

mass transport over a characteristic probing length much larger than the terrace width, the diffusing species will have to overcome not only the Schwoebel barrier at one side of the step but also the trapping potential well at the other side of the step. Therefore, determination of both the Schwoebel barrier and the additional trapping energy requires independent knowledge in addition to diffusion. For diffusion parallel to steps, at least two channels in parallel exist, one is on terrace and the other along step edges. If the adsorbate distribution is not under equilibrium in the direction perpendicular to steps in addition to the coverage gradient created parallel to steps, a net cross-channel mass transport can also result. One such scenario is that the molecules will diffuse from a terrace area with high coverage to a step and move along the step edge for a while and then return back to another terrace area with low coverage at a later moment.

In this article, we present our study of CO diffusion on stepped Pt(111) surfaces, for both perpendicular and parallel to step directions. For the former, a slow-down of diffusion by the trapping sites was observed. For the latter, an enhancement was measured at high temperature for all A-, B-, and AB-type stepped surfaces. At low temperatures, an enhancement was found only for parallel to A- and AB-type steps. Most of our results were published elsewhere [6,7], we here include them with additional measurements to show a complete picture of step effects on surface diffusion.

## 2. Experiment

The key requirements for this experiment are (1) samples with known and well characterized steps (both density and orientation), and (2) ability to measure diffusion along a given direction. For requirement (1), we prepared six samples with different miscut off the (111) plane to obtain single atomic height steps [14]: one with a miscut $<0.1^{\circ}$ for terrace diffusion measurement, three with a miscut of $1^{\circ}$, $2^{\circ}$, and $4^{\circ}$, respectively, to provide AB-type steps for step density dependence study, and two with a miscut of $2^{\circ}$ to provide A- and B-type stepped surfaces for step orientation dependence study. While A- and B-type steps are defined as along [110] but (100) and (111) faceted, respectively [11,15], the AB-type step is defined as along $[11\bar{2}]$, perpendicular to A- and B-type steps. For requirement (2), we used a linear optical diffraction method to measure diffusion, which is able to probe diffusion along a given direction.

The experiment was performed in an ultrahigh vacuum (UHV) chamber that had a base pressure of $2.0\times10^{-10}$ torr. Sample cleaning was performed by extensive cycles of $Ar^{+}$ sputtering, oxygen treatment at 1000 K, and high temperature annealing at 1200 K. The routine cleaning procedure of the Pt surface was to $Ar^{+}$ sputter the surface at room temperature for 30 min, followed by annealing at 1200 K for 5 min. Auger spectra of the clean surfaces showed no detectable impurities. Observation of (1x1) LEED patterns from clean Pt(111) surfaces (with splitting for stepped surfaces) and a c(4x2) pattern from 0.5 ML CO on Pt(111) ensured that the surface was well ordered [16,17].

The linear optical diffraction technique for surface diffusion measurement has been described in detail previously [18,19]. Here, we summarize only the key points. First, a uniform CO layer with a prescribed coverage was deposited at 170 K on a vicinal Pt(111) surface. Then a one-dimensional CO grating was created at a diffusion temperature by interfering two laser beams (Nd:YAG laser at 1.06 μm, 10 ns) at the surface via laser-induced thermal desorption (LITD). For diffusion perpendicular to steps, this one-dimensional CO grating was made parallel to the steps. For diffusion parallel to steps, the grating is made perpendicular to steps. In all cases, a coverage modulation of ~0.02 ML was chosen. The smearing of the CO grating via surface diffusion can be detected by the decay of the first-order linear diffraction signal of a He-Ne laser beam with polarization modulation [18,19]:

$$S=S(0)\exp(-t/\tau),\quad \text{with } \tau=s^2/8\pi^2D, \tag{1}$$

where D is the chemical diffusion coefficient and s=4.4 μm is the grating period. In the present study, D was measured as a function of substrate temperature over a range from 120 to 310 K for CO coverages at 0.3 and 0.5 ML.

## 3. Results

The experimental results are shown in Figs. 1 (a) and (b) for 0.5 ML and 0.3 ML CO coverage, respectively, where D is plotted against the reciprocal temperature 1/T for diffusion perpendicular and parallel to steps for a number of miscut samples.

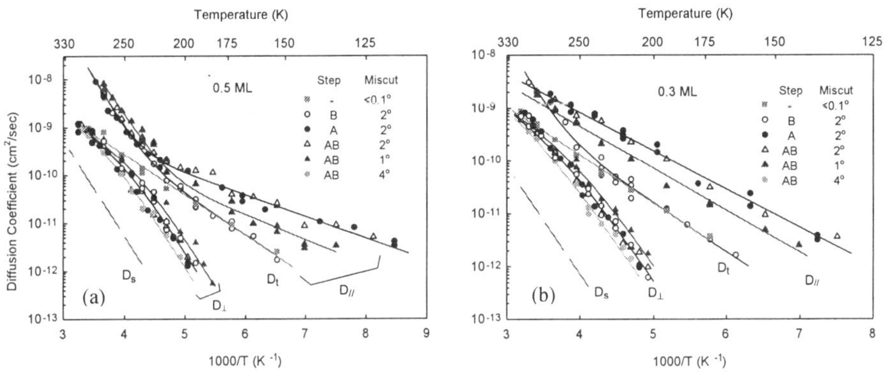


*Figure 1.* (a) CO diffusion coefficient D in the direction of perpendicular or parallel to steps on vicinal Pt(111) surfaces versus reciprocal temperature 1/T in an Arrhenius plot for a number of step orientations and densities at 0.5 ML CO coverage. The step orientations and densities are indicated by the legend. The solid lines are fittings described in the text and the dashed line represents the diffusion coefficient $D_s$ over steps as deduced from the fitting by Eq. (2). (b) Same as for (a) except CO coverage is 0.3 ML.

At 0.5 ML coverage, while the data on the <0.1° miscut sample represent terrace diffusion ($D_t$) with a prefactor $\nu_t a^2 = 6\times10^{-7}$ cm²/sec and an activation energy $E_t$ = 3.9 kcal/mol [20], diffusion perpendicular to steps ($D_\perp$) is found to increasingly slow down as the step density increases. The overall diffusion rate $D_\perp$ is limited by terrace diffusion at high temperatures and by the cross-step diffusion at low temperatures. This makes the diffusion perpendicular to steps not a simple Arrhenius behavior. It is also noted that diffusion perpendicular to A- and B-type steps on the 2° miscut surfaces coincides within the experimental uncertainty with that of AB-type steps at the same step density, indicating no step orientation dependence. This is in strong contrast to the case of diffusion parallel to steps where significant step orientation dependence is found. For diffusion parallel ($D_{//}$) to B-type steps, the diffusion rate is the same as that on terrace at low temperatures (<225 K), but much higher than that on terrace at high temperatures (>225 K). For diffusion parallel to A- and AB-type steps at the same step density ($2^o$ miscut), however, an enhancement over terrace diffusion is also found at low temperatures [6], in addition to the similar enhancement observed on the B-type stepped surface at high temperatures. The step density dependence for diffusion parallel to AB-type steps is also more complicated than that for diffusion perpendicular to steps. In the low temperature regime (<225 K), as the step density increases from $1^o$ to $2^o$ miscut, the enhancement of the diffusion rate increases. This relation does not seem to hold in the high temperature regime (>225 K), where the diffusion rate on the $1^o$ miscut sample is a little faster than that on $2^o$ miscut sample. As the miscut further increases to $4^o$, the enhancement of diffusion disappears in the entire temperature range. Diffusion coefficients similar to those on terrace are measured instead (the data are not shown). At 0.3 ML coverage, all the above features are reproduced, except for diffusion parallel to steps; the on-set of the enhancement of $D_{//}$ occurs at a higher temperature than that at 0.5 ML. This renders the non-Arrhenius behavior for diffusion parallel to A- and AB-type steps as being less pronounce compared to that at 0.5 ML. Because of the similarity, we will only discuss the data at 0.5 ML.

## 4. Discussion

### A. PERPENDICULAR TO STEPS

Since diffusion perpendicular to steps must be subjected to the following processes in sequence, namely diffusing over a terrace, and then over a step edge barrier (i.e., Schwoebel barrier) to reach a trapping site, then hopping out of the trapping site to the adjacent terrace, the diffusion time is additive. Based on the lattice gas model with the nearest neighbor hopping, Merikoski and Ying have analyzed the case in detail [9]. For a Langmuir gas with no adsorbate-adsorbate interaction except site blocking, the overall diffusion coefficient $D_\perp$ at $\theta$>>1/L and L>>1 is given by [9]

$$\frac{1}{D_\perp} = \frac{1}{\nu_t a^2 e^{-E_t/kT}} + \frac{\theta}{L} \frac{1}{\nu_{eff} a^2 e^{-(E_t+\Delta E_s+\Delta E_B)/kT}} \qquad (2)$$

$$\text{with} \quad \nu_{eff} = \frac{\nu_s}{1+(\nu_s/\nu_B)e^{-\Delta E_s/kT}},$$

where $\theta$ is the adsorbate coverage, L is the terrace width in units of lattice constant a. With the activation energies and attempt frequencies defined in Fig. 2, $D_t=\nu_t a^2 \exp(-E_t/kT)$ is the terrace diffusion coefficient and $D_s=\nu_{eff} a^2 \exp[-(E_t+\Delta E_s+\Delta E_B)/kT]$ is the diffusion coefficient over steps.

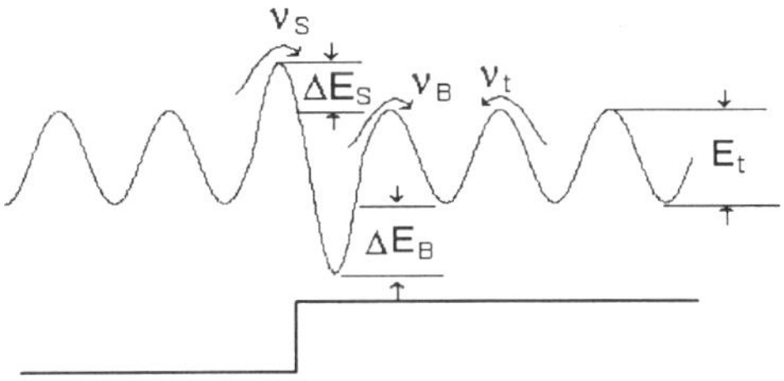


*Figure 2.* Schematics of potential energy near a step for a gas molecule on metal surfaces. The relevant diffusion parameters are defined.

The above equation has two regimes, one with $\nu_s/\nu_B >> \exp(\Delta E_s/kT)$, which leads to the elimination of $\Delta E_s$ in the diffusion coefficient over steps so that $D_s = \nu_B a^2 \exp[-(E_t + \Delta E_B)/kT]$; the other with $\nu_s/\nu_B << \exp(\Delta E_s/kT)$, which leads to a simplification of $\nu_{eff} = \nu_s$. In both regimes, $D_s$ has a simple Arrhenius form to allow a simple fitting procedure. In the intermediate regime, $\nu_s/\nu_B \sim \exp(\Delta E_s/kT)$, however, the situation is more complicated. For a positive $\Delta E_s$, one can always reach to the second regime by going to low temperatures independent of the exact value of $\nu_s/\nu_B$. For a negative $\Delta E_s$, it is likely that the diffusion is in the first regime as long as $\nu_s$ and $\nu_B$ are comparable. We must note that in the first regime, no definitive information on $\Delta E_s$ can be extracted from the fitting. Physically, this is because either due to a lowered step edge barrier with a negative $\Delta E_s$ or due to a large $\nu_s$, diffusion over this step barrier now takes negligible time as compared to terrace diffusion. Thus, the measured diffusion coefficient contains no signature of diffusion over the Schwoebel barrier.

Assuming an Arrhenius-behaved $D_s$, our data can be fitted by Eq. (2) quite well for all step densities (L=48,24,12 for miscut of 1°, 2°, and 4°, respectively) with $E_t = 3.9 \pm 0.2$ kcal/mol [20, 21] and $E_t + (\Delta E_s) + \Delta E_B = 9.3 \pm 0.8$ kcal/mol, where $\Delta E_s$ in the parentheses appears only for the second regime. The corresponding prefactors from the fittings are $D_{t0} = 6 \times 10^{-7}$ cm$^2$/sec, $D_{s0} = 6.4 \times 10^{-4}$ cm$^2$/sec ($D_{s0} = \nu_B a^2$ in the first regime or $\nu_s a^2$ in the second regime of Eq. (2)). Judging from the quality of the fittings, it is safe to say that our measurement does not fall into the intermediate regime. The prefactor across the steps is about three orders of magnitude larger than over the terraces, similar to what has been reported for Ag/Ag(111) [22].

In order to deduce the value for $\Delta E_s$, we must know $\Delta E_B$ first. Fortunately, in the case of CO/Pt(111), $\Delta E_B$ is well studied [14,21,23,24]. While most of the measurements are made by thermal desorption spectroscopy, other types of measurements such as FTIR for the equilibrium population at step and terrace sites also exist [21]. From all the studies, on samples with different step orientations (e.g., for A-, B- and AB-types) and densities (ranges from a terrace width of 28 atoms to 3 atoms), the extra binding energy at the trapping site is found to be $\Delta E_B \sim 7 \pm 1$ kcal/mol at low coverages. As we argued in Ref. [7], $\Delta E_B \sim 7 \pm 1$ kcal/mol should be the lower limit at a finite coverage of 0.5 ML when the effect of CO-CO interaction is included. Now, with the terrace barrier $E_t$ independently measured, we deduce $\Delta E_s \sim -1.6 \pm 1.3$ kcal/mol, where the error bar is mainly caused by the uncertainty in $\Delta E_B$. Knowing $D_{t0}$, $E_t$, $\Delta E_B$ from independent measurements, we can also fit our data with the full Eq. (2) without making any approximation to $\nu_s/\nu_B$. In this case, similar values of $\Delta E_s$ and $\nu_s a^2$ are obtained as before. However, a very large $\nu_B a^2$ (>0.5 $cm^2/sec$) is required in order to compensate the contribution from a large negative $\Delta E_s$ to fulfil the condition $\nu_s/\nu_B \ll \exp(\Delta E_s/kT)$. Comparing to the "normal prefactor" value (~$10^{-3}$ $cm^2/sec$) or $\nu_s a^2$ (~$6.4 \times 10^{-4}$ $cm^2/sec$) from the fitting, this $\nu_B a^2$ value seems to be physically unreasonable. Thus, it is likely that the above value of $\Delta E_s$~-1.6 kcal/mol represents the lower bound of $\Delta E_s$. Any smaller negative $\Delta E_s$ would reduce the required value for $\nu_B a^2$. The fact that we have deduced a large negative $\Delta E_s$ value is possibly due to the large error in determining $\Delta E_B$. From the scattering of the existing data on the adsorption energy of CO on Pt(111), for example from 32 kcal/mol [16] to 27 kcal/mol [23] for adsorption on terrace sites at low coverages, it is not surprising that $\Delta E_B$ might be over-estimated. On the other hand, our results also indicate that $\Delta E_s$ must be small if not negative. Assuming our measurement falls into the second regime of Eq. (2) with $\Delta E_s$~2 kcal/mol, the condition $\nu_s/\nu_B \ll \exp(\Delta E_s/kT)$ can be easily satisfied with comparable and reasonable values of $\nu_s$ and $\nu_B$ except that $\Delta E_B$ would be too small, ~3.2 kcal/mol, beyond the uncertainty of the other independent measurements. If our measurement falls into the first regime of Eq. (2), we would deduce $\Delta E_B$~5.2 kcal/mol, somewhere near the lower bound of the other independent measurements. With $\Delta E_s$~2 kcal/mol in this regime, it would further require $\nu_s/\nu_B \gg \exp(\Delta E_s/kT)$~23 at 320 K and $\nu_s/\nu_B \gg \exp(\Delta E_s/kT)$ ~153 at 200 K. Despite the lack of knowledge for the ratio of $\nu_s/\nu_B$, we feel that $\nu_s$ should not be too much larger than $\nu_B$ and $\Delta E_s$~2 kcal/mol might be an upper limit for CO/Pt(111) even for the first regime. Based on the above argument, we conclude that the Schwoebel barrier for CO on Pt(111) is small if not negative and diffusion perpendicular to steps is dominated only by the step edge trapping well.

In the past, the diffusion over steps was often reported to be controlled by the Schwoebel barrier. For metal-on-metal systems studied by FIM and STM [10-13], it was very natural that this conclusion was reached, since the diffusion process after trapping at the step sites was not followed. There, the Schwoebel barrier was the only one of interest. Diffusion out of the step, i.e., dissociation of atom from steps, was studied as an independent process [11]. For gas atom/molecule diffusion on stepped surfaces studied previously by LITD and FEM methods, however, the additional trapping potential must have been involved but was not appropriately considered. While for Xe on stepped Pt(111) [25], H and D on Rh(332) [26], H and D on W(123) [27], and H and D on W(023) [28], either the lack of anisotropy observed between diffusion parallel to and perpendicular to steps or the complication of adsorption structure renders a determination of the Schwoebel barrier and the additional trapping energy impossible, diffusion of H [3], CO [4] and n-butane [5] on Ru(001) with double atomic height steps measured by LITD provides good examples for comparison. From the anisotropy of diffusion parallel to and perpendicular to the steps, the authors were able to deduce a Schwoebel barrier of $\Delta E_s$=2.7 kcal/mol ($E_t$=3.2 kcal/mol) for H, $\Delta E_s$ =7.7 kcal/mol ($E_t$=9.4 kcal/mol) for CO, and $\Delta E_s$ =2.0 kcal/mol ($E_t$=9.4 kcal/mol) for n-butane, by assuming a negligible $\Delta E_B$. While this assumption for H is consistent with thermal desorption spectroscopy, for CO there exists contradictory evidence from the thermal desorption data that show an additional trapping potential [4]. For n-butane, the strongly coverage-dependent diffusion coefficient [5] also indicates a strong trapping at the steps. Our analysis here includes $\Delta E_B$ in a systematic way and is the most complete one to date. Thus, our observation for CO/Pt(111) is the first example illustrating that the additional trapping potential well at step edges controls diffusion perpendicular to steps.

B. PARALLEL TO STEPS

For diffusion parallel to steps, two diffusion channels must be considered: one is over the terraces and the other is along the steps. Both channels simultaneously contribute to mass transport from high coverage areas to low coverage areas at the length scale of our experiment. In a theoretical work of Merikoski and Ying, it was shown that under conditions L>>1 and $\theta$>> 1/L, the diffusion coefficient parallel to steps is given by [9]:

$$D_{//} = D_t \left( 1 + \frac{\nu_2}{L\theta^2 \nu_t} e^{(E_t - E_2 - \Delta E_B)/kT} \right), \tag{3}$$

where $D_t$ is defined in the previous section and $D_2 = a^2\nu_2 \exp(-E_2/kT)$ is the diffusion rate along the steps. Enhancement over the terrace diffusion rate $D_t$ will result if the second term in the parenthesis is comparable to or larger than unity. If this term is much less than unity, $D_{//}$ reduces to $D_t$, and will have temperature dependence corresponding to an activation energy of $E_t$. In the opposite limit when the second term is much larger than unity, the temperature dependence of $D_{//}$ will have a slope corresponding to an activation energy of $(E_2+\Delta E_B)$. These can readily be compared with the experimental observations for diffusion parallel to B-type steps on the 2° miscut sample. Using $\nu_t a^2 = 6\times10^{-7}$ cm$^2$/sec, $E_t$ = 3.9 kcal/mol, by fitting the data with Eq. (3) we obtain $E_2+\Delta E_B$=10.3 kcal/mol and $\nu_2 a^2 \sim 5$ cm$^2$/sec, about $10^7$ times $\nu_t a^2$. With the established $\Delta E_B$ value of 7 kcal/mol [14, 21,23, 24], we deduce $E_2$=3.3 kcal/mol. The fitting quality is excellent over the entire temperature range of our experiment, spanning close to four orders of magnitude of values of D, as shown by the solid line in Fig. 1(a).

Although mathematically the above fitting would hold, physically, it is a puzzle why the diffusion prefactor along the steps is so much different from that on the terraces. Realizing that Eq. (3) was derived under an assumption of equilibrium coverage distribution between the steps and terraces, which *eliminates* any net cross-channel mass transport between the terraces and steps, we performed a Monte Carlo simulation to lift this assumption [29]. In the simulation, we used the terrace diffusion parameters as given above, $\Delta E_B$=7 kcal/mol, $\Delta E_S$=0, and normal prefactors ($10^{-3}$ cm$^2$/sec) for diffusion across steps and along the steps. The initial condition of the adsorbates was set as a coverage grating with 5% modulation. By adjusting the activation energy $E_2$ to be 2 kcal/mol, we are able to fit our experimental data very well, indistinguishable from the solid fitting line in Fig. 1(a) [29]. The success of our simulation with normal prefactors implies that (i) Eq. (3) is not appropriate for analysing the physics of diffusion under non-equilibrium conditions, where cross-channel mass transport is allowed; (ii) without properly taking into account the inhomogeneity of the surface potential energy and the diffusion conditions, a correct deduction of prefactor is impossible. The physical picture is then the following. At low temperature diffusion mainly takes place on the terrace. Given that there are L (~24) times more terrace sites than step sites, the overall rate is very close to that of the terrace rate. At high temperatures, the activation of hopping out of the step edge sites makes the process for CO diffusion from terrace areas at higher coverage to steps, then along step edges, and finally back to terrace areas at lower coverage effective to the terrace mass transport. Because of the fast diffusion along the step edges due to small $E_2$, this mechanism actually dominates the mass transport for terrace CO. This is why the slope at high temperature is close to $E_t+\Delta E_B$. The simulation indicates that the extremely large $\nu_2 a^2$ obtained by fitting the data to Eq. (3) is fictitious due to non-equilibrium experimental conditions. Here, we must point out that the diffusion coefficient D deduced by Eq. (1) from both the experiment and Monte Carlo simulation results with a constant $\tau$ is only an approximation under non-equilibrium conditions. In principle, the accurate D value must be extracted in the long time regime of the coverage grating decay, which imposes a significant challenge to the experiment.

Our experimental results also contain richer physics than the simple model used in the derivation of Eq. (3). At temperatures above 225 K, data from all three samples (A-, B-, and AB-type steps with 2° miscut) fall on a single line and can be interpreted as due to the same mechanism, i.e., diffusion via step edges. This is consistent with earlier observations that the value $\Delta E_B$=7 kcal/mol [14,21,23,24] appears to be insensitive to the types of steps and the local step morphology. The unexpected surprise is that our diffusion data parallel to A- and AB-type steps clearly follow another trend at low temperature. At temperatures lower than 225K where the above step-edge enhancement channel can contribute little, yet another enhancement channel becomes dominant over the terrace diffusion rate. We see from Fig. 1(a) that our data for both of these samples in this temperature range can be fitted by a single Arrhenius curve with an apparent activation energy of 2.1 kcal/mol and prefactor 2 x $10^{-8}$ $cm^2$/sec. This corresponds to a process with an activation energy substantially less than the terrace value $E_t$ and does not see an extra binding energy $E_B$ of 7 kcal/mol. Following the line of argument in Ref. [9], it is therefore necessary to add another term to Eq. (3) to account for this experimental observation for diffusion parallel to A- and AB-type steps:

$$D_{//} = D_t\left(1+\frac{\nu_2}{L\theta^2\nu_t}e^{-(E_t-E_2-\Delta E_B)/kT}+\frac{\nu_3}{L\theta^2\nu_t}e^{-(E_t-E_3-\Delta E_B^{'})/kT}\right). \quad (4)$$

Fitting the data with Eq. (4) results in $E_3+\Delta E'_B$ =1.9 kcal/mol and $\nu_3 a^2$ = 7 x $10^{-9}$ $cm^2$/sec. We must stress that the use of Eq. (4) is only to point out there must be more than two diffusion channels on stepped surfaces in order to account for our experimental observation. Because of the non-equilibrium conditions of our diffusion measurement, the parameters deduced by Eq. (4) do not have clear physical meanings as already discussed above.

It is believed [21,24,30,31] that the step edge channel for CO diffusion is on the upper step edge. For the origin of the low temperature channel dominating diffusion parallel to A- and AB-type steps, we can *speculate* on possibilities based on the observed energetics and the observed step orientation dependence. The process is undoubtedly related to steps (A- and AB-types) in some manner, and yet any extra binding energy $\Delta E'_B$, if present, appears to be much smaller than $\Delta E_B$ at the step edge sites, since the overall experimental activation energy is only about 1.9 kcal/mol. From this, it appears likely that the path of adsorbate motion might be close to, but not at the step edge. One probable choice is the first available row next to the step on the lower terrace. It is easy to see that, on the upper step edge path, diffusing molecules along A- and B-type will see the same local geometry. However, along the lower path, the local geometry is *different* for the two types of steps, offering the opportunity to account for the observed difference in diffusion rates along A- and B-type steps. We have the additional observed data that this low activation energy path is also effective for AB-type steps. If we restrict ourselves to atop sites near the step edge on the lower terrace, we can see that the path for A- and AB-type steps is farther away from the edge than the path for B-type, as indicated in Fig 3. Naturally, these proposed paths remain speculative in nature; and, further, even if one assumes these paths to be correct, it is still necessary to show how this difference in distances from the step edge can lead to different potentials governing the motion of the CO molecule. This difference in potential might possibly come from intrinsic potential perturbation from the step structure, or from CO–CO interaction due to the presence of CO molecules on the upper step edge. Much future work will be required to check whether these speculations are valid.

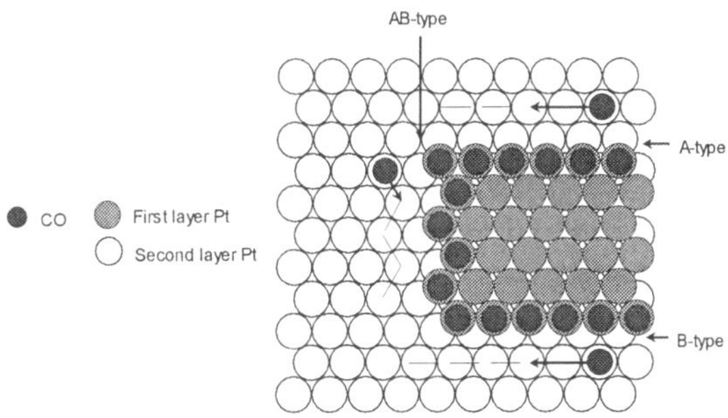


*Figure 3.* The atomic arrangement for the three types steps and the speculated diffusion paths for the low temperature channel along A- and AB-type steps.

A last issue to address is the step density dependence of diffusion parallel to AB-type steps. For the high temperature channel, it does not show a strong dependence. For the low temperature channel, the diffusion seems to increase as the density of steps increases from 1° to 2° miscut (Fig. 1). However, this trend does not continue for 4° miscut sample, on which the enhancement of diffusion disappears in the entire temperature range. The reason for such a behavior might be the following. When the terrace size on the average is large enough, the step-step interaction is negligible. Thus, the modelling of straight step edges seems to hold over the length scale comparable to the grating period in our experiment. Once the terrace width is further reduced, the step-step interaction becomes important and the step morphology may become completely different. At this stage, the diffusion parallel to steps becomes meaningless, since the steps can no longer be treated as straight and parallel. The crossing of steps might terminate the enhancement effect observed for parallel to steps.

## 5. Conclusions

In summary, we have presented the CO diffusion results perpendicular and parallel to steps on vicinal Pt(111) surfaces. Diffusion perpendicular to steps is straightforward and is determined by two parts: terrace diffusion and cross-step diffusion. With increasing step density, the diffusion perpendicular to steps is slowed down dramatically and no step orientation dependence was observed. Based on nearest-neighbor site hopping, we concluded from our diffusion measurement that the Schwoebel barrier is very small and could even be negative. This indicates the observed slow-down of diffusion is likely due to step edge trapping rather than an extra step edge barrier. For diffusion parallel to steps, enhancement was observed for all three types of stepped surfaces at high temperatures, which was identified as activation of the fast step edge diffusion in contributing to the mass transport on terraces. At low temperatures, an enhancement was found for A- and AB- type steps, which requires the assumption of a third diffusion channel parallel to the step edges to explain the data. Further observation shows that the enhancement can only be observed on small miscut samples, since the step may become very wavy if the step density is high, at which the straight step assumption fails and the enhancement along steps terminates.

## Acknowledgments

The author is grateful for the contribution to this work from Jianwei Ma, Lei Cai, M. M. T. Loy, X. R. Wang, and Zhenyu Zhang at various stages, and also thankful to Prof. S. C. Ying for beneficial discussions. We wish to acknowledge the financial support from the Research Grant Council of Hong Kong through Grant No. RGC-HKUST684/96P, and the William Mong Solid State Cluster Laboratory.

## References:

1. C. A. Roulet, *Surf. Sci.* **36**, 295 (1973).
2. N.-J. Wu, H. Yasunaga, and A. Natori, *Suf. Sci.* **260**, 75 (1992).
3. M. V. Area, E. D. Westre, D. E. Brown, J. Kutzner, and S. M. George, *Surf. Sci.* **325**, 151 (1995).
4. E. D. Westre, D. E. Brown, J. Kutzner, and S. M. George, *Surf. Sci.* **302**, 280 (1994).
5. M. V. Area, E. D. Westre, and S. M. George, *J. Chem. Phys.* **96**, 808 (1991); *Surf. Sci.* **261**, 129 (1992).
6. Jianwei Ma, Xudong Xiao, and M. M. T. Loy, *Surf. Sci.* **436**, L661 (1999).
7. Jianwei Ma, Lei Cai, Xudong Xiao and M. M. T. Loy, *Surf. Sci.* **425**, 131 (1999).
8. A. Natori and R. W. Godby, *Phys. Rev. B* **47**, 15 816 (1993).
9. J. Merikoski and S. C. Ying, *Phys. Rev. B* **56**, 2166 (1997).
10. S.-C Wang, T. T. Tsong, *Surf. Sci.* **121**, 85 (1982).
11. Tsu-Yi Fu, Hao-Tse Wu, and T. T. Tsong, *Phys. Rev. B* **58**, 2340 (1998).
12. K. Bromann, H. Brune, H. Roder, and K. Kern, *Phys. Rev. Lett.* **75**, 677 (1995).
13. J. A. Meyer, J. Vrijmoeth, H. A. van der Vegt, E. Vlieg, R. J. Behm, *Phys. Rev. B* **51**, 14,790 (1995).
14. H. R. Siddiqui, X. Guo, I. Chorkendorff, and J. T. Yates, Jr., *Surf. Sci.* **191**, L813 (1987).
15. P. J. Feibelman, S. Esch, and T. Michely, *Phys. Rev. Lett.* **77**, 2257 (1996).
16. G. Ertl, M. Neumann, and K. M. Streit, *Surf. Sci.* **64**, 393 (1977).
17. H. Steininger, S. Lehwald, and H. Ibach, *Surf. Sci.* **123**, 264 (1982).
18. Xu-dong Xiao, X. D. Zhu, W. Daum, and Y. R. Shen, *Phys. Rev. Lett.* **66**, 2352 (1991).
19. Xu-dong Xiao, Yuanlin Xie, and Y. R. Shen, *Surf. Sci.* **271**, 295 (1992).
20. Jianwei Ma, Xudoong Xiao, N. J. DiNardo, and M. M. T. Loy, *Phys. Rev. B* **58**, 4977 (1998).

21. J. E. Reutt-Robey, D. J. Doven, Y. J. Chabal and S. B. Christian, *Phys. Rev. Lett.* **61**, 2778 (1988); *J. Chem. Phys.* **93**, 9113 (1990).
22. K. R. Roos, M. C. Tringides, *Phys. Rev. Lett.* **85**, 1480 (2000).
23. D. M. Collins and W. E. Spicer, *Surf. Sci.* **69**, 85 (1977).
24. J. S. Luo, R. G. Tobin, D. K. Lambert, G. B. Fisher, and C. L. DiMaggio, *Surf. Sci.* **274**, 53 (1992).
25. O. Sneh and S. M. George, *J. Chem. Phys.* **101**, 3287 (1994).
26. G. Hoogers, B. Lesiak-Orlowska, D. A. King, *Surf. Sci.* **327**, 47 (1995).
27. D. S. Choi, C. Uebing, and R. Gomer, *Surf. Sci.* **259**, 139 (1991).
28. C. Uebing and R. Gomer, *Surf. Sci.* **259**, 151 (1991).
29. X. R. Wang, Xudong Xiao, Zhenyu Zhang, submitted to PRL.
30. B. E. Hayden, K. Kretzschmar and A. M. Bradshaw, *Surf. Sci.* **149**, 394 (1985).
31. M. A. Henderson, A. Szabo, and J. T. Yates, Jr., *J. Chem. Phys.* **91**, 7245 (1989).

# ADSORPTION, DESORPTION AND DIFFUSION OF POTASSIUM ON METAL AND OXIDE SURFACES


G. KERNER, I. M. DANZIGER, W. ZHAO[a] and M. ASSCHER

*Department of Physical Chemistry and the Farkas Center for Light Induced Processes, The Hebrew University, Jerusalem 91904, Israel*
*a- Current address: Department of Chemistry and Biochemistry, University of Texas, Austin, U.S.A.*


## ABSTRACT


The adsorption-desorption and diffusion of potassium atoms on Ru(0001), Re(0001) and $Cr_2O_3$(0001) surfaces were studied using work function change ($\Delta\Phi$), temperature programmed desorption (TPD) and optical second harmonic generation (SHG) measurements. Work function and TPD data indicated that on both metallic surfaces the bonding of potassium atoms is typical to alkali metals, namely, highly ionic nature at low coverages that gradually change to neutral layers at coverages approaching monolayer and above. This is revealed by the shift of the desorption peak from near 1000 K at 0.05 monolayer (ML) down to 350 K at 1ML.

Work function change is maximal near 0.5 ML in the case of the metallic surface, where the balance between number of positive dipoles and neutral islands leads to $\Delta\Phi$ decrease of near 4 eV. At higher coverages $\Delta\Phi$ starts to increase, approaching the bulk potassium work function at coverages of 1 ML and above. In the case of the oxide surface, on the other hand, at the completion of the first layer, the potassium atoms are still partially charged. This is indicated by the high desorption temperature of 600 K at 1ML. Higher coverages result in the formation of 2D and 3D islands desorbing at 450 K and 350 K, respectively.

Diffusion of the potassium atoms is studied by the SH diffraction method from coverage grating. Small coverage dependence of the diffusion rate is found on the metal substrates with a barrier for diffusion of 5 kcal/mole on Re(0001) at 1ML. On the oxide surface, the diffraction method followed the descent of potassium atoms from the edges of the 3D islands to the underlying layer of potassium on top of the oxide surface. A barrier for this process of 11 kcal/mole suggests that the effective corrugation on the oxide surface is significantly higher than on the metal substrates, leading to the higher barrier. The origin for these differences between metallic and oxide surfaces in terms of the interaction with alkali metals is discussed.




*M.C. Tringides and Z. Chvoj (eds.),*
*Collective Diffusion on Surfaces: Correlation Effects and Adatom Interactions,* 129–141.

# 1. Introduction

The chemical interaction of alkali metals on top of metal and semiconductor surfaces has been studied extensively over the past two decades, among others due to their importance as promoters in catalysis [1]. A typical behavior of these systems is the profound work function decrease as the alkali coverage increases, found on metal as well as on semiconductor surfaces. It reflects the significant modification of the charge distribution at the vicinity of an adsorbed alkali metal and as a result the chemistry with co-adsorbates often changes dramatically [2-4]. Such is the case when CO and $CO_2$ co-adsorb with potassium on Re(001) [5,6]. They form a surface compound with the potassium and thus significantly strengthen their bonding. Similar behavior was shown with other co-adsorbates and on other metal surfaces [2].

In contrast to metals and semiconductor surfaces, the chemistry of alkali metals on oxide substrates, was far less investigated over the years. Gradually, however, it became the focus of increasing interest, due to the extremely wide range of applications in which these systems play an important role, both in catalysis and in the semiconductor industry [7-12]. In particular, geometric and electronic structural analysis and the resulting chemistry of small metallic particles on oxide surfaces have been studied [7,9,11]. None of the studies so far have addressed the effect of ionic substrates, such as oxides, on macroscopic diffusion of metals, in general, and alkali metals in particular. The diffusion of metal particles is very important for understanding important phenomenon such as sintering.

In this work we shall attempt to learn from the comparison of results we obtained on the chemical interaction and macroscopic diffusion [13] of potassium on metals, e.g. Ru(001) and Re(001), to that on an oxide surface – $Cr_2O_3$(0001) grown on Cr(110). Work function change and temperature programmed desorption (TPD) measurements were employed for the characterization of bonding and strength of chemical interaction of the adsorbed alkali metal. Coverage grating formation by means of laser induced desorption (LITD) combined with optical second harmonic generation (SHG) – diffraction method was the technique employed to study the macroscopic surface diffusion. This technique was previously introduced and utilized for the study of surface diffusion on metals and semiconductor surfaces [14-17].

## 2. Experimental

The details of the experimental set-up were given elsewhere [18, 19]. Briefly, it is based on a UHV chamber with a minimum pressure below $8{\cdot}10^{-11}$ Torr, equipped with sputter gun, a quadrupole mass spectrometer (QMS) and a Kelvin probe. A *p*-polarized pulsed Nd:YAG laser (Quantel YG585) at a fundamental wavelength of 1.064 μm, having a repetition rate of 10 Hz and pulse duration of 10 nsec (FWHM) was used for both potassium coverage determination and the diffusion measurements. The laser pulse was split into two beams, the high power beam was used for the coverage grating formation via LITD, while the lower power beam was used as a probe beam, based on optical second harmonic (SH) diffraction from the adsorbate grating [14-17].

The preparation and characterization of the metallic and oxide surfaces were described elsewhere in detail [18-20]. The actual structure of these surfaces was verified by low energy electron diffraction (LEED). Potassium was dosed onto the metals and oxide surfaces by passing current (4.5-6.5 A) through a commercial potassium source (SAES Getters) located approximately 20 mm from the sample. This was sufficient to produce a monolayer of potassium on the surface within 3 minutes. Potassium surface coverages were determined by first exposing the surface to ~ 1.5ML of potassium atoms and subsequently annealing the surface to a given temperature (approached at 3K/sec) in the range 300-900K. This way, highly reproducible coverages could be obtained (within 5% of a ML) in the coverage range 0-2.5ML, as determined and calibrated by integrated TPD peak area measurements.

## 3. Results and Discussion

TPD spectra of potassium from metal and the oxide surface ($Cr_2O_3$(0001)) reveal the clear differences between these two surfaces in terms of the nature of potassium bonding. The two TPD spectra are shown as a function of potassium coverage in Figures 1 and 2. The bonding to a metal surface, here to Ru(001), is characterized by an extremely strong repulsion among neighbor K atoms. This is demonstrated by the typical shift of the peak desorption all the way from near 1000 K down to 350 K as coverage increases from near zero to 1ML. The change of the TPD spectra with coverage could be nicely expressed by an electrostatic model based on dipole-dipole repulsion [3]. The vast shift of the desorption peak to lower temperatures has been explained in terms of a gradual change of the potassium adsorbate from mostly ionic in character to a neutral metallic adsorbate [1,3,21]. On Ru(001) at one monolayer, potassium forms an ordered structure of $\sqrt{3}\times\sqrt{3}R30°$, which corresponds to $5\cdot10^{14}$ potassium atoms per $cm^2$. At coverages higher than a monolayer, the only change in the TPD spectra is the growth of the peak at 350 K, reflecting the enthalpy of vaporization of K atoms from bulk of potassium.

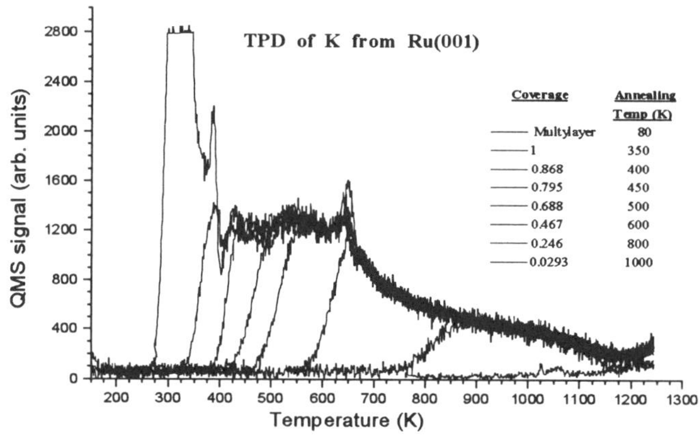


*Figure 1*: TPD of potassium from Ru(001) as a function of coverage. The potassium coverage continuously increases as the TPD curves shift from high to low temperatures, as indicated. Heating rate was 6 K/sec.

The TPD spectra of potassium from the oxide surface are clearly different, as shown in Figure 2. Here the desorption of potassium at a monolayer coverage peaks near 600 K (compared to 350 K on the metal), a significantly higher temperature which reflects a much stronger binding of potassium to the oxide surface. While there is a shift in desorption temperature on the oxide surface as well ( from near 850 to 600 K), it is far smaller than on the metal. One way to understand the unusually high desorption peak at full monolayer coverage is that potassium remains partially ionic at this coverage. On the oxide surface it means that very strong Coulombic attractive interactions exist between the negative oxygen ions on the surface and the potassium ion. Unlike the case of the metallic substrate, however, at higher coverages two more desorption peaks are seen in the TPD spectra. One near 450 K and the lowest temperature is the typical potassium evaporation peak near 350 K. The peak near 450 K was attributed to 2D potassium islands on top of a monolayer, as revealed by a combination of work function and second harmonic generation study [22]. The lowest temperature desorption peak near 350 K was attributed to 3D islands that, at higher coverages, develop into bulk potassium. Both the TPD and work function change studies suggest that the 3D islands start to form before the completion of the second layer in the form of 2D islands.

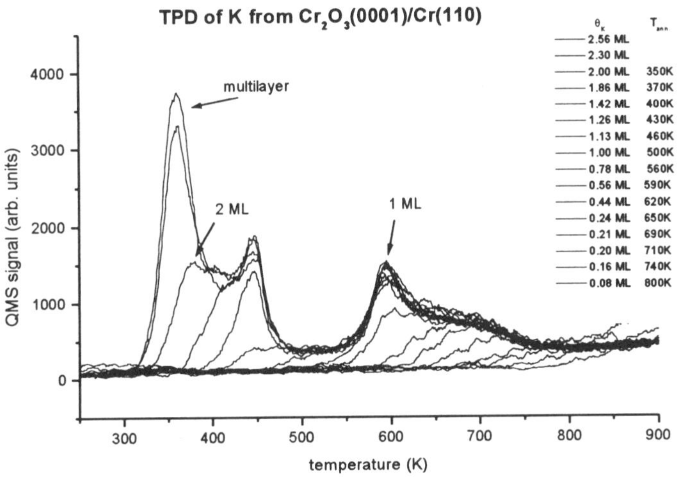


*Figure 2*: TPD of potassium from $Cr_2O_3$(0001)/Cr(110). The potassium coverage continuously increases as the TPD curves shift from high to low temperatures, as indicated. Heating rate was 10 K/sec.

A comparison of the work function change measured as a function of potassium coverage on Re(001) and on the oxide surface of $Cr_2O_3$(0001)/Cr(110) reveals again the difference in the nature of bonding. The maximum change is more than 4 eV on the metal, while only 2.5 eV is found on the oxide at half of a monolayer, where the maximum change is recorded on both surfaces, as shown in Figure 3. This difference can be attributed to the different charge distributions on the potassium adsorbate on the two substrates. It seems as if at coverages up to half of a monolayer the overall charge on the potassium is more positive on the metal than on the oxide, therefore, the change in the work function is larger. But, as coverage increases, the potassium atoms aggregate to form neutral islands on top of the metallic substrate [21,3], therefore the asymptotic value at coverages approaching a monolayer (and above) coincides with the work function of the bulk potassium. In the case of the oxide substrate, however, the effective positive charge near the potassium atom is smaller at half of a monolayer, thus leading to a smaller work function change. At higher coverages, on the other hand, a faster increase of the work function is found as coverage increases above half of a monolayer, to values higher than the bulk potassium work function. As indicated from the TPD data, at one monolayer the high binding energy of potassium suggests that it is still partially ionic. An independent assignment of the ionic character of potassium on chromium oxide at one monolayer was provided also from XPS measurements [23]. The steep increase of the work function above half of a monolayer could qualitatively be explained by a rearrangement of the potassium ions in such a way that they form a surface compound with the oxygen ions on the oxide surface [22]. In this compound the potassium ions adsorb below the oxygen ions, thus forming an opposite dipole. Alkali metal oxides were characterized on other oxide surfaces using photo-emission spectroscopy. The formation of reversed dipole structures was suggested before to explain the work function *increase* in the coadsorption of potassium with $H_2O$ [2]. The absolute numbers of the work function change on the oxide should be treated semi-quantitatively only though, since the precise number of potassium atoms at a monolayer on the oxide surface has not yet been determined.

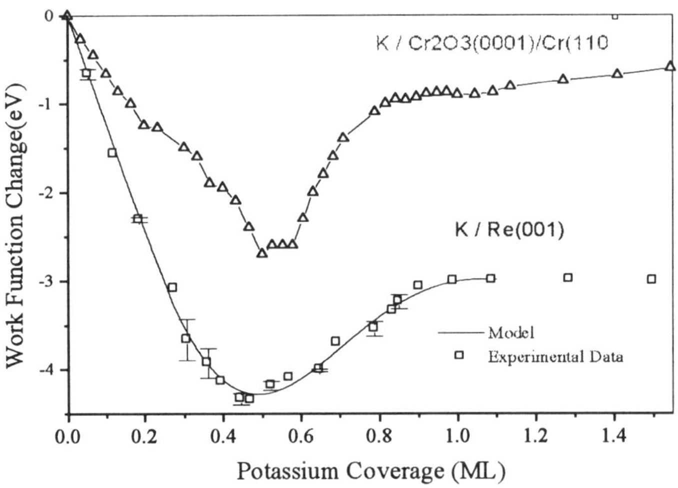


*Figure 3*: Work function change as a function of potassium coverage on $Cr_2O_3(0001)/Cr(110)$ (upper curve) and on Re(001) (lower curve).

Diffusion measurements were performed on the two metal surfaces, Re(001) and Ru(001) and also on the surface of chromium oxide. These measurements were performed using the laser induced thermal desorption (LITD) technique to create coverage grating of potassium on these surfaces. The first order second harmonic (SH) diffraction was then used as the monitor for the presence of a periodic coverage modulation. As the first order diffraction peak decayed at higher surface temperature, it followed in real time the diffusion process as it progressed and smeared out the periodic coverage modulation [14-17]. The initial overages were chosen to be compatible with laser pulse energies to avoid damage of the substrate. This limited the LITD process to desorption temperatures near 350-400 K. In the case of potassium desorption from the metal substrates, this temperature range meant a coverage in the range > 0.8ML. In the case of the oxide, however, the coverage was higher and as will be shown below it was limited to the 3D islands that desorb near 350 K. To demonstrate the actual grating formed on the surface, the diffraction pattern obtained from potassium grating on $Cr_2O_3$(0001)/Cr(110) is shown in Figure 4.

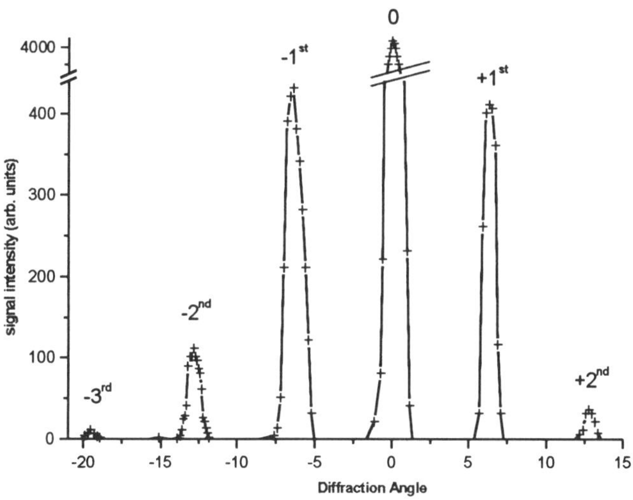


*ure 4*: SH diffraction spectrum from K coverage grating on $Cr_2O_3$(0001)/Cr(110). The ial potassium coverage was 2.5 ML.

The diffraction pattern shown in figure 4 represents a relatively deep modulation into the initial potassium coverage of approximately 3ML by using laser power of 15 $MW/cm^2$. In order to follow surface diffusion, we typically use a lower laser power that results in shallower grating troughs. The decay of a shallow modulation (estimated here to be 10-20% of the coverage at the 3rd layer) is expected to be very near an exponential decay and approximately representing the diffusion at the initial coverage [16,17].

Typical decays of the first order SH diffraction peaks while monitoring the diffusion of potassium on Re(001) at different temperatures at initial coverage of 1ML are shown in Figure 5. Similar decays are observed when the grating is formed on an initial potassium coverage of 3.1ML on the oxide surface. This is demonstrated in Figure 6 for different temperatures and initial coverages. The coverage dependence shown in Figure 6 (a) is a demonstration of the strong lateral effects expected to influence the diffusion process when partially charged species are interacting. As discussed above, while the diffusing species are clearly the atomic potassium in direct contact with the metal substrate in both the Re(001) (Figure 5) [17] and Ru(001) (not shown), it is not so simple in the case of the oxide surface. For example, one cannot rule out the effect of grating profile change as coverage increases on the measured decay rate [5,17].

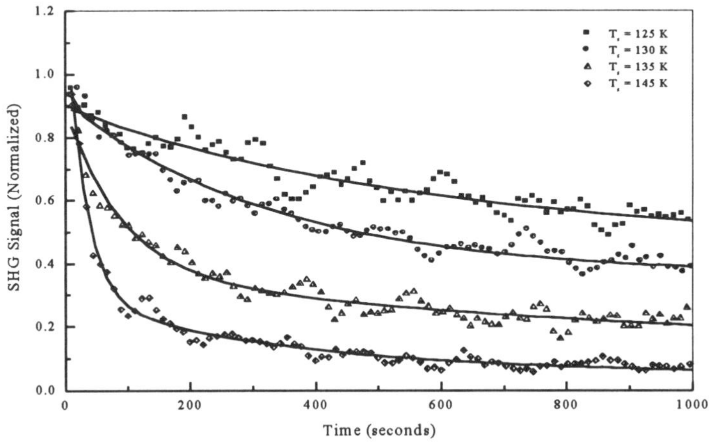


*Figure 5*: Decay of the 1st order SH peak during the diffusion of K on Re(001) at the indicated surface temperatures.

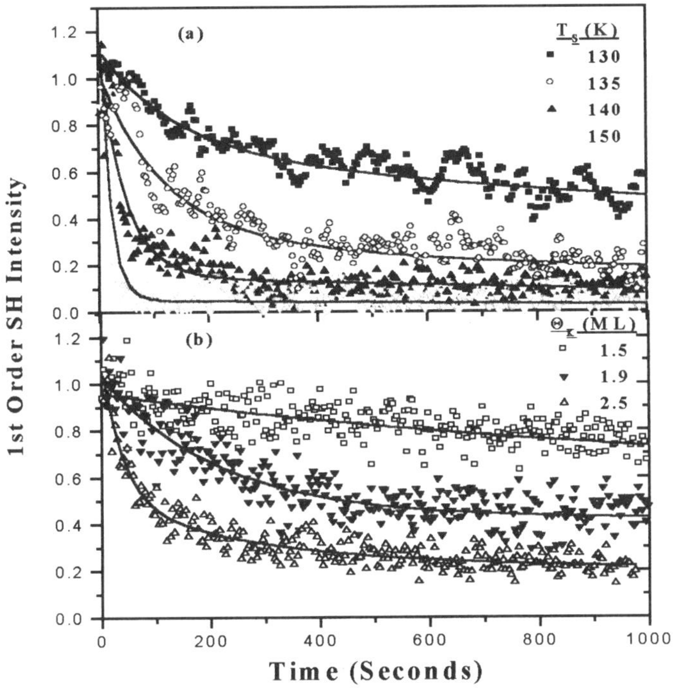


*Figure 6*: Decay of the 1st order SH diffraction peak during the diffusion of K on $Cr_2O_3(0001)/Cr(001)$ at the indicated temperatures (a) and at different initial coverages at 150 K (b).

When potassium diffuses on the oxide surface, one must assign the character of the diffusing species. It needs to be clarified whether potassium atoms diffuse in the first layer that directly interact with the substrate surface ions or in 2D islands where the interaction is with the under laying potassium atoms (ions) in the first layer, or atoms within 3D islands. These are distinguished above by the specific peak desorption within the TPD spectra of each of these groups of atoms. In order to address this question, 10 laser pulses at identical power as that used in the grating formation were used to strike the potassium covered surface without forming a grating (by removing the beam splitter). TPD was then recorded of the remaining potassium atoms on the surface. A spectrum of this kind is shown in Figure 7. One can clearly see the exclusive removal of potassium atoms from the lowest desorption peak near 350 K.

These atoms were thought to belong to 3D islands that form on top of a smooth first layer that spreads over the oxide surface (the first layer atoms desorb near 600 K).

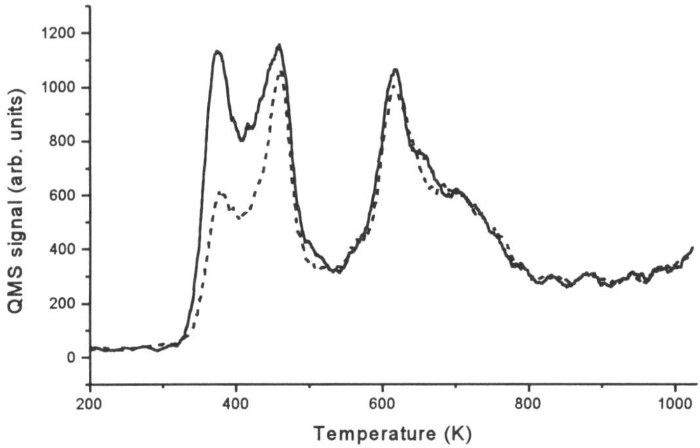


*Figure 7*: TPD of potassium from chromium oxide at initial coverage of 2.5 ML at a heating rate of 10 K/sec before (solid line) and after (dashed line) 10 high power laser strikes.

We conclude from Figure 7 that coverage modulation is created by the LITD method within the distribution of 3D islands that are formed near 3 monolayers. The diffusion process we monitor, therefore, follows potassium atoms that descend from the edge of these 3D islands and add-up to the edges of 2D island on top of the under laying homogeneous first layer of the potassium atoms. The temperature dependence of this process has been analyzed as an Arrhenius plot to extract the apparent activation energy for this process [22]. A barrier of 11 kcal/mole was obtained at the coverage range between 2-3ML. Above this coverage it started slightly to descend to 9 kcal/mole. Based on our conclusion that the rate limiting diffusion we follow represents the process of atoms descending or hopping over the 3D islands edges, we assign the activation energy for diffusion to the Schwoebel-Ehrlich barrier for atoms to descend from metal islands to the smooth layer underneath. Other processes that take place on 3D islands, such as breaking from kinks and edges should not affect the coverage modulation formed by the LITD method, therefore, we cannot be sensitive to them. Comparison of the barrier for diffusion measured on the metallic surfaces of about 5 kcal/mole, suggests that the Schwoebel-Ehrlich barrier for atom descent is almost twice the barrier for diffusion on the smooth metal. This relatively large difference may, however, be specific for the case here where the descent is from metallic 3D islands to partially ionic first layer of potassium on chromium oxide.

## 4. Summary

The chemical bonding, work function change, and diffusion of potassium atoms were measured on two hexagonal closed packed metallic surfaces of Re(001) and Ru(001) and on top of an oxide surface, $Cr_2O_3$(0001)/Cr(110). The results were compared and a very clear effect of the substrate was found. On the metallic surfaces, the typical strong lateral repulsion among neighbor adsorbates stretched the TPD spectra as a function of coverage all the way from 1100 K down to 350 K, where the desorption from the first monolayer coincided with the bulk potassium enthalpy of vaporization. Atoms near the completion of one monolayer are believed to be practically neutral potassium atoms. On the oxide surface, on the other hand, the peak desorption observed from a monolayer covered surface was found near 600 K. The strong binding energy implied by this high desorption temperature, suggested that potassium is partially positively charged even at the completion of the first monolayer, which leads to a strong attraction with the negatively charged oxygen ions at the surface. Work function change measurements imply to the formation of surface compound between potassium and the surface oxygen which may explain the fact that asymptotic values do not coincide with those of bulk potassium, as is the case in the metallic substrates.

Diffusion measurements, performed using the optical SH diffraction form potassium coverage grating, indicated that the diffusing species at coverages near one monolayer on the metallic substrates were neutral atoms, directly interacting with the substrate surface atoms, as concluded from the relatively small barrier for diffusion of ca. 5 kcal/mole. On the oxide surface of $Cr_2O_3$(0001)/Cr(110), we have assigned the rate limiting mechanism for surface diffusion, to potassium atoms descending from the edges of 3D islands to the underlying first layer of potassium atoms in direct contact with the oxide surface. This Schwoebel-Ehrlich barrier was found to be two times higher than the barrier for diffusion on the smooth metallic surface.

## Acknowledgments

This work has been partially supported by the Israel Science Foundation and the Ger Israeli Foundation. The Farkas Research Center for Light Induced Processes is supporte the Minerva Gesellschaft für die Forschung, GmbH, München.

## References

1. Bonzel, H. P., Bradshaw, A. M., and Ertl, G., (eds.), (1989) Physics and Chemistry of Alkali Metal Adsorption, in *Materials Science Monographs*, 57, Elsevier, Amsterdam.
2. Bonzel, H. P., and Pirug, G., (1993) Coadsorption, Promoters and Poisons, in King, D. A. and Woodruff D. P. (eds.), *The Chemical Physics of Solid Surfaces, Vol 6,* Elsevier Science Amsterdam.
3. Verhoef, R., W., Zhao, W., and Asscher, M., (1997), *J. Chem. Phys.*, **106**, 9353, and references therein.
4. Song, K., J., Heskett, D., Dai, H., L., Liebsch, A., and Plummer, E., W., (1988), *Phys. Rev. Lett.*, ***61***, 1380.
5. Zhao, W., and Asscher, M., (1999), *Surf. Sci.,* **429**, 1
6. Zhao, W., Chacham, I., Vackart, Y., and Asscher, M. (1999), *Isr. J. of Chemistry*, **38(4)**, 399.
7. Noguera, C., (1996) *Physics and Chemistry at oxide surfaces*, Cambridge University Press, Cambridge.
8. Street, S. C.,Xu, C., and Goodman, D. W., (1997), *Ann. Rev. Phys. Chem.*, **48**, 43.
9. Freund, H.-J., (1997) *Angew. Chem. Int. Ed. Engl.*, **36**, 452.
10. Campbell, C. T., (1997) *Surf. Sci. Rep.*, **27**, 1.
11. Valden, M., Lai, X., and Goodman, D. W., (1998) *Science*, **281(**5383), 1647.
12. Valden, M., and Goodman, D. W., (1999) *Isr. J. of Chem.*, 38(4), 285.
13. Gomer, R., (1990) *Rept. Prog. Phys.* **53**, 917.

14. Zhu, X. D., Rasing, Th., and Shen, Y. R., (1988) *Phys. Rev. Lett.* **61**, 2883.

15. Xiao, X.-D., Xie, Y., and Shen, Y. R., (1992) *Surf. Sci.* **271**, 295

16. Zhu, X. D., (1992) *Mod. Phys. Lett.* B **6**, 1217.

17. Zhao, W.,Verhoef, R. W., and Asscher, M., (1997) *J. Chem. Phys.*, **107**, 5554.

18. Rosenzweig, Z., and Asscher, M., (1992) *J. Chem. Phys.* **96**, 4040.

19. Livneh, T., and Asscher, M., (1997) *J. Phys. Chem.*, **B 101**, 5705.

20. Rohr, F., Bäumer, M., Freund, H.-J., Mejias, J. A., Staemmler, V. Müller, S., Hammer, L., and Heinz, K., (1997) *Surface Science* **372**, L291.

21. Neugebauer, J., M., Scheffler, M., (1993) *Phys. Rev. Lett.* **71**, 577.

22. Zhao, W., Kerner, G., Asscher, M., Wilde, M., Al-Shamry, K., Freund, H.-J., Staemmler, V., Wieszbowska, M., (2000) *Phys. Rev. B* **62(11),** 7527.

23. Wilde, M., Beauport, I., Stuhl, F., Al-Shamery, K., Freund, H.-J. (1999) Phys. Rev. **B 59**, 13401.

# DYNAMICS OF INTERACTING ADPARTICLES UNDER NON-EQUILIBRIUM CONDITIONS

Z.Chvoj
*Institute of Physics, Academy of Sciences of the Czech Republic*
*Cukrovarnická 10*
*162 53 Prague 6*
*Czech Republic*

The generalized (wave-vector $\vec{q}$ dependent) diffusion coefficient $D(\vec{q})$ is determined for a system of interacting adsorbed particles under non-equilibrium conditions, characterized in particular by the strong gradient of coverage. Our contribution is based on calculations of dynamic structure factor, with Mori's projection operator method being used to determine $D(\vec{q})$. It was shown that the diffusion coefficient depends on the state of the system, even within the limit $\vec{q} \to 0$. The deviations of the Green - Kubo formula $D_c^K$ (derived for the relaxation of fluctuations around the equilibrium state) from the non-equilibrium results reflect the influence of the gradient of coverage on the collective behavior of adparticles. The diffusion coefficient under non-equilibrium conditions can be higher or less than $D_c^K$, depending on the character of atom interaction. In the case of repulsive interaction and the increasing role of the ordering mechanism, these deviations increase with interaction energy or with a decreasing of temperature.

## 1. Introduction

Surface diffusion influences many processes affecting surface technologies, e.g., growth of thin layers, transport of mass, catalysis or chemical processes, etc., and hence, it has been intensively investigated in recent years. Even if many questions were answered in connection with the particular experimental problems, many fundamental problems remain unsolved. One of these is the understanding of diffusion processes during the formation of ordered phases, i.e., the relaxation of a system from a state far from equilibrium to the ordered equilibrium phases. In ordered phases experimentally determined diffusion coefficients usually show local extrema (maxima or minima). The diffusion processes were simulated by Monte Carlo calculations, the molecular dynamics method or by methods based on the master equation. The standard analytical approach for calculations of collective diffusion coefficient is the dynamical mean-field theory (DMF) [1], in which we



*M.C. Tringides and Z. Chvoj(eds.),*
*Collective Diffusion on Surfaces: Correlation Effects and Adatom Interactions,* 143–155.

neglect memory effects. This theory determines collective diffusion coefficient, $D_c$, as a time correlation of density fluctuations in a system in equilibrium multiplied by the thermodynamical factor. The roles played by memory effects in tracer and collective diffusion coefficients were studied by means of Monte Carlo simulations [2]. It was shown that in strongly interacting cases (where order is present), the memory effects play an important role, considerably influencing the effective energy barrier for diffusion. In this case, the results obtained from Monte Carlo simulation differ substantially from DMF theory results.

Even analytical theories gave us the answers to many questions, as they study systems in or near equilibrium conditions and describe the relaxation of an adatom system to equilibrium within the hydrodynamic limit. But in many experiments we meet with non-equilibrium conditions, with the non-equilibrium distribution of adatoms, which in a finite time period forms, for example, the ordered equilibrium phase. Monte Carlo simulations of the dynamics of adatoms in such conditions [3] show the time-dependence of the diffusion coefficient. The large gradients of coverage also represent systems far from equilibrium. We encounter large gradients on the boundaries of islands, steps or another nanostructures, during the deposition of atoms in strip shape, etc. It is problematic to describe the dynamics of such systems by methods which consider the relaxation of fluctuations around the equilibrium state. For a greater understanding of the dynamics of adatoms under non-equilibrium conditions, we follow adatoms within the very simple lattice gas model. We will describe the dynamics of interacting adatoms by the time correlation functions of the distribution of atoms, i.e., by the dynamic structure factor. We use the projection operator method described by Kawasaki [4].

Extracting a generalized diffusion coefficient from the dynamic structure factor, we obtain several interesting results. First we studied the system in its disordered phase. The diffusion coefficient obtains the imaginary part, which disappears within the limit $\vec{q} \to 0$. The real part depends on the gradient of surface coverage and its limit at $\vec{q} \to 0$ can be higher or less than this coefficient under a homogeneous distribution of atoms on surface. The value depends on the character of the atom interaction.

We expect more interesting results in systems near or even under the critical temperature of phase transition, under temperatures where the role of ordering increases. Such a system was studied by Chvoj [5]. It was shown in this work that the deviations of the diffusion coefficient under non-equilibrium conditions from its equilibrium value is more pronounced near the phase boundary of order-disorder phase transition.

## 2. Non-equilibrium Dynamics

Since we are investigating the basic features of the dynamics of adsorbed atoms in non-equilibrium conditions, we select a simple model which enables us to give a clear interpretation of our results. The dynamics of an adatom system is studied within a lattice gas model with square symmetry. The non-equilibrium conditions are characterized by the gradient of coverage $\Theta$, whose direction coincides with that of the lattice axe. We suppose jumps to occur only into the nearest-neighbor positions. The interaction of atoms is restricted to the nearest-neighbor sites and we focus on the study of one-dimensional diffusion in the direction of grad $\Theta$. Let the positions on the square lattice be indicated by the pair of indexes $(i,j)$. Index $i$ runs across the positions in a direction parallel with the gradient of coverage. Index $j$ runs across lattice positions perpendicular to the coverage gradient. In order to proceed with the determination of the diffusion coefficient, we introduce some simplifications. First, we assume local equilibrium in the lines of atoms perpendicular to the gradient of coverage. This condition enables us to introduce a mean value of coverage in these lines, which is a function of lattice site $(i,j)$ - $\Theta(i)$ and changes only in the direction of grad $\Theta$.

Let us take a surface with $N*N$ adsorption positions. $p_{i,j}$, $i,j = 1,2,...N$ represents the occupation number of $(i,j)$th position on the square lattice. $p_{i,j} = 1$ if the $(i,j)$th position is occupied and $p_{i,j} = 0$ if it is empty. The mean value of $p_{i,j}$ is equal to the mean coverage $\Theta(i)$, which is the relative occupancy of adsorption sites.

Information about the dynamics of atoms on the surface is obtained in the time correlations of the adatom distributions - the Fourier transforms of which is the well known dynamic structure factor. This is important not only from a theoretical point of view but, owing to the progress of experimental techniques, it allows us to measure dynamic structure factor directly [6]. We study a system far from equilibrium and thus these correlations depend on the initial time $t$ from which they are calculated. This means that the dynamic structure factor will depend on time - $S(\vec{k},\omega,t)$. We shall suppose the relaxation of the system to equilibrium to be the slowest process with its relaxation time much longer than the characteristic time of the correlation functions decay. This means that the mean coverage is constant during the evaluation of correlation functions and we omit the variable $t$ where no confusion arises. We start our analysis from the definition of $S(\vec{k},z,t)$:

$$S(\vec{k},\tau,t) = \frac{1}{n} < \delta A(-\vec{k},t+\tau)\delta A(\vec{k},t) > \tag{1}$$

where $n$ is the total number of atoms and

$$\begin{aligned} \delta A(\vec{k},t) &= \sum_{i,j}(p_{i,j}(t) - \Theta(\vec{r}_{i,j},t))\exp(-i\vec{k}\vec{r}_{i,j}) \\ A(\vec{k},t) &= \sum_{i,j} p_{i,j}(t)\exp(-i\vec{k}\vec{r}_{i,j}) \end{aligned} \tag{2}$$

$\vec{r}_{i,j}$ is the position vector of $(i,j)$-th lattice site, $\Theta(\vec{r}_{i,j},t)$ is mean coverage at positions $(i,j)$ and at time $t$. The question arises as to what the average value $< A, A >$ means. In experiments and also in Monte Carlo simulations, this mean value is evaluated as an ensemble mean value over the observed area. In theoretical analysis, e.g. within the linear response theory, this mean value is usually substituted by the average ensemble value of the system in equilibrium and the deviations are supposedly small. But now we will interest ourselves in situations arbitrarily far from equilibrium and will have to use an extended theory which respects the non-equilibrium distribution of atoms. The basis of our following analysis is found the theory published by Kawasaki [4].

We consider a classical system whose microscopic state is given by the point in the phase space $X$. The normalized microscopic distribution function at a time $t$ is denoted by $F(t,X)$. The time evolution of the system is given by the Liouville equation:

$$\frac{\partial F(t,X)}{\partial t} = LF(t,X) \tag{3}$$

where $L$ is the Liouville operator. If the system is in a non-equilibrium state, the mean values of dynamic variables in $F(t,X)$ are different from their equilibrium values and they are functions of time. It is very difficult to determine the distribution function $F(t,X)$ and thus the reduced distribution function is introduced, which has a more clear form and the appropriate mean values are identical with the mean values in $F(t,X)$. This reduced distribution function describes the distribution of the fluctuations around the non-equilibrium state in the same form as the canonical distribution of fluctuations around the equilibrium state. We introduce the canonical local equilibrium state, which is generally described by the distribution function $F^c(t,X)$:

$$\begin{aligned} F^c(t,X) &= \exp[\beta H^c(t,X)] & (4)\\ H^c(X) &= -H(\{p_{i,j}\}) + V(G,\{p_{i,j}\}) & (5) \end{aligned}$$

where $\beta = 1/k_BT$, $k_B$ is the Boltzmann constant, and $T$ temperature $\{p_{i,j}\}$ represents the set of occupation numbers which corresponds to the particular distribution of atoms on the surface. $H(\{p_i\})$ is the Hamiltonian of the system in configuration $\{p_i\}$, $G$ represent the 'external field' and $-V(G,\{p_{i,j}\})$ the energy of the adatoms in this field. $G$ is so chosen as to reproduce the same average values of $p_{i,j}(t)$ as in $F(t,X)$ as in $F^c(t,X)$. This means that these average values correspond to the non-equilibrium distribution of atoms.

The mean values and correlation of occupation numbers are then given by the relations:

$$\begin{aligned} < p_{i,j} >_{ne} (t) &= \int p_{i,j} F^c(t,\{p_{i,j}\}) dp_1 ... dp_{N*N} & (6)\\ < p_{i,j} ... p_{k,l} >_{ne} (t) &= \int (p_{i,j} ... p_{k,l}) F^c(t,\{p_{i,j}\}) dp_1 ... dp_{N*N} & (7) \end{aligned}$$

We suppose a dominating role of the gradient of coverage to occur, along with the lesser role of the interaction of atoms on their distribution. The modification of position correlations due to atom interaction was studied by Chvoj [5]. Within the framework of this model, we shall suppose the above mean values in the form:

$$\begin{aligned} < p_{i,j} >_{ne} &= \Theta(\vec{r}_{i,j}) = \bar{\Theta} + g(i) & (8)\\ < p_{i,j} p_{k,l} >_{ne} &= \Theta(\vec{r}_{i,j}) \ \ (k,l) = (i,j) & (9) \end{aligned}$$

$$< p_{i,j}p_{k,l} >_{ne} \quad = \quad \Theta(\vec{r}_{i,j})\Theta(\vec{r}_{k,l}) \ \ (k,l) \neq (i,j) \tag{10}$$

$$< p_{i,j}p_{k,l}...p_{m,n} >_{ne} \quad = \quad \Theta(\vec{r}_{i,j})\Theta(\vec{r}_{k,l})...\Theta(\vec{r}_{m,n}) \ \ (i,j) \neq ... \neq (m,n) \tag{11}$$

where $\bar{\Theta}$ is the overall mean value of coverage, and the mean value of coverage at position $\vec{l}_{i,j}$ is given by $\bar{\Theta}$ modified by the function $g(i)$, which varies only in the direction of the gradient of coverage. This approximation signifies:
a) in the direction perpendicular to the gradient of coverage the mean value of occupation numbers is supposed to be constant. It is acceptable if the relaxation of fluctuations in this direction is faster than the relaxation of the gradient of concentration in the system. This assumption is the main assumption of the whole contribution, since we suppose the static structure factor does not depend on time during the decay of correlation of coverage fluctuations.
b) the mean value of occupation number changes according to the mean coverage, it means there are no ordered phases in the system. The dominant role of the 'external field' which constituents grad $\Theta$ is the main assumption. The influence of the adatom interaction and thus the correlation of simultaneous occupation are depressed. The role of interaction will be analyzed later.

Now we introduce a time-dependent projector onto $F^c(t, X)$, which is defined by ($f$ - arbitrary function)

$$P_t f(\vec{k},t) =< f(\vec{k},t) >_{ne} + \frac{< \delta A(-\vec{k},t) f(\vec{k},t) >_{ne}}{< \delta A(-\vec{k},t)\delta A(\vec{k},t) >_{ne}} \delta A(\vec{k},t) \tag{12}$$

The Mori's projection operator method allowed us to rewrite the dynamic structure factor in a form more suitable for further analysis:

$$\begin{aligned} \frac{dS(\vec{k},\tau,t)}{d\tau} &= \frac{S(\vec{k},\tau,t)}{nS(\vec{k},t+\tau)} < \delta A(-\vec{k},t+\tau) L A(\vec{k},t+\tau) >_{ne} - \\ &- \int_t^{t+\tau} S(\vec{k},s,t) M(\vec{k},t+\tau-s) ds + \qquad (13) \\ &+ \int_t^{t+\tau} < \delta A(-\vec{k},t) \exp(sL)(\partial_s P_s) R(t+\tau,s) >_{ne} \ ds \\ \partial_s P_s R(t+\tau,s) &= \delta A(\vec{k},s) < \delta A(-\vec{k},s) R(t+\tau,s) \delta A(\vec{k},s) >_{ne} \\ & \quad (nS)^{-2}(\vec{k},s)(\partial_s < A(\vec{k},s) >_{ne}) \end{aligned}$$

$M(\vec{k}, \tau - s)$ is the memory kernel (the rather complex expression is given in [3]), $R(\tau, s)$ is the random force acting upon the collective diffusion, $S(\vec{k}, t)$ is the static structure factor

$$S(\vec{k},t) = \frac{1}{n} < \delta A(-\vec{k},t)\delta A(\vec{k},t) >_{ne} \tag{14}$$

The last term of (13) represents a coupling of the drift of mean value $< A(\vec{k},t) >_{ne}$ and $A(\vec{k},t)$ through the random force $R(\tau,s)$, which is reminiscent of the mode-mode coupling mechanism. We shall now analyze the dynamic equation (13). First, we focus on the first term, i.e., at this stage we neglect memory function, and second the influence of drift terms on the dynamic structure factor. (In future investigations we hope to return to looking into the role of these terms.) Within the framework of such an approach, we obtain for the Laplace transform of $S(\vec{k}, \tau, t)$

$$S(\vec{k}, z, t) = \frac{S(\vec{k}, t)}{z + \Omega(\vec{k}, t)} \tag{15}$$

where

$$\Omega(\vec{k}, t) = -\frac{< \delta A(-\vec{k}, t) L A(\vec{k}, l) >_{ne}}{n S(\vec{k})} \tag{16}$$

The collective diffusion coefficient is determined as a hydrodynamic pole of the dynamic structure factor.

## 3. Dynamics of Atoms on a Surface

This paragraph is concerned with the determination of the dynamics of an atom on the surface. We use the theory commonly used for lattice gas systems. The dynamics of the whole system obey a master equation for the time evolution of the $p_{i,j}$:

$$\frac{dp_{i,j}}{dt} = L p_{i,j} \tag{17}$$

In our model we are interested in the component of the diffusion coefficient tensor which is parallel to the grad $\Theta$, so we proceed to the one-dimensional model. In such a description the Liouville operator $L$ is defined by the following relation:

$$L p_{i,j} = p_{i-1,j} P_{i-1 \to i} + p_{i+1,j} P_{i+1 \to i} - p_{i,j} (P_{i \to i-1} + P_{i \to i+1}) \tag{18}$$

$P_{i \to k}$ is the probability of the jump of an atom from position $i$ to site $k$ per unit time in the direction of the coverage gradient, as determined by Chvoj [5]. We now briefly shortly repeat its derivation. The notation corresponds to Fig. 1.

The probability of a single atom jump, whose nearest positions are empty, is supposed to be of the Arrhenius form:

$$W_0 = \nu \, \exp(-\frac{E_0}{k_B T}) \tag{19}$$

where $\nu$ is the frequency factor, and $E_0$ is the basic value of the activation energy of diffusion.

We then introduce the changes in the energy barrier for diffusion as a result of the interaction of ad-atoms. The total change in the energy barrier consists of two contributions. The first part results from the change in equilibrium energy of the particle due to an interaction with atoms in adjacent positions. The energy barrier is changed by $\delta E$ if an adjacent site is occupied and this change is additive. So, for this particular configuration, the energy barrier for diffusion $E$ is:

$$E(i) = E_0 - i \delta E \tag{20}$$

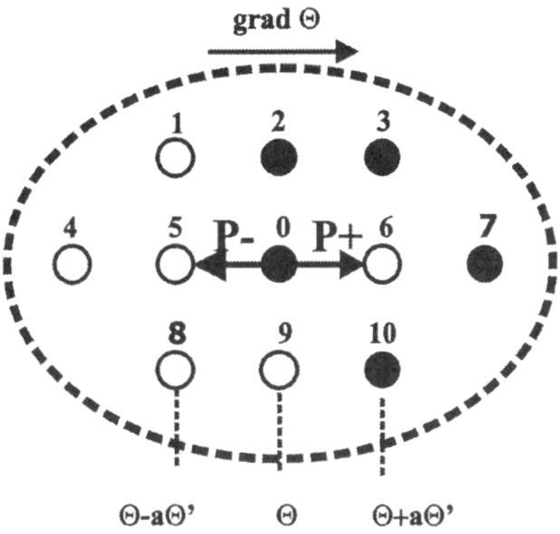


Figure 1: Microscopic model of the probability of jumps in a particular configuration, showing all adjacent sites which influence the dynamics of the atom.

if $i$ adjacent sites are occupied. $\delta E > 0$ for the repulsive interaction.

In addition to these changes in the energy barrier for diffusion, the effect of lateral interaction at the saddle point position is taken into account. Such models were also studied by Danani et al. [7], where the importance of this type of interaction on the dynamics of adatoms was demonstrated. This can accelerate or retard the diffusion. In our microscopic model the additive change of the saddle point energy increases by $k\ \Delta E$, where $k$ represents the number of atoms which influence the energy barrier in the saddle point. We assume that particles at positions 2,3,9,10 (Fig. 1), as well as at positions 5,7, will have the same effect on the saddle point (between 0 and 6) energy. Within this approximation the total energy barrier $E$ takes the form:

$$E = E_0 - i\delta E + (i + j)\Delta E = E_0 - i(\delta E - \Delta E) + j\Delta E \tag{21}$$

$i$ - is the number of atoms in positions adjacent to the initial position and $j$ is the number of atoms in positions adjacent to the final position of the jumping atom.

The resulting change in the energy barrier can be interpreted as resulting from the occupation of sites adjacent to the initial position (the change of energy is characterized by $\delta E' = \delta E - \Delta E$) and is due to the occupation of the adjacent sites of the final position (change of energy $\Delta E$). To describe this feature of adatom dynamics we introduce the parameters:

$$v_1 = \exp(\frac{\delta E'}{k_B T}) \tag{22}$$

and

$$v_2 = \exp(-\frac{\Delta E}{k_B T}) \tag{23}$$

$\Delta E > 0$ for the repulsive interaction. The probability of a jump from an $i$ to $j$ position of an adatom per unit time $P_{i\to j}$ is given in this model by the relation:

$$P_{i\to j} = W_0(1-p_j)\prod_{k\neq j}^{NNi}(1+p_k(v_1-1))\prod_{l\neq i}^{NNj}(1+p_l(v_2-1)) \tag{24}$$

$NNi$ is the summation over the nearest neighboring positions upon $i$. To determine the function $\Omega(\vec{k})$, we need to calculate the following mean values:

$$< p_i P_{i\to j} >_{ne} = W_0 < p_i(1-p_j)\prod_{k\neq j}^{NNi}(1+p_k(v_1-1))\prod_{k\neq i}^{NNj}(1+p_l(v_2-1)) >_{ne} \tag{25}$$

Relation (25) obtains the correlation of the simultaneous occupation of clusters up to eight sites. Within our model we can write for the jumps between positions 0 and 6:

$$< p_0 P_{0\to 6} >_{ne} = \Theta(0) < P_{06} >_{ne} \tag{26}$$

Following the method given by Chvoj [5], the linear dependence of the probabilities of jumps per unit time on the gradient of coverage can be expressed by the relation:

$$\begin{aligned} < P_{06} >_{ne} &= < P_{06} >^0_{ne} [1 + \Delta\ B] && (27)\\ \Delta &= a\mathrm{grad}\Theta && (28)\\ B < P_{06} >^0_{ne} &= -(1+\Theta(0)(v_1-1))^2(1+\Theta(0)(v_2-1)) \\ &\quad \{(1+\Theta(0)(v_2-1))^2(1+\Theta(0)(v_1-1)) \\ &\quad +(1-\Theta(0))[(v_1-1)(1+\Theta(0)(v_2-1))^2 - \\ &\quad 4(v_2-1)(1+\Theta(0)(v_1-1)(1+\Theta(0)(v_2-1))]\} && (29) \end{aligned}$$

$a$ is the lattice constant.
In the relation (29) $< P_{06} >^0$ is the mean jump rate in a system with a homogeneous distribution of atoms on the surface, i.e., with a constant coverage which is equal to $\Theta(0)$. This is given by the relation:

$$< P_{06} >^0_{ne} = W_0(1-\Theta(0))(1+(v_2-1)\Theta(0))^3(1+(v_1-1)\Theta(0))^3 \tag{30}$$

For repulsive interaction is $v_1 > 1$, $v_2 < 1$, and $B < 0$. The preferential jumps occur in the opposite direction to grad $\Theta$. In the case $v_1 = v_2 = 1$ is $< P_{06} >_{\acute{n}e} = W_0((1-\Theta) - a\Theta')$. In this case the change of jump probability with gradient of coverage does not depend on $\Theta$. It can be shown that because of this simple dependence of $< P_{06} >_{ne}$ on $\Theta'$ the diffusion coefficient does not depend on coverage gradient and our results are in keeping with the linear response theory. Thus, the coverage dependence of the coefficient $B$ is important.

For jumps in the opposite direction (see Fig. 1) we obtain:

$$< P_{05} >_{ne} = < P_{06} >^0_{ne} [1 - \Delta B] \tag{31}$$

To simplify the notation, we introduce for the equilibrium mean value

$$P^0(\Theta(i)) =< P_{i,i+1} >^0 /(1-\Theta(0)) \tag{32}$$

## 4. Statical Structure Factor

The statical structure factor $S(\vec{k})$ is determined in our 1D model by the relation:

$$S(\vec{k}) = \frac{1}{N\bar{\Theta}} \sum_i \sum_k [< p_{i,j} p_{k,j} >_{ne} - \Theta(i)\Theta(k)] \exp[-i\vec{k}(\vec{r}_{k,j} - \vec{r}_{i,j})] \tag{33}$$

Within the framework of our linear model, we arrive at the relation:

$$S(\vec{k}) = 1 - \frac{\Theta^2}{n} \tag{34}$$

$$\Theta^2 = \sum_i \Theta^2(i) \tag{35}$$

## 5. Dynamical Structure Factor and Diffusion Coefficient

The function which reflects the dynamics of adatoms in the dynamical structure factor is the correlation:

$$\begin{aligned} \Omega_1(\vec{k}) &= - < \delta A(-\vec{k}, t) L A(\vec{k}, t) >_{ne} \\ &= - \sum_{i,l} < p_{\vec{r}_{i,j}} L p_{\vec{r}_{l,j}} >_{ne} \exp(-i\vec{k}(\vec{r}_{l,j} - \vec{r}_{i,j}) \end{aligned} \tag{36}$$

Within the framework of linear approximations, we obtain the final expression for correlations $\Omega_1$:

$$\begin{aligned} \Omega_1(\vec{k}) &= 2\sum_j^N P^0(\Theta(j))[(\Theta(j)(\Theta(j)-1)(1-\cos(\vec{k}\vec{a})) - \\ & \quad i\Delta(\Theta(j))\Theta(j)B(\Theta(j))(\Theta(j)-1)\sin(\vec{k}\vec{a})] \end{aligned} \tag{37}$$

$\vec{a}$ is the lattice vector in grad $\Theta$ direction. We input this relation into Eq. (16) and we obtain for the function $\Omega(\vec{k})$:

$$\Omega(\vec{k}) = \Omega_R(\vec{k}) + i\Omega_I(\vec{k}) \tag{38}$$

$$\Omega_R(\vec{k}) = 2\sum_j P^0(\Theta(j))\frac{\Theta(j)(1-\Theta(j))(1-\cos(\vec{k}\vec{a}))}{n-\Theta^2} \tag{39}$$

$$\Omega_I(\vec{k}) = 2\sum_j \frac{\Delta(\Theta(j))\Theta(j)P^0(\Theta(j))B(\Theta(j))(\Theta(j)-1)\sin(\vec{k}\vec{a})}{n-\Theta^2} \tag{40}$$

In the limit case $grad\Theta = 0$ we obtain:

$$\Omega(\vec{k}, grad\Theta \to 0) = 2P^0(\bar{\Theta})(1-\cos(\vec{k}\vec{a})) \tag{41}$$

This is the familiar relation taken from former lattice gas theories.

If we return back to the diffusion processes, the chemical diffusion coefficient is determined from the parabolic dependence of $\Omega$ at the hydrodynamic limit $\vec{k} \to 0$. It is evident that our function $\Omega$ equals 0 at $\vec{k} = 0$. If we determine the second derivative of $\Omega$ at this limit, we obtain:

$$D_c = 2\frac{\sum_j P^0(\Theta(j))\Theta(j)(1-\Theta(j))}{n-\Theta^2} \tag{42}$$

We denote $D_0$ the value of $D_c$ in the limit grad$\Theta = 0$. This corresponds to the linear response theory.

Returning now to the wave vector dependence of the dynamic structure factor, the new result which we obtain is due to the presence of the imaginary part of the function $\Omega(\vec{k})$. This imaginary part may correspond to the energy of some excited state. The real part of $\Omega$ represent the inverse relaxation time of this excited state. This result is not surprising, since the non-equilibrium state is the state with the higher free energy. Not only is the entropy lower, but even the inner energy of such a state is higher in comparison with the equilibrium state. The relaxation process is the diffusion process combined with the ordering process in the case of low temperatures. Both these processes are described by the real part of function $\Omega(\vec{k})$.

In Figs. 2 and 3 diffusion coefficient $D_c$ is drawn as a function of grad$\Theta$ and the different values of interaction energies and temperature characterized by parameters $v_1, v_2$. The value of parameter $v_1 = 20$ corresponds with the interaction energy $\delta E = 0.08\ eV$ and temperature $T = 300\ K$; $v_2 = 0.2$ corresponds with the value $\Delta E = 0.04\ eV$ at the same temperature. We will express the value of grad$\Theta$ in relative values. The value grad$\Theta = 1$ represents the maximal value of the gradient of coverage on the observed area. It is evident that the diffusion coefficient depends on the state of the system. This dependence is a consequence of the break of the symmetry of jumps of atoms in or in the opposite direction to the gradient of coverage. Such a break in the symmetry is a result of the mechanism of atom interaction.

In Figure 3 the influence of the short-range correlation of the occupation of atoms on the collective diffusion coefficient in non-equilibrium conditions is presented. The short-range correlation is a result of the repulsive interaction of adatoms at low temperatures and it modifies the correlations (8-11). The details are described by Chvoj [5]. Strong deviations from the equilibrium theory

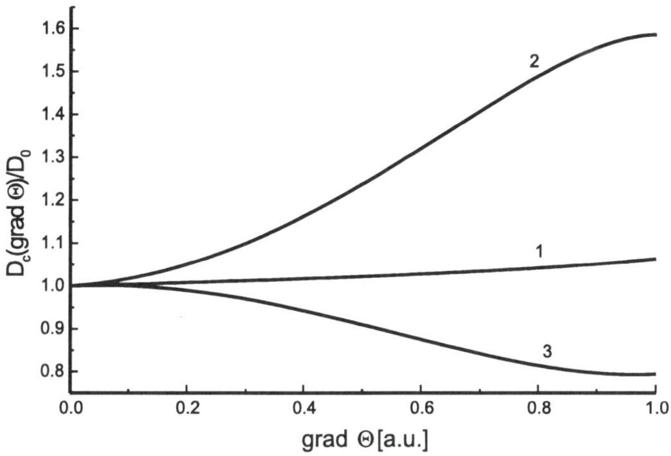


Figure 2: Dependence of the relative value of the diffusion coefficient $D_c/D_0$ on relative value of the gradient of coverage. The following different models of atom interaction were chosen: curve 1 - $v_1 = 1$, $v_2 = 1$ ($D_0 = 1$), curve 2 - $v_1 = 20$, $v_2 = 1$ ($D_0 = 1160$), curve 3 - $v_1 = 20$, $v_2 = 0.2$ ($D_0 = 250$).

appear, even for low values of interaction energy, and in comparison with the 'disordered system' (Fig. 2) the effect of non-equilibrium conditions is much more pronounced.

## 6. Conclusions

In this paper we have analyzed the role of the non-equilibrium state of the system of adatoms on its relaxation process to equilibrium. Even though we used only a very simple model and focused only on the linear terms, it was shown that the deviation of the system from equilibrium plays an important role. Not only does the correlation function $\Omega(\vec{k})$ change with the state of the system, but even the diffusion coefficient in the hydrodynamic limit has a different value from the Green-Kubo formula, i.e., from the value of $D_c$ calculated at equilibrium conditions. Such a deviation is sometimes connected with 'memory effects'. This difference is more pronounced at low temperatures, where ordered phases form. Since the state of the system depends on time, the collective diffusion coefficient is also a time-dependent value.

We can conclude that close attention must be paid to collective diffusion under

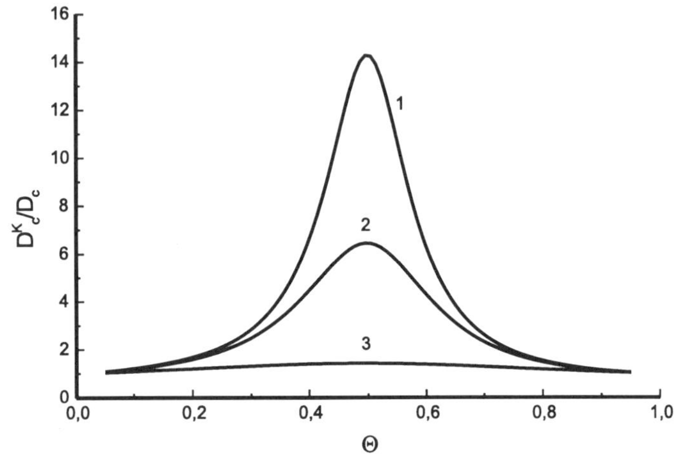


Figure 3: Ratio of the collective diffusion coefficient determined from the Green-Kubo formula, $D_c^K$ and $D_c$ for different values of the interaction energy: curve 1: $v_1 = 1.3$, $v_2 = 0.8$; curve 2: $v_1 = 1.1$, $v_2 = 0.7$; curve 3: $v_1 = 1.1$, $v_2 = 0.9$, grad$\Theta$=1.

non-equilibrium conditions and study the memory effects in these systems. Interesting results can be expected at low temperatures and on phase boundaries.

The investigation was supported by the Grant Agency of the Academy of Sciences of the Czech Republic (Grant No. A1010718).

## References


1 Ferrando, R., Scalas, E. and Torri, M. (1994) Projection-operator route to the generalized Darken equation. *Phys. Let. A* **186,** 415-418.

2 Vattulainen, I., Ying, S.C., Ala-Nissila, T. and Merikoski, J. (1999) Memory effects and coverage dependence of surface diffusion in a model adsorption system. *Phys. Rev. B* **59,** 7697 - 7707.

3 Nikunen, P., Vattulainen, I. and Ala-Nissila, T. (2000) Non-equilibrium Effects in Profile Evolution Measurements of Surface Diffusion. *Surf. Sci.* **447,** L162.

4 Kawasaki, K., (2000) Theoretical methods dealing with slow dynamics. *J.Phys.: Condens. Matter* **12,** 6343-6351

5 Chvoj, Z. (2000) Dynamics of adsorbed atoms under non-equilibrium conditions, *J. Phys.: Condens. Matter* **12,** 2135-2151.

6 Conrad, E.H., Menzel, A., Kiriukhin, S. and Tringides, M.C. (1998) Observations of Surface Temporal Fluctuations by Low Energy Electron Diffraction. *Phys. Rev. Lett.* **81,** 3175-3178.

7 Danani, A., Ferrando, R., Scalas, E. and Torri, M. (1998) Collective surface diffusion on triangular and square interacting lattice gases. *Surf. Sci.* **409,** 117-131.

# STATISTICAL-MECHANICAL DESCRIPTION OF NONEQUILIBRIUM PROCESSES IN INTERACTING LATTICE GASES

V.S. VIKHRENKO, G.S. BOKUN, AND YA.G. GRODA
*Belarussian State Technological University*
*13a Sverdlova Str., Minsk, 220050, Belarus*

## Abstract

The nonequilibrium statistical ensemble method due to D. Zubarev is used to develop a theory of nonequilibrium processes (diffusion, thermal diffusion, heat conduction, structural relaxation) in interacting lattice gases. Deviations of particle and energy densities from their equilibrium values obey a system of Langevin-type equations that allow one to deduce microscopic expressions for different transport coefficients. All the expressions consist of two parts, one is proportional to a static correlation function and the other to the time integral of a time correlation function. The latter describes specific statistical memory effects, which are important, for example, in highly ordered states. In many cases, however, these memory effects can be disregarded. Then the transport coefficients are represented by the chemical diffusion coefficient at a zero concentration limit and equilibrium lattice gas characteristics, namely, its chemical potential, lattice concentration, temperature, and distribution functions, which are calculated within the selfconsistent diagram approximation.

## 1. Introduction

Mass transfer in solids and on their surfaces plays a key role in numerous phenomena, e. g., chemical reactions, crystal growth, diffusion, thermal diffusion, structural relaxation [1–6]. Many of these processes are accompanied by energy redistribution or take place at a nonuniform temperature distribution. Thus, the mass transport must be considered in combination with other transport processes.

Lattice models are frequently used for simplified description of equilibrium and nonequilibrium systems [1–22]. These relatively simple models of interacting many-body systems allow one to clarify different aspects of complex phenomena. Thus, different approaches for their investigation are developed. Main efforts have been spent for understanding lattice gas diffusion by Monte Carlo (MC) simulations [3,6,9–12,15–17] and, to a much lesser extent, by statistical-mechanical [4,8,13,18–22] methods.

In this paper we consider lattice gas diffusion, thermal diffusion, thermal conduction,



*M.C. Tringides and Z. Chvoj (eds.),*
*Collective Diffusion on Surfaces: Correlation Effects and Adatom Interactions,* 157–168.

and structural relaxation (the mutual particle redistribution from one equilibrium state to the other after a temperature change) in the framework of the non-equilibrium statistical ensemble method (NESEM) due to Zubarev [23–25] and derive microscopic expressions for the respective transport and relaxation coefficients. After appropriate approximations these expressions are represented via lattice gas equilibrium characteristics, such as the chemical potential and two-, three- and four- site distribution functions, and the Langmuir (noninteracting) lattice gas transition rates. These equilibrium characteristics are estimated on the basis of recently proposed [26] a selfconsistent diagram approximation (SCDA).

## 2. Langevin-type Equations for Particle and Energy Densities

### 2.1 THE MODEL, DYNAMICAL VARIABLES AND THEIR FLUXES

We consider an interacting lattice gas system of $n$ particles distributed over $N$ sites of a $d$-dimensional lattice. The microscopic density of particles (occupation numbers)

$$\hat{n}_i = \begin{cases} 0 & \text{if site } i \text{ is vacant} \\ 1 & \text{if site } i \text{ is occupied} \end{cases} \tag{1}$$

and energy

$$\hat{e}_i = u\hat{n}_i + \frac{1}{2}\sum_{j=1}^{N} \Phi_{ij}\hat{n}_i\hat{n}_j\,, \tag{2}$$

where $u<0$ and $\Phi_{ij}$ are external site potentials and the interaction energy of particles on sites $i$ and $j$, respectively. Occupation numbers $\hat{n}_i$ are considered as the dynamical variables (designated by hats) characterizing the *microscopic* system state. There is no interaction of a particle with itself ($\Phi_{ii} = 0$).

It is supposed that the lattice gas evolution obeys the master equation [3,5,27]

$$\frac{d\mathbf{P}(t)}{dt} = \mathbf{W}\cdot\mathbf{P}(t)\,, \tag{3}$$

where $\mathbf{W}$ is a transition matrix and $\mathbf{P}(t) = \sum_\alpha P_\alpha(t)\vec{\alpha}$ is a probability distribution vector with components $P_\alpha(t)$ defining the probabilities of particular microscopic system states designated by $\alpha$ at time $t$. An orthonormalized set of vectors $\vec{\alpha}$ represents all states of the system.

A full set of probability components $P_\alpha(t)$ defines the *macroscopic* system state. The mean value of a function $f_\alpha(\hat{n}_i)$ of dynamical variables that can be considered as a vector $\vec{f} = \sum_\alpha f_\alpha\vec{\alpha}$ is determined as a scalar product in the usual way

$$f(t) = \vec{f}\cdot\mathbf{P}(t) = \sum_\alpha f_\alpha P_\alpha(t)\,. \tag{4}$$

The transition matrix must satisfy the detailed balance condition

$$\mathbf{W} \cdot \mathbf{P}_{\text{eq}} = 0 , \tag{5}$$

where subscript „eq“ designates the equilibrium distribution.

Equations (1) through (5) describe the lattice gas model under consideration. It is not necessary to specify a particular dynamics (Kawasaki, hopping, etc.) at this stage.

An important step is the definition of fluxes of dynamical variables. The particle flux can be determined in accordance with the master equation as follows

$$\dot{\hat{n}}_i = \sum_{j=1}^{N} \left[ w_{ij} \hat{n}_j (1 - n_i) - w_{ji} n_i (1 - n_j) \right], \tag{6}$$

where $w_{ij}$ is a transition rate for particle jumps from site $j$ to site $i$. Although the lhs of Eq. (6) cannot be considered as a time derivative, it is instructive to formally calculate the time derivative of Eq. (2) and combine it with Eq. (6) for determining the microscopic energy flux. The result can be written as

$$\dot{\hat{e}}_i = u \dot{\hat{n}}_i + \frac{1}{2} \sum_{j=1}^{N} \Phi_{ij} \left( \dot{\hat{n}}'_{i(j)} \hat{n}_j - \hat{n}_i \dot{\hat{n}}'_{j(i)} \right), \tag{7}$$

where primes at occupation number time derivatives mean that in Eq. (6) the term corresponding to the site indicated in the subscript parentheses must be omitted, because after a particle jump from site $i$ to $j$ or in the opposite direction, no change occurs in the interaction energy between sites $i$ and $j$ (equal to zero because one of the sites is vacant either before or after the jump).

## 2.2 THE EVOLUTION EQUATIONS

Thermodynamical parameters conjugated to the microscopic densities introduced in the previous section are the inverse temperature $\beta_i = (k_{\text{B}} T_i)^{-1}$, where $k_{\text{B}}$ is the Boltzmann constant, and the chemical potential $\mu_i$ rescaled by the inverse temperature, i.e., $\beta_i \mu_i$.

For the statistical mechanical description of non-equilibrium processes the NESEM [23–25] can be used in a convenient way.

The relevant (or quasiequilibrium) probability distribution is defined as

$$P(\alpha, t) = \exp\left\{ -\sum_{i=1}^{N} \beta_i \left[ \Psi_i + \hat{e}_i^{\alpha} - \mu_i \hat{n}_i^{\alpha} \right] \right\}, \tag{8}$$

where quantities calculated for a particular microscopic lattice gas state $\alpha$ are directly indicated, $\Psi_i$ is a normalization constant

$$\exp(\beta_i \Psi_i) = \sum_{\alpha} \exp\left\{ -\sum_{i=1}^{N} \beta_i \left[ \hat{e}_i^{\alpha} - \mu_i \hat{n}_i^{\alpha} \right] \right\},$$

and summation in the last equation is carried out over all system states $\alpha$ (over all possible distributions of occupation numbers over lattice sites).

The system of Langevin-type evolution equations

$$\frac{\partial(\delta\zeta_i)}{\partial t}-\sum_{j=1}^{N}\sum_{\xi=\beta,\mu}\Omega_{ij}^{\zeta\xi}\delta\xi_j=-\sum_{j=1}^{N}\sum_{\xi=\beta,\mu}\int_0^t dt'\,\Theta_{ij}^{\zeta\xi}(t')\delta\xi_j(t-t') \tag{9}$$

is written for deviations of the quasiequilibrium mean values of the dynamical variables from their equilibrium values,

$$\delta\zeta_i=\left\langle\hat{\zeta}_i\right\rangle_{\mathrm{q}}-\left\langle\hat{\zeta}_i\right\rangle,\qquad \zeta=n,e, \tag{10}$$

where $\langle\cdots\rangle_{\mathrm{q}}$ and $\langle\cdots\rangle$ designate mean values over quasiequilibrium and equilibrium distributions, respectively. $\xi$ is a thermodynamic parameter conjugated to the dynamical variable $\hat{\xi}$.

Transport properties are determined by the mixed static correlation matrix of the dynamical variables and their fluxes

$$\Omega_{ij}^{\zeta\xi}=\left\langle\dot{\hat{\zeta}}_i\hat{\xi}_j\right\rangle \tag{11}$$

and the time correlation matrix of subtracted fluxes $J_i^{\zeta}$

$$\Theta_{ij}^{\zeta\xi}=\beta\left\langle J_i^{\zeta}\mathrm{e}^{-it\tilde{L}}J_j^{\xi}\right\rangle, \tag{12}$$

$$\tilde{L}=QLQ,\qquad Q=1-P,\qquad J_i^{\zeta}=Q\dot{\hat{\zeta}}, \tag{13}$$

where the evolution operator $L$ is defined by the master equation (3). The Mori's projection operator $P$ acts on a function $\hat{b}_i$ of the dynamical variables as [24,25,28]

$$P\hat{b}_i=\left\langle\hat{b}_i\right\rangle+\beta\sum_{j,k}\sum_{\zeta,\xi}\left\langle\hat{b}_i\delta\hat{\zeta}_j\right\rangle\chi_{jk}^{\zeta\xi}\delta\hat{\xi}_k\,. \tag{14}$$

The susceptibility matrix can be calculated inverting the static correlation matrix of the dynamical variables fluctuations

$$\chi_{ij}^{\zeta\xi}=\beta^{-1}\left\langle\delta\hat{\zeta}_i\delta\hat{\xi}_j\right\rangle^{-1}. \tag{15}$$

In turn, the latter determines thermodynamic relations between variations of mean values of the dynamical variables and thermodynamic parameters

$$\delta\xi_j=\sum_{k=1}^{N}\sum_{\eta}\chi_{jk}^{\xi\eta}\left\langle\delta\hat{\eta}_k\right\rangle. \tag{16}$$

Equation (9) can be used to derive exact expressions for the frequency and wave vector dependent energy and mass transport coefficients in analogy with the approach developed in Refs. [29,30] for the diffusion coefficients. The static correlation matrix $\Omega$ gives rise to spatial dispersion of the transport coefficients, whereas the memory matrix $\Theta$ contributes to their spatial as well as temporal dispersion.

As was shown earlier [30], in the hydrodynamic (low frequency long wavelength) limit the contribution of matrix $\Theta$ to the diffusion coefficient can be neglected in many cases. As an approximation, here we disregard its contribution to the other coefficients as well. The way to write expressions for the transport coefficients is the representation of the second term in the lhs of Eq. (9) as a second finite difference with respect to lattice site positions. Below, this technique is demonstrated calculating transport coefficients for Langmuir gas and the thermal diffusion coefficient for a lattice gas with nearest-neighbor interactions.

## 3. Transport Coefficients for Langmuir Lattice Gas

The model of traps [14], i.e., thermally-activated hopping transition rates to nearest-neighbor sites at constant (zero) barrier level

$$w_{ji} = \nu \exp(\beta_i E_i), \qquad E_i = u + \sum_{k=1}^{N} \Phi_{ik} \hat{n}_k \tag{17}$$

is used to specify the lattice gas dynamics. $\nu$ is a frequency determining the time scale of transport processes in the lattice gas. The transition rate from site $i$ to a nearest-neighbor site $j$ is determined by the well depth, $E_i$, at site $i$ and the inverse temperature $\beta_i$ at this site. Hence, the transition rates are modulated by the temperature distribution in the system. The effective dimensionless well depth $\beta_i E_i$ is inversionally proportional to temperature at a given site.

For Langmuir (noninteracting) lattice gas, $\Phi_{ij} = 0$. If the temperature distribution is nonuniform, the transition rates depend on the position of site $i$,

$$w_{ji} = \nu \exp(\beta_i u). \tag{18}$$

Simple intuitive arguments show that at uniform particle distribution, the particle flux (thermal diffusion) arises in the direction opposite to the temperature gradient because the effective well depth decreases with temperature increase.

The dynamical variables fluctuation matrix elements can readily be calculated

$$\left\langle \delta\hat{n}_i \delta\hat{n}_j \right\rangle = c(1-c)\delta_{ij}, \qquad \left\langle \delta\hat{e}_i \delta\hat{e}_j \right\rangle = u^2 c(1-c)\delta_{ij},$$

$$\left\langle \delta\hat{n}_i \delta\hat{e}_j \right\rangle = \left\langle \delta\hat{e}_i \delta\hat{n}_j \right\rangle = -uc(1-c)\delta_{ij}, \tag{19}$$

where $c = \langle \hat{n}_i \rangle$ is a lattice concentration of particles ($c=n/N$, $n$ is a total number of particles). This is a degenerate matrix because for a Langmuir gas the microscopic particle and energy densities are proportional to each other [see Eq. (2)]. However, thermodynamic equations of state can be written without making use of Eq. (16)

$$\beta\mu = \ln[c/(1-c)], \qquad e = uc\,, \tag{20}$$

where $e$ is a mean system energy per lattice site. Thus,

$$d(\beta\mu) = dc / c(1-c)\,, \tag{21}$$

and the successability matrix is also degenerate.

For the Langmuir gas, the subtracted fluxes as well as the Θ-matrix are equal to zero and transport properties are completely represented by the Ω-matrix. Its elements

$$\Omega_{ij}^{nn} = \left\langle \dot{\hat{n}}_i \hat{n}_j \right\rangle = \begin{cases} -zwc(1-c) & i=j \\ wc(1-c) & i \neq j, \end{cases}$$

$$\Omega_{ij}^{ne} = \Omega_{ij}^{en} = \left\langle \dot{\hat{n}}_i \hat{e}_j \right\rangle = u\Omega_{ij}^{nn}, \qquad \Omega_{ij}^{ee} = \left\langle \dot{\hat{e}}_i \hat{e}_j \right\rangle = u^2\Omega_{ij}^{nn}, \tag{22}$$

where $z$ is a number of nearest neighbors and $w = \nu\exp(\beta u)$ is an equilibrium transition rate for a single particle.

To prove Eqs. (22), one can use Eq. (2) for calculating static mean values

$$\left\langle \dot{\hat{n}}_i \hat{n}_j \right\rangle = w\sum_{k=1}^{z} \left\langle \left[\hat{n}_k(1-\hat{n}_i) - \hat{n}_i(1-\hat{n}_k)\right]\hat{n}_j \right\rangle, \tag{23}$$

where $k$ runs over $z$ nearest-neighbors of site $i$. In the last sum, any term is equal to zero if $j{\neq}i,k$ due to that mean values of occupation numbers on different lattice sites are independent of each other. Due to identities $\hat{n}_j^2 = \hat{n}_j$ and $\hat{n}_j(1-\hat{n}_j) = 0$ only the term at $k{=}j$ survives if $i$ and $j$ are nearest-neighbors. All $z$ terms contribute to Eq. (23) with equal values $-w\langle \hat{n}_i(1-\hat{n}_k)\rangle = -wc(1-c)$ when $j{=}i$. Thus, matrices $\Omega_{ij}^{nn}$ and $\Omega_{ij}^{en}$ have nonzero elements if indices $i$ and $j$ coincide or represent NN sites only.

For calculating the thermal diffusion coefficient, we consider a uniform particle $[n_i{=}\langle \hat{n}_i \rangle_q{=}const,\ \delta n_i{=}0]$ and nonuniform temperature $\delta\beta_j$ distributions. The second term in the lhs of Eq. (9) is represented as

$$-\sum_{j=1}^{N}\sum_{\xi=n,e}\Omega_{ij}^{\zeta\xi}\delta\xi_j = -\left[\sum_{j=1}^{z}\Omega_{ij}^{ne}\delta\beta_j + \Omega_{ii}^{ne}\delta\beta_i\right]. \tag{24}$$

Taking into account Eqs. (22) and using the two-index chess-board notations for lattice sites, we rewrite the last expression for, e.g., a two-dimensional square lattice as

$$-\sum_{j=1}^{N}\sum_{\xi=n,e}\Omega_{ij}^{\zeta\xi}\delta\xi_j = -uwc(1-c)\left[\delta\beta_{p+1,g} + \delta\beta_{p-1,g} + \delta\beta_{p,g+1} + \delta\beta_{p,g-1} - 4\delta\beta_{p,g}\right], \tag{25}$$

where $\delta\beta_{p,g} \equiv \delta\beta_i$ and $p$ and $g$ denote positions along horizontal and vertical axes, respectively. The second finite difference of inverse temperature is represented in the rhs of Eq. (25) that is equivalent to the Laplasian $(\Delta\beta)$ for a continuous system. Thus,

$$a^2 \frac{\partial \rho}{\partial t} = uwa^2 c(1-c)\Delta\beta \,, \tag{26}$$

where the number density $\rho = c / a^2$ is introduced and $a$ is a lattice parameter. From the last equation the density current can be written as follows

$$J_\rho = -uwc(1-c)\mathrm{grad}\beta = \left[uwc(1-c) / k_\mathrm{B} T^2\right]\mathrm{grad}T \,, \tag{27}$$

and the thermal diffusion coefficient

$$D_{nT} = -uc(1-c)w / k_\mathrm{B} T^2 = -u\rho D_\mathrm{J} / k_\mathrm{B} T^2 \tag{28}$$

appears to be proportional to the jump diffusion coefficient $D_\mathrm{J} = wa^2(1-c)$. For negative values of $u$ the particle flux is opposite to the temperature gradient.

In the same way other fluxes can be calculated that leads to transport equations of nonequilibrium thermodynamics [31]

$$J_\rho = -D\mathrm{grad}\rho - D_{nT}\mathrm{grad}T \,,$$

$$J_\rho = -D_{Tn}\mathrm{grad}\rho - D_e \mathrm{grad}T \,, \tag{29}$$

where transport coefficients

$$D = wa^2 , \qquad D_{Tn} = uD, \qquad D_e = uD_{nT} \,. \tag{30}$$

Contrary to Eqs. (28) and (30), the matrix of transport coefficients is symmetric in the original thermodynamic variables $\beta$ and $\beta\mu$. It is interesting to note that the thermal conduction coefficient $D_e$ is negative and the energy flux coincides with the positive direction of the temperature gradient because the temperature is an external parameter and there is no mechanisms in the model that allows the temperature to relax. In accordance with Eq. (20) the energy flux is completely determined by the particle flux. In another definition of the heat conduction coefficient the energy flux due to the particle flux is subtracted from the total energy flux. In the latter case, the heat conduction coefficient of the Langmuir lattice gas is equal to zero.

The same results for the particle and energy fluxes and transport coefficients can be obtained, calculating the corresponding fluxes through a plain orthogonal to the chemical potential or temperature gradient for the stationary distribution function

$$P_N = const \cdot \exp\left[ -\sum_{i=1}^{N} \beta_i (u - \mu_i)\hat{n}_i \right]. \tag{31}$$

It means this distribution is consistent with the master equation.

## 4. Thermal Diffusion Coefficient for Lattice Gas with NN Interactions

Although interparticle interactions complicate calculations, the latter essentially repeat the steps of the previous section. As the thermal conduction coefficient requires rather tedious calculations, we restrict our consideration by the thermal diffusion coefficient, which is of considerable practical interest.

The Ω-matrix is composed of components

$$\left\langle \dot{\hat{n}}_i \hat{e}_j \right\rangle = \sum_{k=1}^{z} \left\langle \left[ w_{ik} \hat{n}_k (1-\hat{n}_i) - w_{ki} \hat{n}_i (1-\hat{n}_k) \right] \left( u\hat{n}_j + \frac{J}{2} \sum_{l=1}^{z} \hat{n}_j \hat{n}_l \right) \right\rangle, \tag{32}$$

where the transition rates are defined by Eq. (17). This expression can be split on two parts one of which is proportional to the external potential $u$ and the other to the interaction constant $J$.

The first part is similar to Eq. (23) except for more complicated transition rates. Nevertheless, any term with different indices $i$, $k$ and $j$ is equal to zero due to the detailed balance condition Eq. (5). Thus, the terms of Eq. (32) contribute to this part only when $j$ coincides with $i$ or $k$, i.e., $i=j$ or $i$ and $j$ are nearest-neighbors by $uw_1$ and $-zuw_1$, respectively. The mean transition rate $w_1$ was calculated in Refs. [29,30]

$$w_1 = c^{-1} \left\langle w_{ij} \hat{n}_j (1-\hat{n}_i) \right\rangle = c^{-1} \exp(\beta\mu) F(0,0) = D_J / a^2, \tag{33}$$

where $F(0,0)$ is a probability for two NN sites to be vacant. This expression reproduces Zhdanov's result for the jump diffusion coefficient [13].

Having defined these contributions, we can calculate the corresponding contribution to the particle flux,

$$-\rho u D_J \, \mathrm{grad}\beta . \tag{34}$$

The second part has more complicated four-site structure

$$\frac{J}{2} \sum_{k=1}^{z} \left\langle \left[ w_{ik} \hat{n}_k (1-\hat{n}_i) - w_{ki} \hat{n}_i (1-\hat{n}_k) \right] \sum_{l=1}^{z} \hat{n}_j \hat{n}_l \right\rangle . \tag{35}$$

Due to the condition of detailed balance any term in Eq. (35) with four different indices is equal to zero. However, nonzero contributions are possible if sites $i$ and $j$ coincide or are nearest, next nearest ($nn$) or next next nearest ($nnn$) neighbors.

If $i=j$, there are $z(z-1)=12$ nonzero terms $\left\langle -w_{ki} \hat{n}_i (1-\hat{n}_k) \hat{n}_l \right\rangle$ of two types: when links $ik$ and $il$ are orthogonal or of opposite directions. If $l=k$ the corresponding term is equal to zero due to the occupation numbers algebra.

When $i$ and $j$ are nearest-neighbors, the terms with $k=j$ or $l=i$ contribute to the sum by $\left\langle w_{ij} \hat{n}_j (1-\hat{n}_i) \hat{n}_l \right\rangle$ or $\left\langle -w_{ki} \hat{n}_i (1-\hat{n}_k) \hat{n}_j \right\rangle$. There are 3 terms of each type for each out of 4 nearest-neighbors from site $i$. Again, the configuration of three sites $ijl$ or $kij$ is either straight line or rectangular.

Two rectangular/one linear configurations at $k=l$ contribute by $\left\langle w_{ik} \hat{n}_k (1-\hat{n}_i) \hat{n}_j \right\rangle$ to the sum for each $j$ out of 4 *nn*/*nnn* neighbors of site $i$, respectively.

Again, the sum of all these contributions is equal to zero. After a straightforward algebra the corresponding term in Eq. (9) is represented as a second finite difference of the inverse temperature. It allows to derive the following expression for the thermal diffusion coefficient

$$D_{nT} = -\left(u + 2J\frac{F_{\text{lin}}(0,0,1) + F_{\text{rec}}(0,0,1)}{F(0,0)}\right)\frac{\rho D_{\text{J}}}{k_{\text{B}}T^2}, \tag{36}$$

where $F_{\text{lin}}(0,0,1)$ and $F_{\text{rec}}(0,0,1)$ are probabilities for two vacancies and a particle to occupy three nearest-neighbor sites of linear or rectangular configuration, respectively.

The superposition type approximation [32]

$$F(0,0,1) \cong c(1-c)^2 g(0,0) g(0,1) \tag{37}$$

simplifies Eq. (36) to

$$D_{nT} = -\left[u + 4cJg(0,1)\right]\frac{\rho D_{\text{J}}}{k_{\text{B}}T^2}, \tag{38}$$

where $g(0,1) = F(0,1)/c(1-c)$ and $g(0,0) = F(0,0)/(1-c)^2$ are similar to the radial distribution functions of liquid state theories [32]. In Eq. (37) the three-site distribution functions originally contain additional $g$-functions for *nn* (rectangular configuration) or *nnn* (linear configuration) neighbor sites, which are approximately equal to one.

The thermal diffusion coefficient is proportional to the jump diffusion coefficient. The expression in square brackets in Eqs. (36) and (38) is a heat of transfer, which consists of the external field ($u$) and interparticle interaction ($J$) contributions. As mentioned above, the external field potential is negative while the interaction constant $J$ is negative for attractive and positive for repulsive interactions. Thus, the thermal diffusion is enhanced by attractive and suppressed by repulsive interactions.

## 5. Structural Relaxation

Interacting lattice gases obey short range order, which is characterized by distribution functions. Changes of a mutual distribution of particles over lattice sites with temperature can be considered as a structural relaxation. The interaction energy depends on the short range order and can be used to investigate the structural relaxation by means of Eq. (9). Considering the inverse temperature change $\delta\beta_i = \delta\beta$ as a small quantity independent on the lattice site position we rewrite Eq. (9) as

$$\frac{d(\delta e)}{dt} = -k_{\text{s}}\delta e, \tag{39}$$

where the structural relaxation rate

$$k_{\text{s}} = \sum_{j=1}^{N}\left\langle \delta\dot{\hat{e}}_i \delta\hat{e}_j \right\rangle \Big/ k_{\text{B}}T^2 C, \tag{40}$$

$C$ is a heat capacity per lattice site that is calculated through the energy density fluctuations according to Eq. (15). Contributions for sites $i$ and $j$, which are neighbors up to the sixth order have to be taken into account. After some algebra, the result is represented via four-site distribution functions and the final expression is simplified by the superposition-type approximations to the form

$$k_S = 3J^2 D_J c^3 g^2(0,1) / a^2 C k_B T^2 . \tag{41}$$

The last expression shows that the structural relaxation is controlled by the jump diffusion and its rate is inversely proportional to the heat capacity. In the near critical region the heat capacity growths leading to slower structural relaxation. Taking into account the jump diffusion coefficient behavior, one can conclude that the structural relaxation is strongly enhanced by repulsive and suppressed by attractive interactions.

## 6. Calculation Results and Discussion

As calculation results for diffusion coefficients were represented elsewhere [29,30] we concentrate our attention on thermal diffusion and structural relaxation characteristics taking into account attractive NN interactions for a gas on a square lattice. The SCDA, that was shown to be quite accurate for lattice gas equilibrium quantities [26], is used for calculating the chemical potential and distribution functions.

The thermal diffusion coefficient as a function of temperature for different concentrations is shown in Fig. 1. In the left panel it is given in units of the Langmuir gas chemical diffusion coefficient while in the right panel units of the interacting lattice gas jump diffusion coefficient are used. The multiplier $a^2T$ reduces the ratio of the thermal and chemical or jump diffusion coefficients to the dimensionless form. The significance of thermal as compared to chemical diffusion can be understood by comparing Figs. 1a and 2a, because the thermodynamic factor $\chi_T$ reflects the ratio of the chemical and jump diffusion coefficients. At middle and high concentrations and temperatures close to critical, the thermodynamic factor attains very low values and thermal diffusion can be considered as a main mass transfer mechanism. The ratio of the thermal and chemical diffusion coefficient is depicted in Fig. 2b. Points above the horizontal line (drawn at 1) correspond to system states when thermal diffusion is a preferable transport mechanism.

It is necessary to note that according to Eq. (38), the main contribution to the thermal diffusion coefficient comes from the external field potential. This contribution (which significance is defined by the ratio $u/J$) is not taken into account here. At the same time, the bare relaxation rate $w$ strongly decreases when the external potential well depth increases. But the latter equally influences the jump, chemical and thermal diffusion coefficients.

In Fig. 3a the structural relaxation rate is given in units of the bare (isolated particle) transition rate $w$. It weakly depends on density and grows with temperature.

The relaxation time, which is inverse to the relaxation rate ($\tau = k_S^{-1}$), is shown in Fig. 3b. At higher temperatures (above $2T_C$) the relaxation time is rather short and does

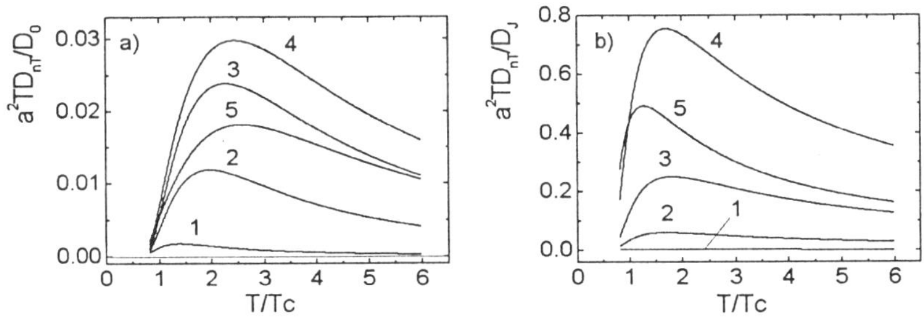


*Figure 1*. Thermal diffusion coefficient versus temperature in units of the chemical diffusion coefficient at zero concentration limit (a) and jump diffusion coefficient (b) at concentrations $c$=0.1 (curve 1), 0.3 (2), 0.5 (3), 0.7 (4) and 0.9 (5). The values corresponding to curves 5 are divided by 5.

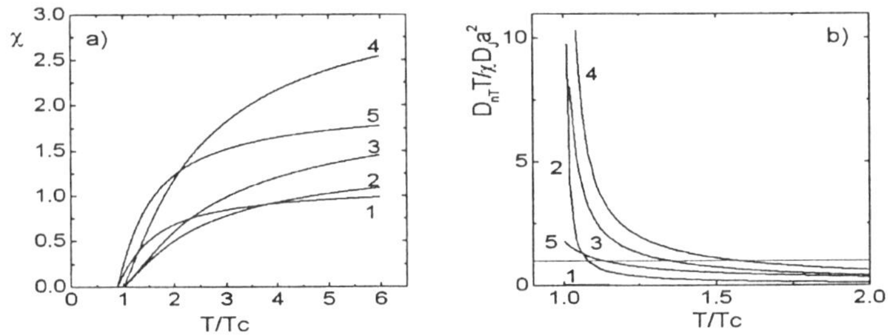


*Figure 2*. Thermodynamic factor (a) and ratio of the thermal and chemical diffusion coefficients (b) versus temperature. The curves are numbered as in Fig. 1. The values corresponding to curve 5 in (a) are divided by 5.

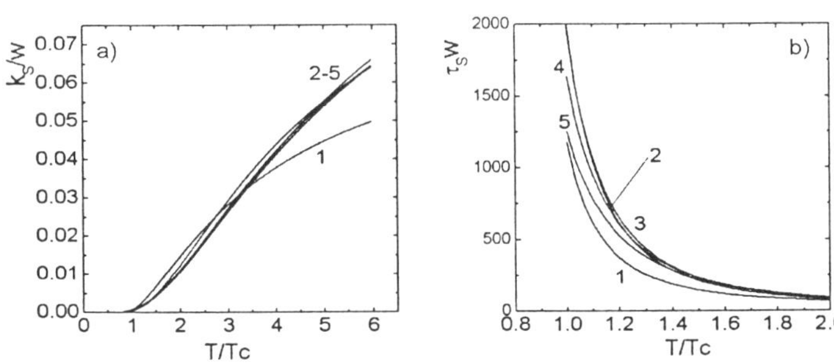


*Figure 3*. Structural relaxation rates (a) and times (b). The curves are numbered as in Fig. 1.

not exceed several tens of MC steps, although at near critical temperatures it attains hundreds or even thousands MC steps. Thus, in systems with attractive interactions the equilibration process is quit fast.

**Acknowledgments:** VVS and GYaG gratefully acknowledge the hospitality of Prof. Panos Argyrakis and Aristotle University of Thessaloniki where the paper was finally prepared. This work was supported by the NATO Fellowship Programme (VVS), INTAS Fellowship grant for young scientists YSF-004154 (GYaG), and Ministry of Education of Belarus.

## References

1. Manning, J.R. (1968) *Diffusion kinetics for atoms in crystals*, Van Nostrand, Princeton, New York.
2. Philibert J. (1991) *Atom movements: diffusion and mass transport in solids*, Editions de Phys., Les Ulis.
3. Gomer, R. (1990) Diffusion of adsorbates on metal surfaces, *Rep. Progr. Phys.* **53**, 917.
4. Zhdanov, V.P. (1991) *Elementary Physicochemical Processes on Solid Surfaces*, Plenum, New York.
5. Allnatt, A.R. and Lidiard A.B. (1993) *Atomic Transport in Solids*, Cambr. Univ. Press, Cambridge.
6. Tringides, M.C., Ed. (1997) *Surface Diffusion. Atomic and Collective Processes*, Plenum, New York.
7. Fisher, M.E. (1965) *The Nature of Critical Points*, Univ. of Colorado Press, Colorado.
8. Chumak, A.A., and Tarasenko, A.A. (1980) *Surf. Sci.* **91**, 694.
9. Murch, G.E. (1981) *Phil. Mag.* A **43**, 871.
10. Kutner, R., Binder, K., and Kehr, K.W., (1983) *Phys. Rev.* B **28,** 1847.
11. Tringides, M.C., and Gomer, R. (1984) *Surf. Sci.* **145,** 121.
12. Tringides, M.C., and Gomer, R. (1985) *Surf. Sci.* **155,** 166.
13. Zhdanov, V.P. (1986) *Surf. Sci.* **102**, L13.
14. Haus, J., and Kher, K.W. (1987) *Phys. Reps.* **150**, 263.
15. Uebing, C., and Gomer, R. (1991) *J. Chem. Phys.* **95,** 7626, 7641, 7648, 7636.
16. Tringides, M.C., and Gomer, R. (1992) *Surf. Sci.* **265,** 283.
17. Uebing, C., Pereyra, V., and Zgrablich, G. (1997) *J. Chem. Phys.* **106**, 9320.
18. ZaluskaKotur, M., and Turski, L.A. (1994) *Phys. Rev.* B **52**, 16102.
19. Scalas, E., and Ferrando, R. (1995) *Phys. Rev.* E **49**, 513.
20. Danani, A., Ferrando, R., and Scalas, E. (1997) *Int. J. Mod. Phys.* B **11,** 2217.
21. Ying , S. C., Vattulainen, I., Merikoski, J., Hjelt, T., and Ala-Nissila, T. (1998) *Phys. Rev.* B **58**, 2170.
22. Bokun, S., Uebing, C., Vikhrenko, V.S., and Zhuk, V.A. (1999) *Solid State Ionics*, **119**, 331.
23. Zubarev, D.N. (1974) *Nonequilibrium Statistical Thermodynamics*, Consultants Bureau, New York.
24. Zubarev, D., Morozov, V., and Repke, G. (1996) *Statistical Mechanics of Nonequilibrium Processes*, Vol. 1, Akademie Verlag, Berlin.
25. Zubarev, D., Morozov, V., and Repke, G. (1997) *Statistical Mechanics of Nonequilibrium Processes*, Vol. 2, Akademie Verlag, Berlin.
26. Bokun, S., Groda, Ya.G., Belov, V.V., Uebing, C., and Vikhrenko, V.S. (2000) *Eur. Phys. J.* B **15**, 297.
27. Kawasaki K. (1973) In: *Phase Transitions and Critical Phenomena*. Domb, C., and Green, M.C., eds. Academic Press, New York.
28. Grabert, H. (1982) *Projection Operator Techniques in Nonequilibrium Statistical Mechanics*, Springer Tracts in Modern Physics, Vol. 95, Springer-Verlag, Berlin.
29. Bokun, G.S., Groda, Ya.G., Uebing C., and Vikhrenko V.S. (2000) *Tech. Phys.* **45**, 1375.
30. Bokun, G.S., Groda, Ya. G., Uebing C., and Vikhrenko V.S. *Physica* A (submitted).
31. de Groot, S.R. and Mazur P. (1984) *Nonequilibrium Thermodynamics*, Dover Publications, New York.
32. Fisher I.Z. (1964) *The Statistical Theory of Liquids*, University of Chicago, Chicago.

# MODEL STUDIES OF COLLECTIVE DIFFUSION

*Non-Equilibrium Effects in Diffusion Barriers*

I. VATTULAINEN
*Department of Chemistry, Technical University of Denmark*
*Building 207, DK-2800 Lyngby, Denmark*

## 1. Introduction

Since the pioneering surface diffusion experiments about 80 years ago [1], the field of surface diffusion has matured such that we have nowadays a fairly good understanding of how the adparticles migrate along the surface, and what principles govern diffusion processes [2–5]. This is possibly best illustrated by the low-temperature limit where adatoms often move by a sequence of jumps from one adsorption site to another. In these cases one often assumes that each jump is thermally activated and characterized by a well-defined activation barrier. Consequently, one also assumes that the diffusion coefficient can be described by an Arrhenius form, thus yielding some activation barrier for the diffusion process. Despite its common use, however, this approach is problematic for a number of reasons. Recent studies have demonstrated, for example, that interactions with other adatoms affect the potential energy surface felt by the tagged adatom [6, 7]. Therefore, even in the low-coverage limit [7], the diffusion barrier extracted from an Arrhenius analysis is some complex average of various activation barriers rather than a well-defined activation energy of some individual process. Another concern is the fact that there are many experimental techniques whose results for similar surface systems are not consistent. Instead, the results may strongly depend on the technique used, a fact whose reasons are not thoroughly understood. Obviously the differences may be due to variations in sample preparation or surface contamination, or due to the fact that some techniques such as scanning tunneling microscopy may interfere with the true diffusion process [8, 9]. On the other hand, the differences can also be explained by the fact that some methods probe the diffusion process strictly in equilibrium, while other techniques operate under far-from-equilibrium conditions. Indeed, various experimental studies have proposed



*M.C. Tringides and Z. Chvoj (eds.),*
*Collective Diffusion on Surfaces: Correlation Effects and Adatom Interactions*, 169–178.

[2, 10–13] that non-equilibrium measurements may yield results which are clearly different from the equilibrium ones.

In the present work we consider two surface diffusion processes where non-equilibrium effects play a significant role. In both cases we focus on the temporal behavior of effective diffusion barriers as the system evolves towards an equilibrium state. The time-dependent results [14–16] based on Monte Carlo simulations for a lattice-gas model of O/W(110) are then compared with equilibrium results [17–19], which allows a systematic study of non-equilibrium effects. As it turns out, non-equilibrium conditions can indeed play a prominent role in diffusion behavior and lead to barriers which deviate considerably from their equilibrium limits.

## 2. Model for the O/W(110) System

Due to its simplicity, the lattice-gas description is an attractive and commonly used approach for studies of adsorption systems. In the case of the O/W(110) system, this approach is particularly attractive due to various experimental studies that have revealed oxygen atoms to have well-defined adsorption sites [20] on a surface which does not reconstruct below about 1600 K [21], where desorption starts to play a role. Furthermore, at low temperatures one has found a rich phase diagram, including the $p(2 \times 1)$ phase around the coverage $\theta = 0.50$ below a temperature $T_c \approx 710$ K, the $p(2 \times 2)$ phase around $\theta = 0.75$, and a number of additional coexistence phases [20–22]. Therefore, the O/W(110) system serves as a good candidate for model studies of adatom diffusion.

The main features of the phase diagram have also been obtained by a lattice-gas Hamiltonian [17, 23] that includes pair interactions up to fifth nearest neighbors and certain three-body interactions important at high coverages. This Hamiltonian leads to a phase diagram [17] in close agreement with the experimental observations [20–22]. The good agreement with regard to phase behavior suggests that the lattice-gas approach is realistic for studies of oxygen diffusion on the W(110) surface. However, one must recall that in Monte Carlo studies of any dynamical property such as diffusion, the specification of the interaction parameters in the lattice-gas Hamiltonian is not alone sufficient. The results also depend on the choice of transition probabilities which determine the dynamics in a given system [24]. In all studies presented here, we have used the Transition Dynamics Algorithm [17, 25], where single-particle jumps proceed by two successive steps via an intermediate "saddle point" or transition state, whose height depends on the adatom-adatom interactions as well as the adiabatic substrate potential. Therefore, our approach mimics the idea of thermally-activated diffusion. Further details of this approach can be found in Ref. [17].

In the present work we use the same lattice-gas model to study non-equilibrium properties of this system in two different cases. We first study surface diffusion during a non-equilibrium ordering process where the system evolves from a random initial configuration towards an ordered equilibrium state, and second, we use the Boltzmann-Matano technique to examine non-equilibrium effects in collective diffusion. In both cases, we focus on non-equilibrium effects in diffusion barriers.

## 3. Diffusion During a Non-equilibrium Ordering Process

As a first case, we consider diffusion under far-from-equilibrium conditions after a sudden temperature quench. Main results of this approach have been described in Ref. [14]. Here we discuss only those aspects that are essential for this work, and furthermore complement Ref. [14] with additional (unpublished) results to gain a deeper understanding of the activation barriers.

### 3.1. METHODS

For a study of any transport process in non-equilibrium, one first must face an important problem of how to define the transport coefficients. In the present case for surface diffusion, we consider an approach which yields equilibrium values in the limit of long times, while for shorter times the results reflect effects due to non-equilibrium conditions.

First, to analyze the ordering dynamics in terms of quasi-equilibrium concepts, we divide the total ordering period into different time regimes. These slices of time obviously must be different at different temperatures because of the change in the rate of ordering. For this purpose, we first calculate the time-dependence of the excess energy of the system, namely $E_x(T,t) = E(T,t) - E(T,\infty)$, after an instantaneous quench at time $t = 0$ from a completely random state to a temperature $T < T_c$. Here $E(T,t)$ is the energy of the system at temperature $T$ and time $t$. We then introduce the normalized excess energy $F(T,t) = E_x(T,t)/E_x(T,0)$, which has the maximum $F(T,0) = 1$. The equivalent time regimes at different temperatures are chosen as intervals between times $t_n(T)$ which satisfy $F(T, t_n(T)) = \exp(-n)$, where the integer $n \geq 0$.

For tracer diffusion in non-equilibrium, we then consider the quantity

$$\chi^{(n)}(\delta t) = \frac{1}{4N} \sum_{k=1}^{N} \langle |\vec{r}_k(t_n(T) + \delta t) - \vec{r}_k(t_n(T))|^2 \rangle, \tag{1}$$

where the sum is over $N$ particles to improve statistics, and $\vec{r}_k(t)$ is the position of particle $k$ at time $t$. $\langle \ \rangle$ is used to denote configuration averaging. Within each slice of time, the time difference $\delta t$ obeys $0 \leq \delta t <$

$t_{n+1}(T) - t_n(T)$. The tracer diffusion coefficient $D_T^{(n)}$ [26] is then defined as the effective slope of $\chi^{(n)}$ within the given time regime $n$. In equilibrium and for $\delta t \to \infty$, Eq. (1) is just the tracer diffusion coefficient $D_T$ times $\delta t$. Therefore, at long times when the system approaches equilibrium, $\chi^{(n)}/\delta t$ tends towards $D_T$.

For a study of collective diffusion in non-equilibrium, we first introduce the time-dependent density-fluctuation autocorrelation function

$$S^{(n)}(\vec{r}, \vec{r}\,', \delta t) = \langle \delta\rho(\vec{r}, t_n(T) + \delta t)\delta\rho(\vec{r}\,', t_n(T))\rangle, \tag{2}$$

where $\vec{r}$ and $\vec{r}\,'$ denote position vectors on a lattice and the density fluctuations are given by $\delta\rho(\vec{r}, t) = \rho(\vec{r}, t) - \langle\rho(\vec{r}, t)\rangle$. In the limit of long times and in the hydrodynamic regime, the Fourier transform of this correlation function decays as $S(\vec{k}, \delta t) = S(\vec{k}, 0)\exp(-k^2 D_C\, \delta t)$ (shown here for an isotropic case) and yields the equilibrium value for the collective diffusion coefficient $D_C$. In a non-equilibrium situation, we utilize the same idea and consider $\ln S^{(n)}(\vec{k}, \delta t)$ over the given regime $n$ and define its effective slope divided by $k^2$ as the collective diffusion coefficient $D_C^{(n)}$ [26].

Finally, by looking at individual transition rates during the ordering process, we determine the temporal evolution of the average single-particle transition rate $\Gamma(T, t)$.

## 3.2. RESULTS

We consider diffusion under far-from-equilibrium conditions by starting from a completely random initial state at $\theta = 0.45$. We then make a sudden temperature quench and let the system evolve towards the equilibrium ordered $p(2 \times 1)$ phase at various temperatures ranging from 465 to 635 K (below $T_c$). At each temperature, the temporal behavior of the tracer and collective diffusion coefficients as well as the average transition rate are followed up to times, where the system is close to equilibrium. This approach allows us to extract the time-dependence of the effective diffusion barriers (as extracted from an Arrhenius analysis) and discuss their behavior during a non-equilibrium ordering process.

Results for the effective diffusion barriers are shown in Fig. 1(a). Their temporal behavior can be compared with the time evolution of the order parameter of the $p(2 \times 1)$ phase [15, 16] shown in Fig. 1(b), which characterizes how far the system is from its true equilibrium state. Three interesting features emerge. At very early times, the adsorbate is in a disordered configuration and the adatom-adatom interaction contributions to the activation barriers largely cancel out. Thus, all barriers have values close to the intrinsic barrier $\Delta$ [14, 17], which in the present model mimics the adatom-substrate potential in the low-coverage limit $\theta \to 0$. At intermediate times, all barriers increase rapidly due to the attractive interactions that dominate the system under present conditions, while at long times the barriers approach their equilibrium limits.

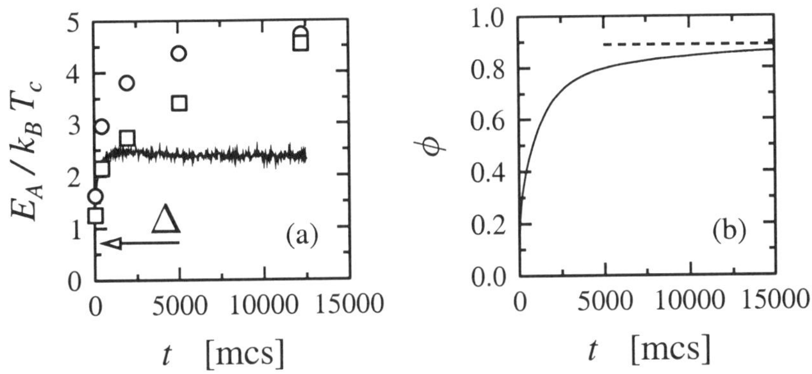


*Figure 1.* (a) Results for the effective diffusion barriers during an ordering process from a completely random initial state to a $p(2\times1)$ ordered state. Barriers shown are for tracer diffusion, $E_A^T$ (circles), collective diffusion, $E_A^C$ (squares), and the average transition rate, $E_A^\Gamma$ (full line). Equilibrium limits of $E_A^T$ and $E_A^C$ are about $4.8\,k_BT_c$. Time is given in units of one Monte Carlo step (mcs). The value of the intrinsic barrier $\Delta$ (see the text) is shown by an arrow. (b) Temporal behavior of the order parameter of the $p(2\times1)$ phase, $\phi$, after a quench to $T = 465$ K. The dashed line shows the equilibrium limit of $\phi$.

In a previous work [14], we have shown that our results for the effective barrier of collective diffusion are in agreement with experiments by Tringides *et al.* [11, 12]. This conclusion was based on a study of ordering kinetics in terms of the average domain size, in a similar fashion as in the original work by Tringides and his collaborators. The main finding of this comparison was that the approach used in Refs. [11, 12] seems to be valid and yields a diffusion barrier that represents the barrier for the dominant diffusion process in the intermediate time regime during an ordering process. In the present work, we analyze the data from a different perspective. We recall that within a lattice-gas description in equilibrium, the tracer and collective diffusion coefficients can be written as [4, 27, 28]

$$D_T(\theta,T) = \frac{a^2}{4}\,\Gamma(\theta,T)\,f_T(\theta,T) \tag{3}$$

and

$$D_C(\theta,T) = \frac{a^2}{4}\,\Gamma(\theta,T)\,f_C(\theta,T)\,\xi(\theta,T), \tag{4}$$

where $a$ is the jump length, $\Gamma$ is the average single-particle transition rate, $f_T$ is the dynamic correlation factor for the motion of a single tagged particle, and $f_C$ is the corresponding correlation factor for the center-of-mass motion. The last term $\xi = 1/k_B T \theta \kappa_T$ is the thermodynamic factor, which is related to the compressibility of the adlayer $\kappa_T$. If these descriptions are analyzed by the Arrhenius form, one easily finds that the effective barriers of $D_T$ and $D_C$ have a number of competing components. For tracer diffusion, we find an effective barrier

$$E_A^T = E_A^\Gamma + E_A^{f_T}, \tag{5}$$

where the first term on the RHS arises from the activated nature of single-particle jumps via $\Gamma$, and the second one is due to the temperature dependence of dynamical correlations. For collective diffusion, the situation is more complex in the sense that then there are three competing barriers,

$$E_A^C = E_A^\Gamma + E_A^{f_C} + E_A^\xi, \tag{6}$$

where the second and third terms on the RHS are not similar to those in Eq. (5). The term $E_A^{f_C}$ now characterizes the temperature dependence of dynamical correlations in the center-of-mass motion, while $E_A^\xi$ results from the temperature dependence of particle number fluctuations.

Using the lattice-gas description for effective diffusion barriers, we can now analyze the results in Fig. 1(a) in a bit more detail. First, we find that $E_A^\Gamma$ levels off more rapidly than $E_A^T$ or $E_A^C$. Furthermore, it is clear that $E_A^\Gamma$ converges to its long-time limit at times when the system is still ordering, as demonstrated by the order parameter in Fig. 1(b). This result is consistent with an idea that, soon after the quench, there is rapid formation of small ordered domains within which the transition rates are approximately equal to their equilibrium counterparts. The tracer diffusion barrier $E_A^T$, on the other hand, is found to increase monotonously and in close connection with the order parameter, demonstrating the hydrodynamic nature of tracer diffusion. The difference between $E_A^T$ and $E_A^\Gamma$ is due to dynamical correlations, whose temperature dependence gives rise to a significant contribution in $E_A^T$. In previous equilibrium studies under similar conditions, the contribution $E_A^{f_T}$ has been shown to be almost 50% of $E_A^T$ [18]. In the current study in non-equilibrium, the role of $E_A^{f_T}$ is not significant at very small times but increases rapidly as ordered domains are formed, suggesting that the memory contribution in $E_A^T$ is most important under equilibrium conditions and especially within ordered phases. This conclusion is consistent with previous equilibrium studies [18, 19].

The behavior of the collective diffusion barrier $E_A^C$ is more difficult to interpret. The difference between $E_A^C$ and $E_A^\Gamma$ is a mixture of two components, namely $E_A^{f_C}$ and $E_A^\xi$. The first of these two quantities characterizes the temperature dependence of memory effects in the center-of-mass motion. Based on a recent equilibrium study for the same model system, the role of $E_A^{fc}$ is negligible in disordered phases, and even in low-temperature ordered phases its significance is typically less than 10% of $E_A^C$ [18]. This suggests that the role of $E_A^{fc}$ is not noteworthy but the difference between $E_A^C$ and $E_A^\Gamma$ is mostly due to $E_A^\xi$, supporting the idea that particle number fluctuations are strongly dependent on the state of the system during a non-equilibrium process. This idea is related to Boltzmann-Matano studies described below, thus we will come back to this point later.

## 4. Diffusion During a Non-equilibrium Spreading Process

As a second case, we use the well-established Boltzmann-Matano (BM) method [2, 29] to extract the temporal behavior of the collective diffusion coefficient $D_C(\theta)$ as the system approaches equilibrium. Again, we focus on the non-equilibrium effects on diffusion barriers. A more complete description of the results is available in Refs. [15, 16].

### 4.1. METHODS

We consider the spreading of a coverage profile $\theta(x,t)$ in a semi-infinite system, which ranges from $-\infty$ to $+\infty$ in the $x$ direction, and whose width $L_y$ is typically 30–1000 lattice units in the $y$ direction. Periodic boundary conditions are employed in the $y$ direction. The coverage profile is initially an ideal step function at $x = 0$ [$\theta(x,0) = 1$ for $x < 0$ and $\theta(x,0) = 0$ for $x > 0$], which evolves in time $t$ in the $+x$ direction. Note that the coordinate $x$ acts as the dividing line $x = 0$, which accounts for particle number conservation via $\int_0^1 x(\theta')\,d\theta' = 0$.

To determine the coverage dependence of the collective diffusion coefficient $D_C(\theta)$, we use the Boltzmann-Matano method [2, 29, 30], where $D_C(\theta)$ can be determined from scaled density profiles as

$$D_C(\theta) = -\frac{1}{2t}\left(\frac{dx}{d\theta'}\right)\bigg|_\theta \int_0^\theta x(\theta')\,d\theta'. \tag{7}$$

Note that the Boltzmann-Matano method is based on the assumption that, in the long-time limit, the coverage profiles $\theta(x,t)$ collapse to a single scaling function when expressed as $\theta(x/\sqrt{t})$. If this condition is truly satisfied, then $D_C(\theta)$ obtained from Eq. (7) corresponds to the actual diffusion coefficient in equilibrium. Otherwise, the effective $D_C(\theta)$ in Eq. (7) is a non-equilibrium quantity and depends on a time regime chosen for an analysis.

## 4.2. RESULTS

In Fig. 2(a), we show the scaled coverage profiles $\theta(x/\sqrt{t})$ at three different time regimes during the profile evolution process (10 000–12 000, 50 000–60 000, and 250 000–300 000 MCS). The coverage profiles in Fig. 2(a) clearly evolve in time and tend towards some universal curve at late times. The late-time behavior is particularly interesting, since experimentally obtained coverage profiles of similar kinds have been used to analyze $D_C(\theta)$ by the BM technique [31, 32]. However, as temporal studies of the order parameter of the $p(2 \times 1)$ phase have indicated [15, 16], even the longest time scales shown in Fig. 2(a) do not correspond to true equilibrium behavior.

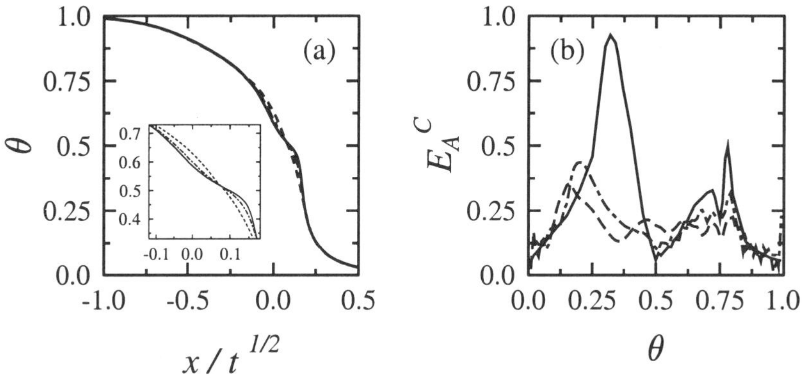


*Figure 2.* (a) Scaled coverage profiles $\theta(x/\sqrt{t})$ at $T \approx 590$ K and at three different times during the profile evolution process. In the data, several profiles from the time regimes 10 000–12 000 (dashed lines), 50 000–60 000 (dash-dotted), and 250 000–300 000 MCS (full lines) have been collapsed to obtain the scaled curves. The inset shows details of the profiles within the developing $p(2\times1)$ phase. (b) Effective activation barriers $E_A^C(\theta)$ as they arise from an Arrhenius analysis of $D_C$. Results shown are for times 10 000–12 000 MCS (dashed lines), 50 000–60 000 MCS (dash-dotted), and in equilibrium (full lines). In this case, the error bars for BM simulations are about $\pm10\%$.

Using Eq. (7) to extract $D_C(\theta)$, we analyzed each of these curves. A similar study was made at a number of temperatures in the vicinity of 590 K, thus allowing us to determine the Arrhenius barrier $E_A^C$ of $D_C$ at different times. As results in Fig. 2(b) now demonstrate, the barriers $E_A^C$ determined from BM studies can be significantly different from the corresponding equilibrium results. This is mainly due to particle number fluctuations, since the barriers $E_A^\Gamma$ studied separately in Ref. [15] cannot explain the differences in Fig. 2(b), and as was pointed out in Section 3.2, the temperature-dependence of $f_C$ accounts typically for only about 10% of the total barrier [18, 19].

These results highlight the important role of non-equilibrium effects played in the BM experiments. They furthermore suggest that particle number fluctuations are particularly sensitive to deviations from equilibrium conditions, and can lead to diffusion barriers that deviate considerably from their equilibrium values. These deviations are strongest within the ordered phases and in the vicinity of phase boundaries [15]. This result may, in part, explain some experiments where it has been observed that the barriers found in BM studies [2, 32, 33] are not fully consistent with barriers extracted from measurements with other techniques [2].

## 5. Summary

Surface diffusion plays a major role in a number of important surface processes such as surface growth and surface reactions. In many cases, one is interested in the activation barriers related to the diffusion processes. Therefore, such measurements have been conducted for a variety of systems using a number of different techniques [2], and perhaps surprisingly, the barriers found are at times inconsistent with each other.

In this work, we have considered one aspect that can partly explain these differences. We have shown how the effective diffusion barriers can be very different under equilibrium and non-equilibrium conditions. This finding is probably most useful for future Boltzmann-Matano experiments, a technique which is one of the most commonly used methods for studying collective diffusion. In this case, we have shown how the non-equilibrium results for effective diffusion barriers may differ significantly from their equilibrium limits. This is mainly due to particle number fluctuations whose equilibration takes place at very long time scales, and especially so within ordered phases and close to phase boundaries where deviations from equilibrium behavior are strongest. Finally, it is worthwhile pointing out that here we have considered an ideal system where the effects due to various kinds of surface defects are not accounted for. Further work remains to be done to clarify their role in macroscopic profile evolution experiments, where defects are in many cases unavoidable.

*Acknowledgments* — The author kindly acknowledges various discussions and collaborative projects with Tapio Ala-Nissila, See-Chen Ying, Juha Merikoski, and Petri Nikunen.

## References

1. Von M. Volmer and I. Estermann, *Z. Phys.* **7**, 13 (1921); I. Langmuir and J. B. Taylor, *Phys. Rev.* **40**, 463 (1932).
2. R. Gomer, *Rep. Prog. Phys.* **53**, 917 (1990).
3. T. Ala-Nissila and S. C. Ying, *Prog. Surf. Sci.* **39**, 227 (1992).

4. A. Danani, R. Ferrando, E. Scalas, and M. Torri, *Int. J. Mod. Phys. B* **11**, 2217 (1997).
5. *Surface Diffusion: Atomistic and Collective Processes*, edited by M. C. Tringides (Plenum, New York, 1997).
6. S. Renisch, R. Schuster, J. Wintterlin, and G. Ertl, *Phys. Rev. Lett.* **82**, 3839 (1999).
7. M. Ø. Pedersen, L. Österlund, J. J. Mortensen, M. Mavrikakis, L. B. Hansen, I. Stensgaard, E. Lægsgaard, J. K. Nørskov, and F. Besenbacher, *Phys. Rev. Lett.* **84**, 4898 (2000).
8. M. Bott, M. Hohage, M. Morgenstern, T. Michely, and G. Comsa, *Phys. Rev. Lett.* **76**, 1304 (1996).
9. U. Kürpick and T. S. Rahman, *Phys. Rev. Lett.* **83**, 2765 (1999).
10. M. C. Tringides, in: *The Chemical Physics of Solid Surfaces and Heterogeneous Catalysis: Phase Transitions and Adsorbate Restructuring of Metal Surfaces*, Vol. 7, Eds. D. A. King and D. P. Woodruff (Elsevier, Amsterdam, 1994) Ch. 6.
11. M. C. Tringides, P. K. Wu, and M. G. Lagally, *Phys. Rev. Lett.* **59**, 315 (1987).
12. P. K. Wu, M. C. Tringides, and M. G. Lagally, *Phys. Rev. B* **39**, 7595 (1989).
13. M. C. Tringides and R. Gomer, *Surf. Sci.* **265**, 283 (1992).
14. I. Vattulainen, J. Merikoski, T. Ala-Nissila, and S. C. Ying, *Surf. Sci.* **366**, L697 (1996).
15. P. Nikunen, I. Vattulainen, and T. Ala-Nissila, *Surf. Sci.* **447**, L162 (2000).
16. P. Nikunen, I. Vattulainen, and T. Ala-Nissila, submitted (2000).
17. I. Vattulainen, J. Merikoski, T. Ala-Nissila, and S. C. Ying, *Phys. Rev. B* **57**, 1896 (1998).
18. I. Vattulainen, *Surf. Sci.* **412/413**, L911 (1998).
19. I. Vattulainen, S. C. Ying, T. Ala-Nissila, and J. Merikoski, *Phys. Rev. B* **59**, 7697 (1999).
20. K. E. Johnson, R. J. Wilson, and S. Chiang, *Phys. Rev. Lett.* **71**, 1055 (1993).
21. C. R. Brundle and J. Q. Broughton, in: *The Chemical Physics of Solid Surfaces and Heterogeneous Catalysis: Chemisorption Systems*, Vol. 3A, Eds. D. A. King and D. P. Woodruff (Elsevier, Amsterdam, 1990) Ch. 3.
22. G.-C. Wang, T.-M. Lu, and M. G. Lagally, *J. Chem. Phys.* **69**, 479 (1978).
23. D. Sahu, S. C. Ying, and J. M. Kosterlitz, in: *The Structure of Surfaces II*, Eds. J. F. van der Veen and M. A. van Hove (Springer-Verlag, Berlin, 1988) p. 470.
24. I. Vattulainen, J. Merikoski, S. C. Ying, and T. Ala-Nissila, *Europhys. Lett.* **51**, 361 (2000); and references therein.
25. T. Ala-Nissila, J. Kjoll, and S. C. Ying, *Phys. Rev. B* **46**, 846 (1992).
26. Strictly speaking, diffusion is well-defined only in equilibrium. Thus here we should make a clear difference between *diffusion* (in equilibrium) and *mobility* (in non-equilibrium). However, for consistency, we follow here the practice used in the Boltzmann-Matano method: despite its non-equilibrium character, one speaks of collective diffusion.
27. D. A. Reed and G. Ehrlich, *Surf. Sci.* **102**, 588 (1981).
28. T. Hjelt, I. Vattulainen, J. Merikoski, T. Ala-Nissila, and S. C. Ying, *Surf. Sci.* **380**, L501 (1997); *Surf. Sci.* **402–404**, 253 (1998).
29. C. Matano, *Jpn. J. Phys.* **8**, 109 (1933).
30. J. Crank, *The Mathematics of Diffusion* (Oxford, Oxford, 1995).
31. M. Šnábl, M. Ondřejček, V. Cháb, Z. Chvoj, W. Stenzel, H. Conrad, and A. M. Bradshaw, *J. Chem. Phys.* **108**, 4212 (1998).
32. A. T. Loburets, A. G. Naumovets, and Yu. S. Vedula, *Surf. Sci.* **399**, 297 (1998).
33. E. Suliga and M. Henzler, *J. Phys. C: Solid State Phys.* **16**, 1543 (1983).

# STUDY OF NON-EQUILIBRIUM SURFACE DIFFUSION FROM AN INITIAL STEP-CONCENTRATION PROFILE WITH A NEW PROBABILITY-BASED CONTINUUM METHOD

M.R. DUDEK
*Institute of Physics, Pedagogical University*
*65-069 Zielona Góra, Poland*

*Institute of Theoretical Physics, University of Wrocław,*
*50-204 Wrocław, Poland*

AND

M.C. TRINGIDES
*Department of Physics and Astronomy, and Ames Laboratory*
*Iowa State University, Ames, IA 50011 USA*

**Abstract.** We have studied the non-equilibrium diffusion coefficient in an adsorbate system which is initially in a state of step-concentration profile to understand the role of adsorbate-adsorbate interactions. We use a new kinetic algorithm based on the probability $p(t,x)$ to find a particle at site $x$ in time $t$. This probability is a measure of the local concentration. The main advantage of this method over standard Monte Carlo methods is the parallel update of the local concentration over the whole lattice (instead of the sequential updates as in conventional Monte Carlo methods; which is particularly important when the system is in the ordered phase). We have calculated the tracer diffusion coefficient, the diffusion coefficient based on the Boltzmann Matano method, and the diffusion coefficient based on the spreading of the width of the profile (which it is easier to measure experimentally). The system is studied both for the case of repulsive nearest-neighbor and competing (nearest-neighbor and next-nearest-neighbor) interactions. We have also analysed the non-equilibrium growth laws describing the evolution of the domains of the ordered phase for $T < T_c$. We confirm that the $c(2\times2)$ phase grows with the characteristic power law in time $t^{\frac{1}{2}}$ (similar to the growth law observed during the evolution of the ordered phase out of a uniform initial random configuration).

## 1. Introduction

Typical experiments on surface diffusion involve the presence of an initial concentration profile $c(x, t=0)$, which evolves in time to reach a final uniform concentration $c(x,\infty)$. In this case, the diffusion coefficient is calculated by comparing the resulting concentration profile $c(x,t)$ with the one following Fick's law [1]. The widely used Boltzmann-Matano method (BM) [1] – [4] is like that. However, it is not clear until now to what degree



*M.C. Tringides and Z. Chvoj (eds.),*
*Collective Diffusion on Surfaces: Correlation Effects and Adatom Interactions,* 179–189.

the method can be applied for interacting adsorbates. In general, the spreading overlayer may undergo a series of phase transitions and they manifest themselves as plateau regions separated by sharp transition regions in the profile shape $c(x,t)$. This makes the profile analysis complex and the application of the BM method (which is based on determining numerically the derivative of the profile) an open question. Furthermore, the spontaneously generated domains of different ordered phases have an effect on the adsorbate diffusion. Thus, in the case of interactive systems the new type of non-equilibrium experiments are performed (see discussion of them, e.g., in Ref. [5]), where the initial system is prepared at $t = 0$ in a disordered phase, i.e., above the critical temperature ($T > T_c$), and next the system is quenched to an ordered phase ($T < T_c$). The system evolves in time by forming domains of the different ordered phases. These domains compete with each other and their evolution towards equilibrium proceeds by means of dissolution of small domains and growth of the large ones, giving rise to a reduction of the total surface of tension generated by the domain boundaries. Typically, the system develops a single length scale (linear size of a typical domain) that grows with time as

$$L(t) = A(T)t^{\alpha}, \tag{1}$$

where $\alpha$ is the growth exponent and $A(T)$ is a temperature-dependent rate constant. It is assumed that $A(T)$ has an Arrhenius-like temperature dependence, e.g. Ref. [5] and Ref. [6]:

$$A(T) \approx A_0 \exp\left(-W/k_B T\right), \tag{2}$$

where $W$ is an activation energy, $A_0$ is a frequency factor, and $k_B$ is Boltzmann constant. So, in the current study we address two questions, first whether the BM method can be used and second whether we can study the non-equilibrium growth laws obeyed by a system evolving in time to obtain equilibrium.

Usually, the theoretical description of both types of non-equilibrium experiments is based on the Monte Carlo method. However, the method is very efficient only in the case of the systems with the small ground state degeneracy (such as a binary alloy or Ising model) otherwise the time to obtain equilibrium in systems with larger degeneracy grows exceedingly long, e.g., when simulating domain growth in the $q$-state Potts model [7]. Therefore, we introduced a new method of propagating probability to describe diffusing adsorbate, which is complementary to the Monte Carlo method. In our case, the lattice sites are occupied by the probability of finding an adatom at a lattice site after time t and not by the adatoms themselves. This is equivalent to the situation as if all possible trajectories of each adatom were considered simultaneously at time t what is far more realistic then the sequential atom interrogation typical of the standard Monte Carlo methods. The method originates from the exact enumeration method constructed by Havlin and Ben-Avraham [8] to describe diffusion of a single particle. We consider a flux of probabilities representing $N$ interacting adatoms instead and the respective transition rates depend on the local adatom concentration (resultant probability in the neigboring lattice sites). This makes our analysis to be mean-field like. However, we can separate from the resultant probability information corresponding to a single adatom and we can follow its time evolution in a rigorous way. The single particle under consideration experiences both surface potential and time varying local adatom concentration.

| | | | | | | |
|---|---|---|---|---|---|---|
| 0 | 0 | 0 | 1 | 0 | 0 | 0 |
| 0 | 0 | 1/2 | 0 | 1/2 | 0 | 0 |
| 0 | 1/4 | 0 | 1/2 | 0 | 1/4 | 0 |
| 1/8 | 0 | 3/8 | 0 | 3/8 | 0 | 1/8 |

*Figure 1.* A diffusion process with the first line corresponding to $t = 0$, the second to $t = 1$, etc.

## 2. Model

We construct a microscopic model of adsorbate evolving from an initial step, where we use the exact enumeration method [8] to describe adatom diffusivity. In the method, lattice sites are occupied by adatom probabilities (probability of finding a random walker at lattice site at time $t$) instead of adatoms. The exact enumeration method [8] has been introduced for a single random walker, where at $t = 0$ the lattice site being the origin has assigned the probability equal to 1 and in all other lattice sites the probability is equal to 0. During the following time steps the probability propagates to $d$ neighboring lattice sites according to system-specific transition rates $w_\mu$ ($\mu = 1, 2, \ldots, d$). In Fig. 1, there is presented an example of the probability propagation mechanism on one-dimensional lattice in the case when $w_1 = w_2 = \frac{1}{2}$.

We generalized this single particle method to describe $N$ interacting adatoms. In this case, at each lattice site $k$ ($k = 1, 2, \ldots, L_x \times L_y$) there is a resultant probability $C(k)$ which contributes all adatoms. One can think of the resultant probability as about the local (site dependent) adatom concentration, which percolates between the neighboring lattice sites. This picture resembles very much the mean-field description of the lattice gas model in terms of the master equation [9]. However, in our model we can separate a component corresponding to every single adatom and we can trace its time evolution, whereas it is not possible in the traditional mean-field approach.

In the model, the substrate is represented by a lattice, $L_x \times L_y$. The initial overlayer has a rectangular geometry, $h \times L_y$, where $h$ is its linear size in the $x$-direction. We assume $L_x >> h$ and the reflecting boundary conditions in the $y$-direction. However, we examined also the case of an initially circular adsorbate. All lattice sites in our model are labelled by reals representing probability of finding an adatom in them. At $t = 0$, the lattice sites, which represent the initial stripe of adatoms have a probability equal to 1, whereas the other lattice sites have the probability equal to 0. At $t > 0$, the adatom probabilities propagate to the neighboring lattice sites with the respective transition rates and under the condition that the resultant flux to any site from its surrounding cannot exceed value 1. This is equivalent to the exclusion principle that two adatoms cannot find themselves at the same lattice site. We assume that in each lattice site $k$ the transition rates are the same in each lattice direction, $w(k)_\mu = w(k)$ ($\mu = 1, 2, 3, 4$). They are defined as follows:

$$w(k) = const\ \exp(-(W/k_BT) - (E/k_BT)\sum_{k_{nn}} C[k_{nn}]), \tag{3}$$

where $W$ is the adatom activation energy, E is the interaction energy between the adatoms located at lattice site $k$, and the adatoms from the neighboring lattice sites $k_{nn}$, $C[k_{nn}]$ represents the resultant probability (concentration) at the lattice site $k_{nn}$. Notice, if we had in Eq. 3 that $C[k] = 1$ or $C[k] = 0$, only then the transition rate becomes the same as the one introduced by Bowker and King [2]. From the normalization constant in Eq. 3, we find the following condition

$$const + 4\ w(k) = 1, \tag{4}$$

where we have included a waiting probability and four transition rates corresponding to four lattice directions.

In our model there are two arrays, $C_1[k]$ and $C_2[k]$, to store the resultant probabilities of finding an adatom at a lattice site $k = 1, 2, \ldots, L_x \times L_y$. The first array $C_1[k]$ represents the concentration of adatoms at time moment $t$, whereas $C_2[k]$ stores the adatom concentration at time moment $t+1$. The concentration propagates according to the set of discrete balance equations. The algorithm is the following:

i) For each lattice site $k = 1, 2, \ldots, L_x \times L_y$ define

$$C_1[k](t) = C_2[k](t), \tag{5}$$

ii) For each lattice site $k = 1, 2, \ldots, L_x \times L_y$ define a new adatom concentration, $C_2[k]$, which is the old one increased by a concentration flux into site $k$ from the neighboring lattice sites $k_{nn}$,

$$C_2[k](t+1) = C_2[k](t) + \sum_{k_{nn}} C_1[k_{nn}](t)\ w(k), \tag{6}$$

and decrease, respectively, the value of the concentration in the neighboring sites, i.e.,

$$C_2[k_{nn}](t+1) = C_2[k_{nn}](t) - C_1[k_{nn}](t)\ w(k). \tag{7}$$

The hard core condition for the adatoms is realized in such a way that the updating of the site $k$ takes place only if the total flux into this site from its neighbors does not exceed the value 1. Otherwise, the updating of the concentration at the site $k$ is forbidden.

iii) Increase time $t$ by 1 and go to step $(i)$ of the algorithm unless the specific time interval expired.

During the concentration flow, we always can follow the propagating probability associated with a single adatom. In this case, we have introduced two additional arrays, say $G_1[\,]$ and $G_2[\,]$, which are analogous to $C_1[\,]$ and $C_2[\,]$, but they store information concerning only the chosen adatom. Thus, although the transition rates (Eq. 3) are defined in terms of the concentrations as in the mean-field description, we still can trace the evolving probability for a chosen single adatom. The only limit on the number of adatoms, which are traced simultaneously, is the capacity of the computer memory once we need the arrays $G_1[\,]$ and $G_2[\,]$ for each of them. The results would become exact if we could consider all adatoms independently.

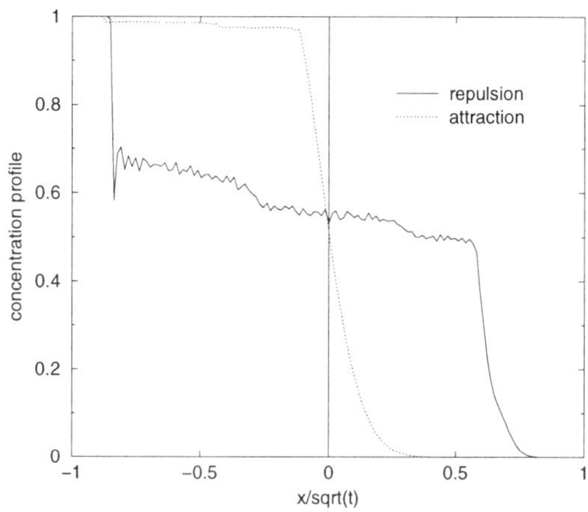


*Figure 2.* Effect of the nearest-neighbor interactions on diffusion profiles: (a) repulsive adatom interactions, (b) the attractive adatom interactions.

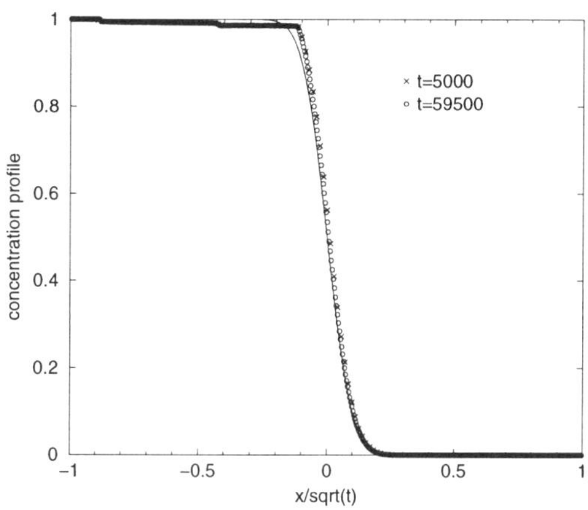


*Figure 3.* Concentration profiles represented by Eq. 8 when $D_c = 0.003$ (solid line) and propagating probability method with the rules determined by Eqs. 6 and 7 at different time moments $t = 5000$ (crosses) and $t = 59500$ (circles). We have chosen $W = 0.5\text{eV}$, $T = 1000\text{K}$, $E = 0$.

## 3. Discussion of the Results

The effect of the lateral interactions in our probability-based continuum method on the shape of the concentration profile has been presented in Fig. 2. The results are consistent with an expectation that repulsion between adatoms produces a flat profile, whereas the attraction favors a tongue-like shape (e.g., see discussion of the profile shapes [4]). If there are only hard-core interactions (the case of free lattice gas) present in the system, then the concentration profile can be represented by an error function [10, 5]:

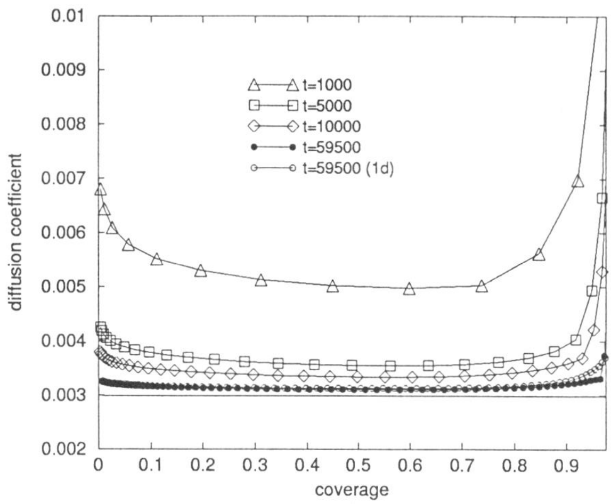


*Figure 4.* Boltzmann-Matano diffusion coefficient vs. coverage calculated for the concentration profiles obtained with the help of our probability-based method, where $W = 0.5$eV, $T = 1000$K, $E = 0$. The geometry of the diffusion run was $h = 300$, $L_y = 300$, $L_x = 5000$ except the case of data represented by the empty circles, which correspond to 1D case, where $h = L_y = 1$. The bottom straight line represents the value of $D_0 \sim 0.003$.

$$c(x,t) = \frac{1}{2}c(0,0)(1 - erf(\frac{x}{2\sqrt{(D_c t)}})), \tag{8}$$

with $c(x,t)$ denoting the concentration profile at position $x$, $D_c$ representing collective diffusion coefficient, and $c(0,0)$ the initial profile concentration. To show this, in Fig. 3 we have plotted our concentration profiles,

$$c(x,t) = \frac{1}{L_y}\sum_y C_1[(x,y),t], \tag{9}$$

in the case when $E = 0$, $W = 0.5$eV, $T = 1000$K and we fitted them their respective error functions. The profiles start to differ from an error function only close to a saturation concentration. However, the scaling applies for them at a whole range of concentrations, i.e., the plots, concentration versus $x/\sqrt{(t)}$, collapse on the same curve.

If we had only a single random walker on the square lattice, then its diffusion coefficient could be determined analytically, i.e.,

$$D_0 = (1/4)a^2(1/t), \tag{10}$$

where $a$ is the lattice constant and $(1/t)$ represents transition rate $w(k)$ (Eq. 3) for the random walker to make a distance $a$ during time $t$. If we put $a = 1$, $E = 0$, $W = 0.5$eV, $T = 1000$K then $D_0 \sim 0.003$, i.e., $D_0 = D_c$. This result for the value of the diffusion coefficient is consistent with the Boltzmann-Matano analysis [1] – [4] of the concentration profiles as one can conclude from Fig. 4. We notice that the Boltzmann-Matano diffusion coefficient does not depend on coverage. The method also weakly depends on the width $L_y$ of the diffusion channel.

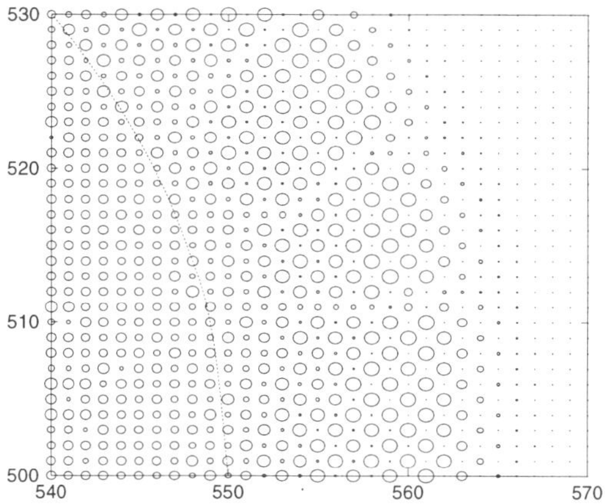


*Figure 5.* Distribution of the resultant probability $C_1(k)$ at $T = 0.36T_c$ when $E = -0.2eV$, $W = 0.5eV$. Their values are represented by circles with the radius equal to $C_1(k)$. The initial overlayer had circular shape (dotted line) with the center at $k_0 = (500, 500)$.

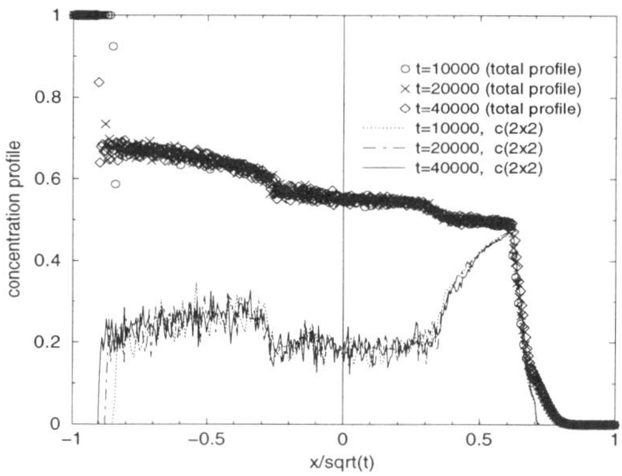


*Figure 6.* Concentration profiles $c(x,t)$ at different times $t = 10000, 20000, 40000$ versus $x/\sqrt{(t)}$ (upper curves). The lower curves correspond to a fraction of the $c(2 \times 2)$ phase contributing to the concentration profiles. $W = 0.5eV$, $T = 0.36T_c$, $E = -0.2eV$.

It is well known that the presence of the short-range adatom interactions makes possible that the ordered states of the adsorbate under consideration can appear. The most discussed case in the literature considers repulsive nearest-neighbor interactions when $c(2 \times 2)$ phase is spontaneously created near the concentration $c \sim 0.5$ at temperature $T < T_c$ (e.g., [10, 11]). Usually, theoretical investigation in the field has been done with the help of the Monte Carlo method, e.g. [2] – [15]. We can compare the results with the case of the repulsive interactions in our probability-based method.

In Fig. 5 we have presented a snapshot of an initially circular adsorbate after $t = 1000$ steps where the various size circles represent the values of the resultant probability $C_1(k)$

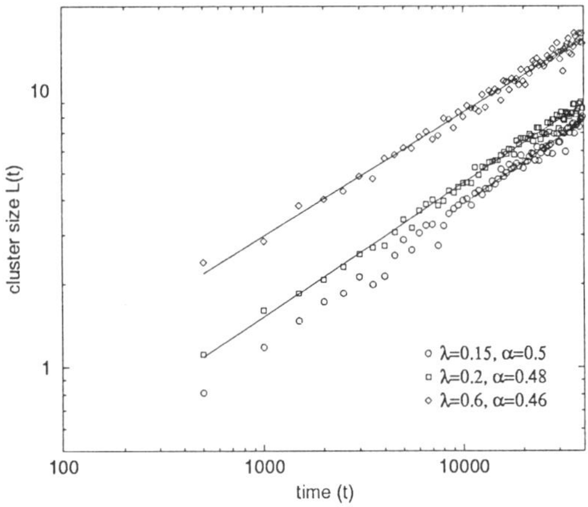


*Figure 7.* Dependence of average linear cluster size on time in non-equilibrium profile experiment for $\lambda = 0.1, 0.2, 0.6$. The increasing cluster size $L(t) \sim t^{\alpha}$ with $\alpha = 0.5$ for $\lambda = 0.15$, $\alpha = 0.48$ for $\lambda = 0.2$, $\alpha = 0.46$ for $\lambda = 0.6$. In all cases $L_x = 5000$, $L_y = 50$, $h = 300$, $W = 0.5\text{eV}$, $T = 0.36T_c$, $E = -0.2\text{eV}$.

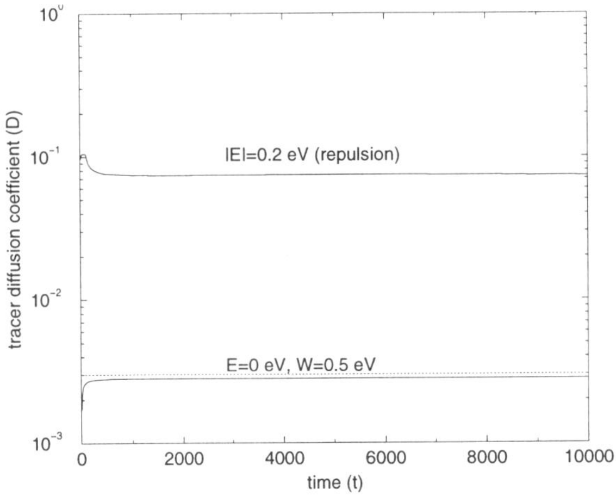


*Figure 8.* Test of the formula Eq. 12 in our model - dependence of tracer diffusion $D$ on time in two cases: free lattice gas ($E = 0$) and repulsive lattice gas ($E = -0.2eV$) at temperature $T = 0.36T_c$. The dotted line represents the value $D_c$ of the phenomenological collective diffusion coefficient (Eq. 8).

at lattice sites $k$. One can observe the domains of the $c(2 \times 2)$ phase near the edge of the evolving overlayer. We have obtained the same results for a diffusion in the rectangular channels where we have performed the majority of our calculations. The corresponding concentration profiles calculated at different time moments, $t$, collapse on the same curve, if they are plotted in terms of $x/\sqrt(t)$, i.e., they have a scaling property. This is shown in Fig. 6.

In the same figure we have plotted a fraction of $c(2 \times 2)$ phase contributing to the respective profiles. We can see that the $c(2 \times 2)$-concentration profiles are also scale invariant. The tooth-like shape of them near the leading edge of the concentration profiles corresponds to the concentration range $c \sim 0.5$, where the $c(2 \times 2)$ phase is the most perfect one. The scaling means that the width $L_W$ of the tooth satisfies the power-like growth law

$$L_W \sim t^{\frac{1}{2}}. \tag{11}$$

Thus, we may think of the size of the $c(2\times2)$ terrace at the leading edge of the concentration profiles as to be a specific kinetic order parameter for this particular type of ordering process. The observation of the power law (Eq. 11) is consistent with the results in Fig. 7 where we presented the time dependence of the average linear size (in the $x$-direction) of the clusters with the structure $c(2 \times 2)$. We defined a cluster as a set of the lattice sites $k$, in the nearest-neighbor relation, in which $C_1(k)$ differs at least by a given factor $\lambda$ from the $C_1(k_{nn})$ in the neighboring lattice sites $k_{nn}$, i.e., when the condition $\lambda C_1(k) > C_1(k_{nn})$ is fulfilled. Thus, parameter $\lambda$ controls fidelity of the local two-sublattice structure. The smaller the value of $\lambda$, the larger the difference between the occupation probabilities of the two sublattices. We can observe a power growth law (Eq. 1) with the growth exponent $\alpha \approx 0.5$ for the clusters which have the $c(2 \times 2)$ structure.

The problem of domain growth is well known in physics. It was stated for the first time by Lifshitz and Slyozov [16]. They constructed a mean-field kinetic theory for the metal grain growth, according to which the average grain radius scales in time as $L(t) \sim t^{1/3}$. Bray and Rutenberg [17, 18] found that in physical systems with a nonconserved order parameter, $L(t) \sim t^{1/2}$ as long as the domain growth is driven only by the curvature of domain boundaries. Thus, we have two types of diffusional processes present in the evolving adsorbate: surface diffusion and diffusional domain growth/dissolving. The domain growth usually is affected by pinning effects at low temperatures, which may suggest a smaller value of the growth exponent $\alpha$.

Once in our model, we can trace the propagation of the probability associated with a single chosen adatom, then, a tracer diffusion coefficient $D$ for the adatom can be determined even in the case of the interacting adatoms:

$$D = \frac{< r^2 >}{4t}, \tag{12}$$

where

$$< r^2 >= \sum r^2 G[r = (x, y), t], \tag{13}$$

and $G(r, t)$ represents the probability of finding the respective random walker representing the chosen adatom at lattice site $r$ after time $t$. In Fig. 8 we have plotted the tracer diffusion coefficient $D$ defined in Eq. 12 versus time in the two cases: the free lattice gas ($E = 0$) and the repulsive lattice gas ($E = -0.2eV$). Notice, that for non-interacting adatoms $D \approx D_c$ according to expectations. To complete all results, in Fig. 9 we have plotted the BM diffusion coefficient dependence on adatom concentration for the repulsive case. One can notice that our approximation works well in the region of low adsorbate concentration.

We can easily generalize the transition rates $w(k)$ (Eq. 3) to include next-nearest-neighbor adatom interactions. In this case, we can consider two scenarios: when the adatom

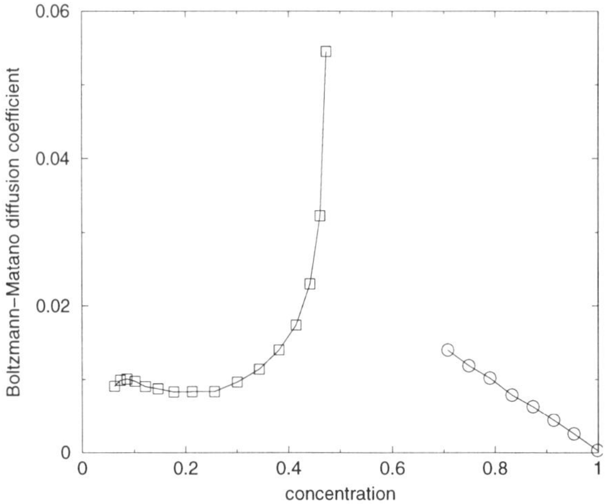


*Figure 9.* Boltzmann-Matano diffusion coefficient vs. coverage when $W = 0.5\text{eV}$, $T = 0.36T_c$, $E = -0.2eV$.

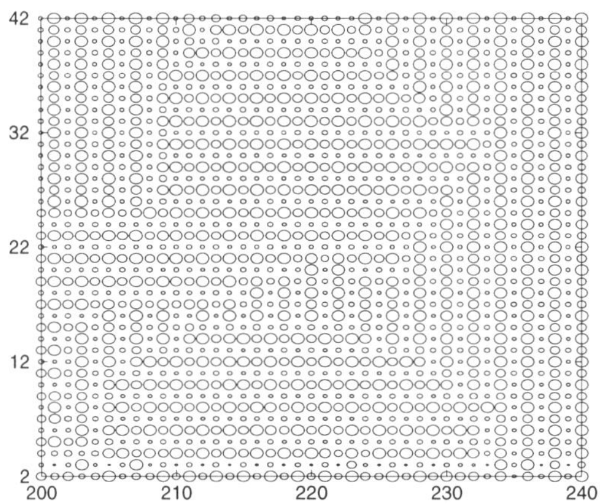


*Figure 10.* Domains with the $(1 \times 2)$ symmetry in the case of the nearest-neighbor and next-nearest-neighbor repulsive interactions in the probability-based continuum method.

concentration propagates only between the nearest-neighbor lattice sites and when the concentration propagates both between the nearest-neighbor lattice sites the next-nearest-neighbor lattice sites. In Fig. 10 we have presented a snapshot for the adsorbate in the latter case. One can observe spontaneously created domains with an ordered $(1 \times 2)$ phase. The ease of inclusion of the further-neighbor interactions in the method of propagating probability makes some perspectives before the method to consider complex behavior of the multi-phase systems. The method can also be adapted to various symmetry substrates.

## 4. Conclusions

We introduced a probability-based method of modeling the adsorbate diffusion in the non-equlibrium experiment. We showed that the model can be complementary to the Monte Carlo methods, which are traditionally used to describe this type of experiments. In the particular case of the repulsive adatom interactions, we showed that the size of the terrace in concentration profile corresponding to $c(2 \times 2)$ phase is scale invariant. This justifies the usage of the phenomenological BM diffusion coefficient in the profile experiment. However, we discussed only the simplified case of the continuous phase transitions appearing in the evolving overlayer. There is still an open question, can we use the BM method when the first order phase transitions occure?

The generalization of our method to consider multi-phase systems is possible, where the various Monte Carlo methods become inefficient.

## References

1. Gomer R. (1990) Diffusion of adsorbates on metal surfaces *Rep. Prog. Phys.* **Vol. 53**, pp. 917-1002
2. Bowker, M. and King, D.A. (1978) Adsorbate Diffusion on Single Crystal Surfaces, *Surface Sci.* **Vol. 71**, pp. 583-598
3. Kellogg, G.L., Tsong, T.T. and Cowan, P. (1978) Direct Observation of Surface Diffusion and Atomic Interactions on Metal Surfaces, *Surface Sci.* **Vol. 70**, pp. 485-519
4. Naumovets, A.G. and Vedula, Yu.S.(1984) Surface diffusion of Adsorbates, *Surface Sci. Rep.* **Vol. 4**, pp. 365-432
5. Tringides, M.C., Gupalo, M., Li, Q. and Wang, X. (1998) *Rep. Prog. Phys.* **Vol. 53**, pp. 917-1002
6. Wu, P.K., Tringides, M.C., and Lagally, M.G. (1989) Ordering kinetics of a chemisorbed overlayer: O/W(110), *Phys. Rev. B* **Vol. 39**, pp. 7595-7610
7. Grest, G.S., Anderson, M.P., Srolovitz, D.J. (1988) Domain-growth kinetics for the Q-state Potts model in two and three dimensions, *Phys. Rev. B* **Vol. 38**, pp. 4752-4760
8. Havlin, S. and Ben-Avraham, D. (1987) Diffusion in disordered media, *Adv. Phys.* **Vol. 36**, pp. 695-798
9. Gouyet, J.-F. (1999) From Microscopic Kinetics to Generalized Allen-Cahn Equations. Application to Adatoms and Intercalation Dynamics, *Lecture Notes in Physics*, **Vol. 519**, pp. 283-297, Springer-Verlag, eds. R. Kutner, A. Pękalski, K. Sznajd-Weron,
10. Reed, D.A and Ehrlich G. (1981) Surface diffusivity and the time correlation of concentration fluctuation, *Surface Sci.* **Vol. 105**, pp. 603-628
11. Uebing, C. and Gomer, R. (1991) A Monte Carlo study of surface diffusion coefficients in the presence of adsorbate-adsorbate interactions, I. Repulsive interactions, *J. Chem. Phys.* **Vol. 95**, pp.7626-7635
12. Natori, A. and Ohtsubo, H. (1986) Surface diffusion and formation of ordered state, *Surface Sci.* **Vol. 171**, pp. 13-44
13. Natori, A. and Ohtsubo, H. (1987) Surface diffusion and formation of an ordered state. II, *Surface Sci.* **Vol. 184**, pp.289-301
14. Uebing, C. and Gomer, R. (1991) A Monte Carlo study of surface diffusion coefficients in the presence of adsorbate-adsorbate interactions, II. Attractive interactions, *J. Chem. Phys.* **Vol. 95**, pp.7637-7640
15. Uebing, C. and Gomer, R. (1991) A Monte Carlo study of surface diffusion coefficients in the presence of adsorbate-adsorbate interactions, III. Repulsive nearest-neighbor and attractive next-nearest-neighbor interactions, *J. Chem. Phys.* **Vol. 95**, pp.7641-7653
16. Lifshitz, I.M. and Slyozov, V.V. The kinetics of precipitation from supersaturated solid solutions (1961) , *J. Phys. Chem. Solids* **Vol. 19**, pp.35-50
17. Bray, A.J. Theory of phase-ordering kinetics (1994) , *Adv. Phys.* **Vol. 43**, pp.357-459
18. Rutenberg, A.D. and Bray, A.J. Energy-scaling approach to phase-ordering growth laws (1995) , *Phys. Rev. E* **Vol. 51**, pp.5499-5513

# DIFFUSION ON AND *IN* SURFACES: THE ATOMIC SLIDE PUZZLE

J.W.M FRENKEN, R. VAN GASTEL, S.B. VAN ALBADA, E. SOMFAI*, AND W. VAN SAARLOOS*

*Kamerlingh Onnes Laboratory and *Instituut-Lorentz, Leiden University, P.O. Box 9504, 2300 RA Leiden, The Netherlands*

**Abstract.** This paper presents first results from scanning tunneling microscopy measurements, which show that a close-packed terrace of a metal surface can be far from static, even at temperatures as low as room temperature. We make the motion visible of the atoms *in* a Cu(001) terrace, by embedding a low density of In atoms in the first Cu layer. The peculiar characteristics of the motion of the In show that the diffusion of surface vacancies is responsible for a continual reshuffling of *all* the (In and Cu) atoms in the first layer. A comparison with model calculations of the statistics of the vacancy-assisted motion of terrace atoms shows there must be an attractive interaction between an embedded In atom and a vacancy, which makes the In atom somewhat more mobile than a Cu surface atom. Such an attraction is indeed found in Embedded Atom Method calculations.

## 1. Introduction

Usually, surface mobility is thought to be restricted completely to steps and kinks on crystal surfaces. These sites form the natural locations for atoms to 'come and go'. The resulting step motion can be observed both on the atomic and on mesoscopic length scales with various types of microscopy, such as, e.g., scanning tunneling microscopy [1-3]. Dynamic step rearrangements can also be responsible for the two-dimensional diffusive motion of adatom and vacancy islands [4]. By contrast, atoms in the close-packed terraces not in the direct vicinity of steps are often considered to be completely static, since they are tightly packed by their neighbors. Although suggestions have been made that also surface vacancies might be responsible for self-diffusion of islands [5], there are no experimental techniques to date, which can follow the diffusion of



*M.C. Tringides and Z. Chvoj (eds.),*
*Collective Diffusion on Surfaces: Correlation Effects and Adatom Interactions,* 191–199.

naturally occurring vacancies in a close-packed metal surface. Available forms of microscopy and spectroscopy either suffer from a lack of spatial resolution, or a lack of temporal resolution, or both.

Here we show that it is possible to use scanning tunneling microscopy to see the effect of diffusing vacancies in a metal surface indirectly, via the motion of impurity atoms. For this purpose, the impurities should have the special property that they embed themselves substitutionally in the outermost crystal layer. One of the surface-impurity combinations known to behave like this, is In on Cu(001) [6,7]. At temperatures just below room temperature, In atoms deposited on Cu(001) are incorporated in substitutional terrace sites. We observe a remarkable mobility of the embedded In atoms. From the statistics of the motion, we conclude that the In atoms are assisted by another 'particle' in their surface diffusion, and we show that the most likely candidate for this assisting particle is a vacancy in the Cu(001) surface. From our observations of the motion of In atoms, we conclude that also the Cu terrace atoms have a surprisingly high mobility, even at room temperature.

In this paper we present preliminary results of this work. A more complete account will be given in a forthcoming publication.

## 2. Experimental

The experiments were performed with a variable temperature scanning tunneling microscope (STM) [8], in an ultrahigh vacuum (UHV) system. A Cu single crystal of 99.999 % purity was mechanically polished parallel to the (001)-plane, and heated in an $Ar/H_2$ atmosphere to remove sulfur traces from the bulk. The Cu surface was further cleaned in UHV by several tens of cycles of sputtering with 600 eV $Ar^+$ ions and annealing to 675 K. The surface was regularly exposed to a few Langmuir of $O_2$ to remove carbon from the surface. STM images showed a well-ordered surface with terrace widths up to 8000 Å. In was deposited in situ from a Knudsen cell, either at cryogenic temperatures, or at room temperature. In this paper we will restrict ourselves to those cases where In was deposited at room temperature.

## 3. Observations

The STM images obtained at room temperature after deposition of In show that the In atoms are introduced into the outermost Cu layer via the steps on the

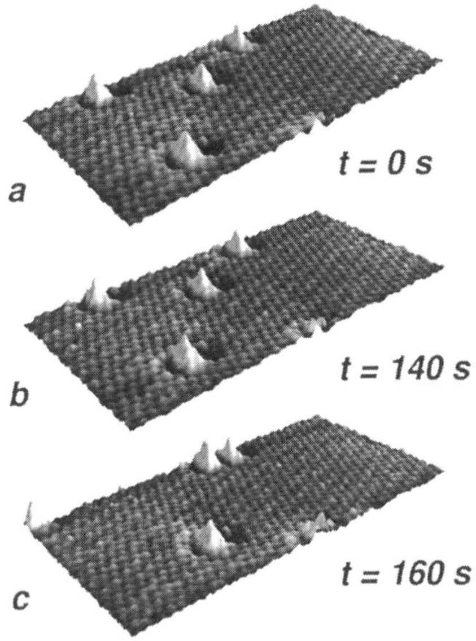


*Figure 1.* Three STM-images of 70×140 Å$^2$ selected from a movie measured at room temperature, illustrating the unusual diffusion of embedded indium atoms. The fine corrugation is due to the Cu(001) lattice. The four larger protrusions are embedded In atoms. In the time interval of 140 s between images (a) and (b), the four imaged In atoms have not moved. During the next 20 s, three of the four In atoms have been displaced over several lattice spacings (c). ($I_t$ = 0.9 nA, $V_t$ = –0.58 V).

Cu(001) surface. We have observed that the embedded In atoms slowly diffuse away from the steps, to homogeneously populate the terraces only after several hours. The observed diffusion distances are orders of magnitude larger than the root-mean-square step fluctuations occurring on the same time scale at room temperature, which implies that the In is not simply incorporated in the first Cu layer as a consequence of the step fluctuations of the Cu surface.

Figure 1 shows three images taken from an STM movie[*] at room temperature, after deposition of 3% of a monolayer of In on the clean Cu(001) surface. The small-scale corrugation corresponds to the atomically resolved Cu surface, while the larger protrusions are individual embedded In atoms. The

* The full movie can be viewed on http://www-lion.leidenuniv.nl/groups/ip

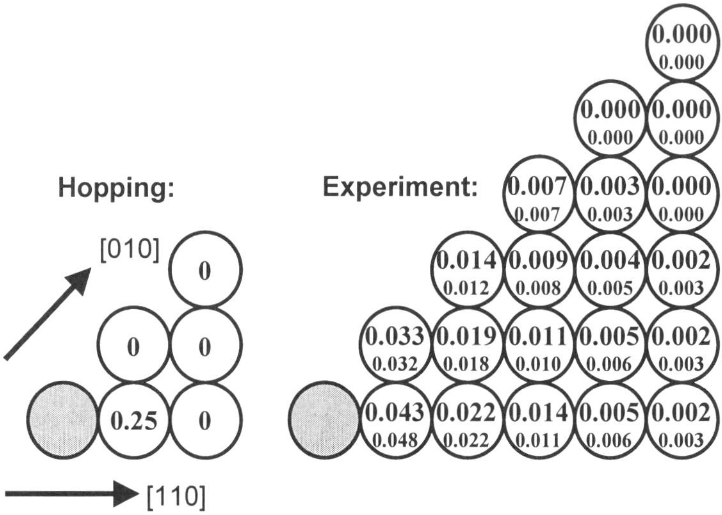


*Figure 2.* Jump vector distributions. The right panel shows the distribution for embedded In atoms in Cu(001). The numbers correspond to the probabilities for jumps of an In atom from its starting position to each of the indicated non-equivalent lattice sites. The upper number on each site is the probability measured from STM-movies at room temperature. The lower number is the result of the model calculation described in the text. For comparison, the jump vector distribution expected for the case of simple hopping is shown in the left diagram.

vertical scale is exaggerated; the In atoms stick out only 0.4 Å above the surface. What is illustrated by Figure 1, and what is observed in all our movies of the motion of In embedded in Cu(001), both at 300 K and at other temperatures, is that the In atoms move via long jumps of more than a single lattice spacing, separated by long time intervals. In addition, the movies show there is a strong tendency for nearby In atoms to jump at the same time.

From the STM movies, we have measured the distribution of jump lengths of the embedded indium atoms, which is displayed in the diagram of Figure 2. The diagram shows the triangle of jump orientations between the [010] and [110] azimuths; the upper number shown for each lattice point corresponds to the average probability measured for the jump vector corresponding to that lattice position and the probabilities for the 3 or 7 equivalent jump vectors in the symmetry related directions that are not shown in the diagram. It is clear that the distribution extends well beyond a single lattice spacing, and that there is even a significant probability for the indium atoms to make jumps as long as five lattice spacings.

## 4. Interpretation

The long jumps and the high probability of nearby indium atoms to jump simultaneously are hallmarks of a diffusion mechanism in which the indium is assisted by another particle, which diffuses so rapidly that it remains invisible to the STM. The simplest scenario is that in which the In moves over a single lattice spacing by changing places with the assisting particle. Even if the assisting particles perform completely random walks, the In has a high probability to move over several lattice spacings as the result of the passage of a *single* assisting particle, simply because the random walk has a high probability to bring the particle back to the In atom after every encounter (see below). The fact that a single vacancy will usually encounter various In atoms, naturally explains the tendency for nearby In atoms to jump at the same time.

Cu adatoms and (surface) vacancies can both be considered as natural candidates for the assisting particle. Figure 3 shows how an encounter of an embedded In atom with a Cu adatom as well as an encounter with a surface vacancy can lead to a displacement of the In atom over a single atomic spacing. On the basis of the displacement statistics of the In atoms, we cannot distinguish between the two scenarios. However, we can decide between the two possibilities by observing the incorporation pathway of In upon deposition on the Cu(001) substrate. As mentioned above, we observe that freshly deposited In atoms cannot invade the Cu(001) terrace directly, but they only reach substitutional terrace sites via the steps. This immediately rules out the possibility of adatom-assisted motion, since the exchange of an In adatom with a terrace Cu atom is a necessary ingredient of such motion (cf. Figure 3).

The mechanism that we end up with is illustrated in Figure 4. The two-dimensional random walk of a single surface vacancy leads to the multiple displacement of an embedded In atom. Note that also many of the terrace Cu atoms undergo similar, large displacements. In the proposed mechanism, the length of the long jumps of the indium atoms depends on the average number of times that a single vacancy changes places with an indium atom, and we

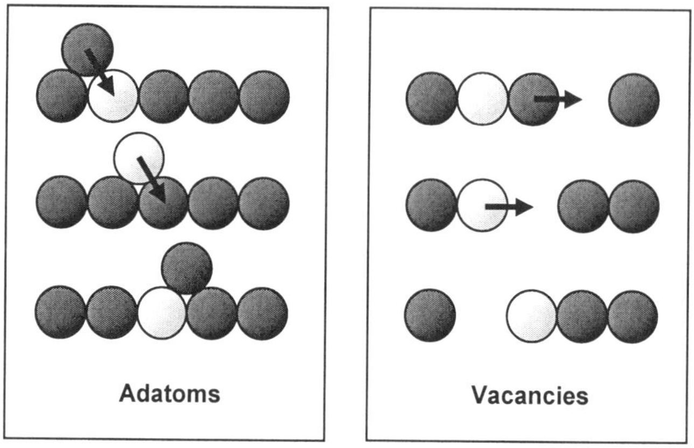


*Figure 3.* Two scenarios for assisted diffusion of embedded In atoms. In the scenario on the left, the In atom moves by an exchange process with a Cu adatom. After the first step, the In atom has temporarily become an adatom itself. A second exchange step embeds the In atom again into the first Cu layer. In the process the In atom is moved over one atomic spacing. In the scenario on the right, the In atom moves as the result of the passage of a surface vacancy.

associate the frequency of the (long) indium jumps with the frequency with which the indium is encountered by new vacancies.

## 5. Model

The simplest model description of the vacancy-assisted motion treats the In atom as a tagged Cu atom. The probability of a vacancy to make single 'jumps' in each of the four directions is simply 25%. Based on these probabilities, it is easy to calculate the four probabilities for a vacancy, after an initial encounter with an In atom, to return to the In atom from each of the four relevant directions: from the same side $P_1$, from the opposite side $P_2$ or from one of the two sideways directions $P_3 = P_4$. If the Cu(001) terrace is of infinite size, these four return probabilities add up to unity: the vacancy will return to the In atom infinitely many times and will therefore cause a diverging displacement of the In atom. However, on a real surface, the vacancy will have a finite lifetime, so that also the displacement of the In atom resulting from the passage of a single vacancy will be finite. The vacancy most probably originates from a step and it will most probably also terminate its 'life' at a step. In our toy model, the finite

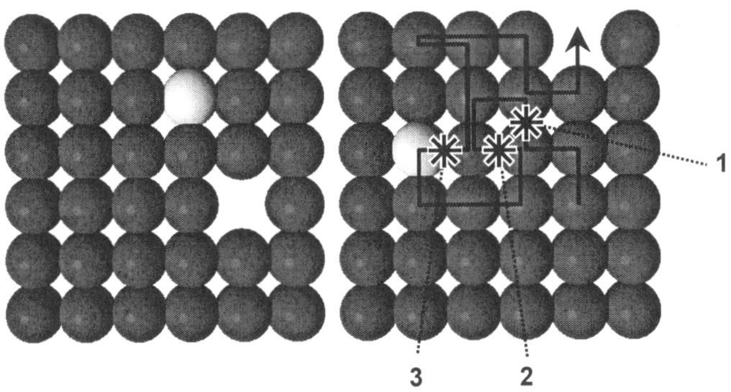


*Figure 4.* Ball model illustrating the vacancy-mediated diffusion mechanism of the In atoms. The random walk of a single vacancy results in the multiple displacement of the In atom. At the three marked locations, the vacancy trades places with the In atom.

lifetime of the vacancy is introduced in the form of a fifth probability $P_{ann}$, which is the probability of the vacancy to be annihilated before it returns once more to the In atom. We treat this probability as a free parameter in an attempt to fit the experimentally determined jump vector distribution. In Figure 2 we compare the theoretical and experimental distributions. The model provides an excellent fit for a value of the annihilation probability of $P_{ann} = 0.027$. The shape of both the measured and the calculated distributions in Figure 2 is significantly different from the two-dimensional Gaussian distribution, expected in the case of ordinary, i.e., non-assisted diffusion. In fact, the shape is very close to that of a modified Bessel function of order zero. As we will demonstrate in a future publication, it is easy to show that this shape is indeed characteristic of the type of assisted diffusion considered here, where the assisting particle has a limited lifetime.

The value expected for $P_{ann}$ for the average step separation observed in the experiment of approximately 400 lattice spacings is 0.23. This is much higher than the value obtained from the fitting described above. The dependence of $P_{ann}$ on the average terrace width is extremely weak (a logarithmic dependence has been derived in a somewhat different context by Brummelhuis et al. [9]). In order to reach the low value of 0.027 found above, the average terrace width would have to be as large as $10^{25}$ atomic spacings, which would be completely unphysical! This result illustrates an important flaw of the simple model, namely the complete neglect of the chemical difference between In and Cu. An increase by an order of magnitude in the average number of returns of the vacancy to the In atom is easily obtained as a consequence of the modifications of the diffusion barriers of the vacancy in the vicinity of the In atom, with respect to the normal barrier for vacancy diffusion in the clean Cu(001) terrace. Indeed, such changes in activation barrier are found in calculations using the Embedded Atom Model (EAM) (we have used the potential described by Breeman et al. [6,7]), which show a noticeable attractive interaction between the vacancy and the (large) In atom. When we modify the model to include the EAM diffusion barriers of the vacancy in the vicinity of the In atom, and if we fix the lifetime of the vacancy via the constraint that the vacancy's random walk terminates once it reaches the nearest step, for a step separation of 400 lattice spacings, we obtain a distribution that is almost identical to the equal-barrier result shown in Figure 2, i.e., with an equally good match to the measured distribution of jump lengths, but this time without any adjustable parameters in the model. Full details of this calculation will be given in a forthcoming publication.

## 6. Conclusion

In summary, STM measurements show that In atoms, embedded in the Cu(001) surface, perform two-dimensional diffusive motion that cannot be explained by step fluctuations. The measurements strongly suggest that the motion is caused by the two-dimensional diffusion of surface vacancies in the Cu surface. These observations imply that not only the In atoms exhibit a high mobility in the surface, but that also all Cu terrace atoms are continually 'on the move'. In this respect we expect that Cu(001) is not very different from other close-packed surfaces of Cu and from other materials, which means that we have to regard all close-packed metal terraces as quite dynamic, even at temperatures as low as room temperature.

## Acknowledgments

This project is part of the research program of the "Stichting voor Fundamenteel Onderzoek der Materie (FOM)," which is financially supported by the "Nederlandse Organisatie voor Wetenschappelijk Onderzoek (NWO)."

**References**

1. Jeong, H.-C. and Williams, E.D. (1999), "Steps on surfaces: Experiments and theory," *Surf. Sci. Rep.* **34**, 171-294.
2. Zhang, Z., and Lagally, M.G. (1997), "Atomistic processes in the early stages of thin-film growth," *Science* **276**, 377-383.
3. Kuipers, L., Hoogeman, M.S., and Frenken, J.W.M. (1993), "Step dynamics on Au(110) studied with a high-temperature, high-speed scanning tunneling microscope," *Phys. Rev. Lett.* **71**, 3517-3520.
4. Morgenstern, K., Rosenfeld, G., Poelsema, B., and Comsa, G. (1995), "Brownian motion of vacancy islands on Ag(111)," *Phys. Rev. Lett.* **74**, 2058-2061.
5. Hannon, J.B., Klünker, C., Giesen, M., Ibach, H., Bartelt, N.C., and Hamilton, J.C. (1997), "Surface self-diffusion by vacancy motion: Island ripening on Cu(001)," *Phys. Rev. Lett.* **79**, 2506-2509.
6. Breeman, M., and Boerma, D.O. (1992), "Sites, mobilities and cluster formation of In atoms on a stepped Cu(100) surface," *Phys. Rev.* B **46**, 1703-1709.
7. Breeman, M., and Boerma, D.O. (1993), "Atomic mobilities on a stepped Cu(100) surface," *Surf. Sci.* **287/288**, 881-885.
8. Hoogeman, M.S., Glastra van Loon, D., Loos, R.W.M., Ficke, H.G., De Haas, E., Van der Linden, J.J, Zeijlemaker, H., Kuipers, L., Chang, M.F., Klik, M.A.J., and Frenken, J.W.M. (1998), "Design and performance of a programmable-temperature scanning tunneling microscope," *Rev. Sci. Instr* **69**, 2072-2080.
9. Brummelhuis, M.J.A.M., and Hilhorst, H.J. (1988), "Single vacancy induced motion of a tracer particle in a two-dimensional lattice gas," *J. Stat. Phys.* **53**, 249-278.

# DECAY OF TWO-DIMENSIONAL ISLANDS ON AG(110)

K. MORGENSTERN[1,2], E. LÆGSGAARD[1], F. BESENBACHER[1]
[1] *Institute of Physics and Astronomy and CAMP, University of Aarhus, DK - 8000 Aarhus C, Denmark*
[2] *Institut für Experimentalphysik, FU Berlin, D - 14195 Berlin, Germany*

**Abstract.** We have investigated the decay of two-dimensional (2D) islands on the anisotropic Ag(110) surface using variable-temperature scanning tunneling microscopy. Contrary to predictions from traditional Ostwald ripening theory, a quasi-one-dimensional (1D) decay mode was observed at low temperatures (175-220 K). A surprisingly sharp transition to the quasi-2D decay mode was observed around 220 K. This transition was accompanied by a fast equilibration of the island shape. These findings have tentatively been rationalized within a simple model to identify the underlying rate limiting atomistic processes.

## 1. Introduction

In recent years the development of fast scanning tunnelling microscopes (STM) triggered the investigation of changes in surface morphologies that were deliberately created far from equilibrium [1–10], as well as thermal fluctuation around equilibrium-shaped structures [11, 12] in high temporal resolution. Model studies of the decay of islands, i.e., structures of monoatomic height consisting of up to several thousand atoms, on metal surfaces [1–10] allow one to predict the limits of stability of atomic systems using a common theory for both metals and semiconductors which are the main components of microtechnology nowadays. In this paper we discuss fundamental laws of the stability of nanostructures on an anisotropic surface, thereby summarizing earlier publications on this subject [13].

Studies of the decay of quasi-isotropic islands have led to an improved understanding of the physical properties and processes involved in their decay. It has been shown that a common theory based on the classic Ostwald ripening theory [14] can be used for island decay on metal and semiconductor surfaces [2–4]. The importance of short-ranged island interaction



*M.C. Tringides and Z. Chvoj (eds.),*
*Collective Diffusion on Surfaces: Correlation Effects and Adatom Interactions,* 201–212.

for Ostwald ripening has been pointed out (Si(100) [2, 3], Ag(111) [10]). Based on this knowledge, energetic barriers have been measured (Si(111) [7] and Ag(111) [9]). Finally, the dominating coarsening mechanisms for different coverages (Ag(100) [5]) and the rate limiting step for ripening between islands on the same terrace (Cu(100) [6]) and on top of each other (Cu(111) [8]) have been determined.

The majority of these studies were carried out on isotropic surfaces, whereas the material of choice of the semiconductor industry, Si(100), has a higher complexity, being anisotropic due to dimerization. The relevant processes on anisotropic surfaces are, however, less well understood, because the only studies on this surface [2, 3, 7] were performed at temperatures well above the predicted crossover temperature from isotropic to anisotropic island growth [15]. Indeed, the anisotropy does not influence the Ostwald ripening in the temperature range investigated [3].

To illucidate the effects of anisotropy on island decay, we investigate islands on an anisotropic surface at lower temperatures. For silver islands on Ag(110) between 155 K and 255 K, we find three different decay behaviors with transitions at $T_l = (175 \pm 5)$ and $T_c = (220 \pm 10)$ K, respectively. Below $T_l$ no island decay is observed. Between $T_l$ and $T_c$, we find a quasi-1D decay mode, contrary to predictions of conventional Ostwald ripening theories [16]. In spite of the high energy for detachment of atoms out of a $[\bar{1}10]$ step [17], a quasi-isotropic decay of the islands sets in already at $T_c$. Above $T_c$ the island shrinks in both dimensions, and the decay can be described within the framework of the Ostwald ripening theory.

## 2. Experimental

The experiments were performed in a UHV system equipped with a home-built, fast-scanning, variable-temperature STM as well as standard facilities for sample preparation and characterization [18]. The Ag(110) surface is prepared by several sputtering (1 keV $Ne^+$) and annealing cycles below the roughening transition temperature [19] (623 K), followed by a slow cool down (10 K/min). This procedure results in a clean, well-ordered surface as shown in Fig. 1 (inset). The anisotropic Ag(110) surface consists of close-packed atom rows with an atomic distance of 0.289 nm running in the $[\bar{1}10]$ direction. The rows are separated in the [001] direction by 0.409 nm.

Recipes were developed to prepare three different types of monatomic high adatom islands on this surface:

(a) Deposition of silver from a tungsten basket with sample tempera-

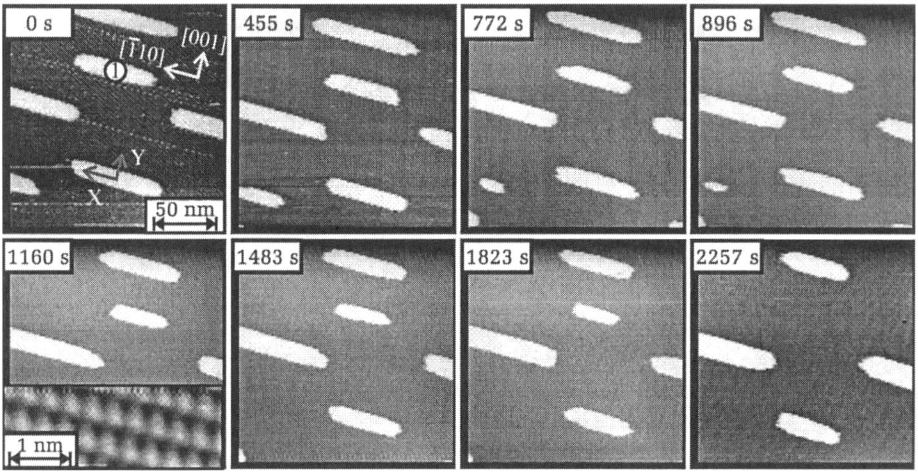


*Figure 1.* Series of STM images recording the development of adatom islands on Ag(110) at 235 K, U = -2.1 V, I = 0.1 nA; Inset: atomic resolution image showing the orientation of the close-packed rows, U = -2.0 V, I = 0.32 nA.

tures of 250-270 K for up to 3 min results in equilibrium-shaped islands (Fig. 1) with an aspect ratio of $R_{eq} := X/Y = 2.9 \pm 0.4$. These islands have their longer axis parallel to $[\bar{1}10]$, i.e., parallel to the direction of fast diffusion. This is in contrast to semiconductor epitaxy on Si(100), where the island elongation driven by anisotropy in bonding is perpendicular to the direction of fast diffusion [20].

(b) Silver deposition at 180-200 K results in islands with an aspect ratio larger than $R_{eq}$. Islands with even larger aspect ratios have been predicted for this temperature range [21].

(c) Sputtering with 1 keV $Ne^+$ ions at 190-200 K for 1-2 sec without any Ag deposition results in islands with $R < R_{eq}$.

After preparation of the desired islands, the sample is transferred to the STM for imaging at the appropriate temperature. With fast-scanning STMs it is possible to visualise the change in surface morphology with high temporal and spatial resolution in consecutive STM images, so-called STM movies, which typically contain a few tens to over a thousand images. When recording the STM movies we employ a dynamic drift compensation mode that practically eliminates thermally-induced drift and allows us to monitor the decay of individual islands over extended periods of time. The position of a non-moving feature (e.g., a kinked step edge) is located in each image using a home-written pattern recognition software routine. During acquisition of the next image, the necessary offset voltages are added to the scanning voltages to keep the feature at the same place in the image.

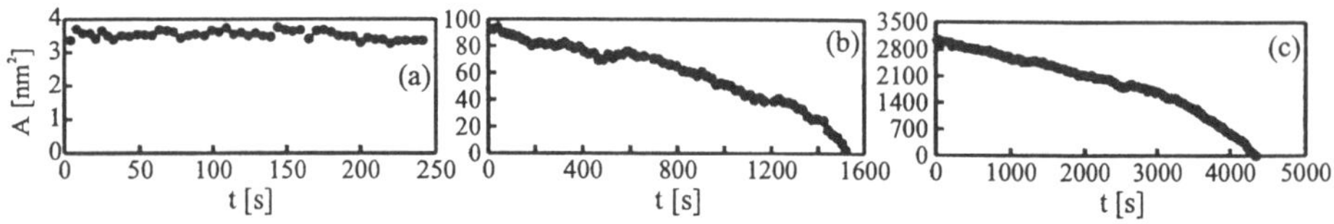


*Figure 2.* Evolution of island area at different temperatures T: (a) T = 169 K (b) T = 212 K (c) T = 255 K.

The achieved stability is typically $\pm 1$ pixel over several hours. Images with $256 \times 256$ pixels are recorded at time intervals between 5 and 70 seconds, according to the decay rate. Special care was taken to insure that the STM data are not influenced by the imaging process [9, 10, 22].

## 3. Results

Excerpts from an STM movie recording Ostwald ripening in an ensemble of islands at 235 K are shown in Fig. 1. Ostwald ripening on a surface refers to the coarsening of an ensemble of differently sized islands via the diffusion of adatoms between them. For equilibrium-shaped islands, small islands have a larger curvature and thus a larger adatom concentration in their vicinity than large islands. This difference in chemical potentials induces a net flux of material from small to either larger islands or pre-existing terrace step edges by diffusion of adatoms through the adatom gas phase.

In Fig. 1, the island marked '1' decays within 40 min. From this type of images, we determined the area $A$ of isolated decaying islands that are situated on terraces so they do not communicate directly with neighboring islands, i.e., usually, the next step in [001] is a terrace step with terrace widths of 200 to 400 nm. In $[\bar{1}10]$ the next islands are typically at a distance of 50 to 200 nm. Examples are shown in Fig. 2. At temperatures below $T_l = (175 \pm 5)$ K, no island decay is observed at 169 K (Fig. 2a). From atom-resolved STM movies, it is concluded that the silver adatoms are simply not able to detach from the island edges at these temperatures. Above $T_l$, adatom detachment from the island is possible and the islands decay (Figs. 2b and c). The decay curves of the islands are not linear in time. At all temperatures the smaller the islands, the faster they decay. Furthermore, the overall decay rate increases with temperature. For the particular islands shown, the decay rate is 0.07 $nm^2/s$ at 212 K (Fig. 2b) and 0.70 $nm^2/s$ at 255 K. The decay rate of islands of the same size should be compared for comparison. For instance, when the island at 255 K has reached the size of 100 $nm^2$, it decays with 1.9 $nm^2/s$.

In the following, we will argue that the island decay is qualitatively

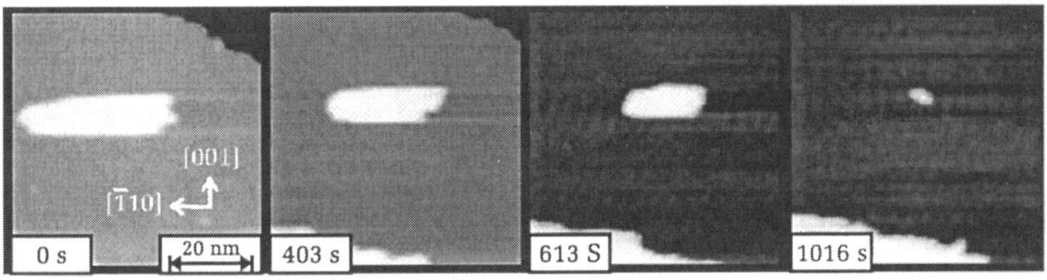


*Figure 3.* Series of STM images recording adatom island decay on Ag(110) at 218 K, U = -0.53 V, I = 0.11 nA.

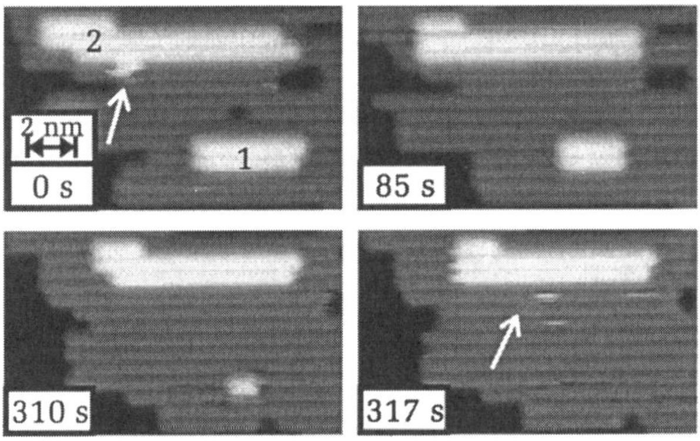


*Figure 4.* Excerpts of a high resolution movie at 186 K showing that adatoms from island '1' diffuse perpendicular to the rows towards island '2'; arrows mark diffusing atoms that are only partly imaged as they diffuse faster than the tip scans an atomic row; U = -0.5 V, I = 1.18 nA.

different below $T_c = (220 \pm 10)$ K (e.g., at 212 K in Fig. 2b) and above $T_c$ (e.g., at 255 K in Fig. 2c). If we compare STM movies of the island decay in both temperature ranges (cf. Figs. 1 and 3) we find that below $T_c$ the island length decreases with constant width, while above $T_c$ both the island's width and length decrease. This is corroborated by measuring the aspect ratio $R$. Indeed, with initial aspect ratios between 1.25 and 10, the aspect ratio changes differently below and above $T_c$. Above $T_c$, the islands equilibrate, i.e., $R$ decreases/increases if the initial value is larger/smaller than $R_{eq}$. Subsequently, $R$ fluctuates within the error bars around $R_{eq}$. In contrast, for islands decaying below $T_c$, $R$ changes towards smaller values regardless of $R_{eq}$. For this 1D decay atoms can only detach from the open step edge. STM movies, however, reveal that the subsequent migration of the released silver adatoms on the Ag(110) terrace is not confined to one specific direction, but occurs isotropically (see Fig. 4).

To analyse the decay quantitatively, we recall that the continuum theory of Ostwald ripening for 2D islands can be used to determine the rate limiting step of the island decay. It states that the area $A$ of the island follows a power law of the form [23]:

$$A \propto \tau^{2\beta}, \qquad (1)$$

$\tau$ being the decay time ($\tau(A=0)=0$), and $\beta$ an exponent, that depends on the ratio of the activation energies for terrace diffusion and reattachment to the island. The island decay is either diffusion-limited with $\beta = 1/3$ or attachment-detachment limited with $\beta = 1/2$ [24]. This continuum theory is generally valid for equilibrium-shaped islands and is *not* restricted to isotropic surfaces as anisotropy manifests itself in the prefactor and leaves the exponent unaltered. For the interface limit Bartelt *et al.* [3] calculated explicitly the prefactor by including the anisotropy in the calculation of the chemical potential. With $X \equiv \sqrt{A * R}$, $Y \equiv \sqrt{A/R}$, the extensions of the islands along the long/short axis parallel to the $[\bar{1}10]/[001]$ directions (see Fig. 1), respectively, we therefore expect for equilibrium-shaped islands, i.e., for $X \propto Y$, throughout the decay:

$$X \propto \tau^{\beta_X}, \qquad Y \propto \tau^{\beta_Y}, \qquad (2)$$

At low temperatures, the island length $X$ decreases continuously, while the island width $Y$ stays approximately constant until the island has almost disappeared (Fig. 5a). At high temperatures, both $X$ and $Y$ decrease continuously (Fig. 5b). Moreover, the decay rate for $X$ is found to be about three times higher than for $Y$, with $R_{eq}$ being conserved throughout the decay. Double logarithmic plots (Fig. 5c) reveal that the parameters $X, Y$ indeed follow the power laws of X, Y in Eq. (2) above $T_c$. Below $T_c$ only $X$ follows the power law scaling, while Y is apparently constant. We determine the exponents $\beta_X$ and $\beta_Y$ by fitting Eq. 2 to the $X(t)$ and $Y(t)$ data (Figs. 5a,b). We find that $\beta_X$ is approximately constant with $0.32 \pm 0.04$ over the whole temperature range; $\beta_Y$ has a similar value above 230 K ($0.27 \pm 0.04$) but drops to a much lower value below 210 K (see Fig. 5d)[1]. We have repeated these measurements for vacancy islands on Ag(110) and find a similar conclusion but with a transition temperature of $(205 \pm 10)$ K.

To summarize our experimental results: below $T_l = 175$ K the island size stays constant. Between $T_l$ and $T_c = 220$ K the islands are not equilibrium-shaped and decay in a preferred direction by adatom detachment from the [001]-oriented step. Above $T_c$ the islands equilibrates and decay two-dimensionally. The decay parameters $\beta$ ($\beta_X \approx 1/3 \approx \beta_Y$) indicate a diffusion-limited decay process qualitatively equivalent to the decay

[1] In principle, it is not very sensible to determine $\beta_Y$ for temperatures below 220 K, since in this case the width $Y$ of the islands stays approximately constant (Fig. 5a). This constancy is, however, not always as obvious as in the example shown. We include the data here for comparison purposes.

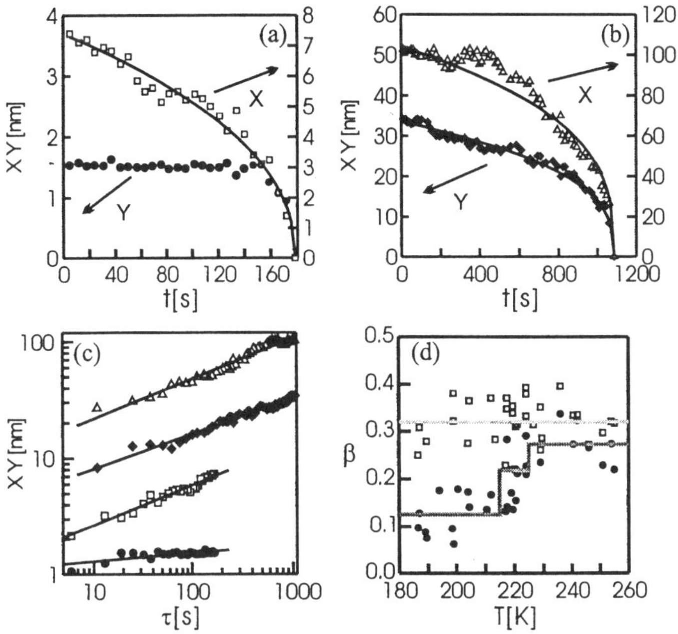


*Figure 5.* (a,b) Evolution of $X$ (open symbols) and $Y$ (filled symbols) extension at different temperatures T: (a) T = 199 K (b) T = 242 K; (c) double logarithmic $X, Y$ - vs. - decay time of (a,b) with linear fit. (d) Exponents $\beta$ - vs. - T determined via eq. 2; open squares $\beta_X$; filled circles $\beta_Y$; horizontal solid lines represent mean values in the regions they are drawn.

on isotropic metal surfaces [4].

At first sight, these experimental findings contradict two theoretical predictions. Supposing a strictly 1D system, Landau-Lifshitz [16] deduced 'the impossibility of the existence of phases in 1D system' within the conventional Ostwald ripening continuum theory, i.e., it should *a priori* not be possible for a 1D island to decay. Furthermore, molecular dynamics calculations predict that a prerequisite for 2D decay, i.e., the detachment of an atom out of a $[\bar{1}10]$ step, should only be possible far above room temperature [25].

## 4. Discussion

To rationalize these results and to understand the atomic processes underlying the two different decay modes, we have performed molecular dynamic calculations of activation energies for a variety of elementary processes (Fig. 6) relevant for the decay of silver islands on Ag(110), using approximate

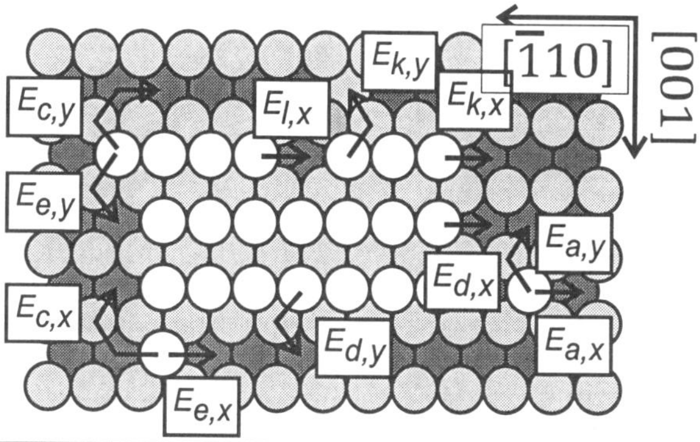


| | a,x | e,x | c,x | a,y | k,x | l,x | d,x | e,y | c,y | k,y | d,y |
|---|---|---|---|---|---|---|---|---|---|---|---|
| $E_{calc}$ [eV] | 0.28 | 0.29 | 0.38 | 0.39 | 0.44 | 0.45 | 0.45 | 0.53 | 0.56 | 0.56 | 0.85 |
| $T_{onset}$ [K] | 110 | 115 | 151 | 156 | 173 | 175 | 178 | 209 | 220 | 220 | 333 |

*Figure 6.* Hard sphere model of an island indicating the processes relevant in island decay, energy barriers calculated with EMT potentials [25], and corresponding onset temperatures. For all processes perpendicular to the close-packed rows ($E_{*,y}$), the exchange process with a substrate atom is found energetically favorable over the hopping over the row. Some of the energies have previously been calculated by Hontinfinde *et al.* [17] based on the RGL potential. Besides process $E_{d,y}$, for which they give the energy for the hopping mechanism (1.05 eV), the energies differ only slightly (in the order of 5%).

effective medium theory (EMT) potentials [25]. With the EMT theory the energy of an atom in a free structure is a simple function of the nearest neighbor (nn) and next-nearest neighbor (nnn) coordination numbers for in this case Ag atoms in the topmost surface layers.

To compare the calculated energies to the experimental results, we recall that calculations based on approximate potentials like EMT tend to underestimate the energy barriers. Agreement with experiments can be achieved by multiplying all energies with the same factor [25]. Under the assumption of equal prefactors for all processes, the calculated energies are therefore proportional to the lowest temperature at which the processes occur. We scale the calculated energies to the experimental onset temperature of $T_l = (175 \pm 5)$ K for island decay, i.e., the adatom detachment from the [001] step ($E_{k,x} = 0.44$ eV and $E_{d,x} = 0.45$ eV). With the resulting calibration factor $C = (378...410)$ K/eV, we calculate the onset temperatures $T_{onset} := C \cdot E_{calc}$ depicted in Fig. 6.

First, we discuss the 2D decay of equilibrium-shaped islands above $T_c$. An effective way for an island to evolve into its equilibrium shape is adatom diffusion along its perimeter. On anisotropic surfaces this implies that the atoms are able to diffuse along the [001] and the [$\bar{1}$10] steps *and* to diffuse around the corners in *both* directions [26]. Adatom migration along the [001]

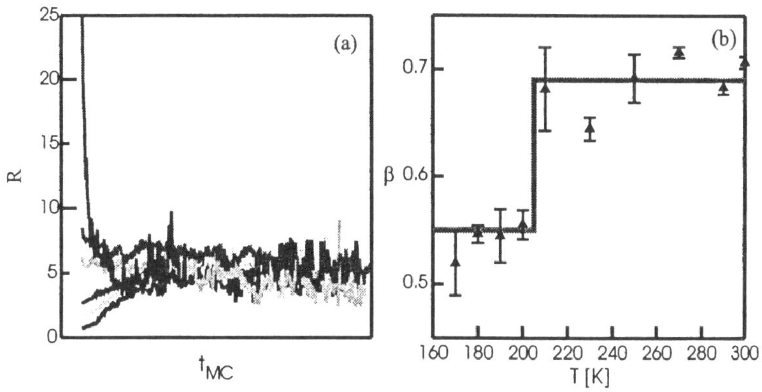


*Figure 7.* kMC calculation: average of 10 runs. (a) Aspect ratio $R$ for islands of 145 atoms each with different initial aspect ratios during their decay at 500 K (b) Exponent $\beta = \beta_X + \beta_Y$.

steps, ($E_{e,y} = 0.53$ eV, 208 K), as well as the corner rounding from [001] steps to [$\bar{1}$10] steps ($E_{c,y} = 0.56$ eV, 220 K) set in around $T_c$, whereas diffusion along the [$\bar{1}$10] step ($E_{e,x} = 0.29$ eV, 115 K) and corner diffusion from [$\bar{1}$10] to [001] steps ($E_{c,x} = 0.38$ eV, 151 K) set in below $T_c$.

The observed 2D island decay behavior should imply that atoms can detach from *both* step edges. The detachment from a close-packed straight [$\bar{1}$10] step ($E_{d,y} = 0.85$ eV) should, however, not occur until about 330 K, and thus this process is not relevant here. Instead, we propose that detachment preferentially occurs from kink sites next to vacancies on the [$\bar{1}$10] step facet. The onset temperature for this process ($E_{k,y} = 0.56$ eV, 220 K) is indeed close to $T_c$. Such a process is, however, only relevant if a sufficient number of kinks exists along the [$\bar{1}$10] step of the islands. The the following three observations makes the existence of a sufficient number of kinks very plausible: (a) a low activation energy for the formation of vacancies in the outermost row ($E_{k,x} = 0.44$ eV, 173 K), (b) a low activation energy for their subsequent diffusion along the [$\bar{1}$10] direction, ($E_{l,x} = 0.45$ eV, 175 K), and (c) the elliptic form of the islands. Based on the atomic arrangement on fcc(110) surfaces, we propose the coincidence of the same onset temperature for edge diffusion (allowing shape equilibration) and detachment out of kink sites (allowing 2D decay) to be a general phenomena. Generally, the difference in energy between $E_{k,y}$ and $E_{c,y}$ should be negligible as the atomic arrangements of both initial and final positions do not differ in the number of their nearest neighbors. Kinetic Monte Carlo (kMC) simulations, based on the calculated EMT energies [25] of islands of identical size but with different aspect ratios, confirm that the islands evolve into their equilibrium shape above $T_c$ (Fig. 7a).

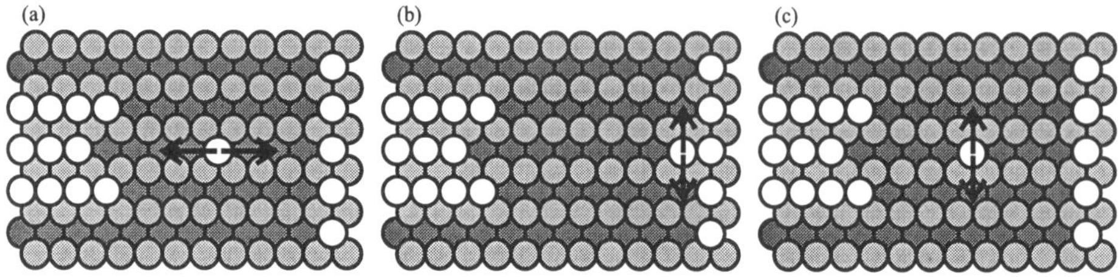


*Figure 8.* Hard sphere model illustrating possibilities for 1D decay.

We will now discuss the situation below $T_c$, where the island width remains approximately constant. The detachment from a $[\bar{1}10]$ step even at a kink site ($E_{k,y} = 0.56$ eV, 220 K) is energetically impossible and atoms can only detach from the [001] step.

Figure 8a shows the situation considered in textbooks [16] for a fcc(110) surface. The adatom migration is strictly 1D, i.e., constrained to the close-packed $[\bar{1}10]$ rows. Within this row the adatoms sense the same chemical potential on both ends, the driving force for ripening is missing, and Ostwald ripening theories predict no decay at all [16]. This is, however, inconsistent with the observed 1D island decay. Between islands there is no potential gradient if all islands have the same dimension in the $Y$-direction, so that the total length of both [001] steps and $[\bar{1}10]$ steps never change when atoms are transferred from one island to another. A chemical gradient, on the other hand, exists between islands of different $Y$-dimension. We exemplify this in the extreme case of an island end being perpendicular to an infinitely long step. When an island looses a complete row of atoms, its perimeter and thereby the total step energy are decreased, whereas atoms attaching at an infinitely straight step do not change the step length. Such an adatom flux becomes possible if the adatoms can diffuse perpendicular to the close-packed rows, either on the terrace (Fig. 8c) or along the island step edge (Fig. 8b). This allows island decay in the case of 1D detachment.

Evidence for 2D diffusion on the terrace is found in atomically-resolved STM images (Fig. 4) at temperatures below $T_c$. In this particular example, the rows in which traces of adatoms are seen are limited by step edges to lower terraces; an unlikely source for adatoms. This is corroborated by the calculated activation energies, where diffusion perpendicular to the close-packed rows, ($E_{a,y} = 0.39$ eV, 160 K) has an onset temperature far below the onset temperature for adatom detachment. (The onset temperature for diffusion along the island edge is 210 K ($E_{e,y} = 0.53$ eV), and thus this effect is negligible.) The 2D diffusivity of the adatoms (8c) is therefore responsible for the decay of the islands, although atoms are released only from the [001] step edge. In this case, the decay rate should be proportional to the open step edge $Y$ (Fig. 5a), and $X$ should decay linearly. Experimentally, however, the decay rate of the $X$-extension increases with smaller island sizes. A non-linear decay in $X$ is possible if atoms are reattaching to the $[\bar{1}10]$ step and afterwards diffuse to the [001] step. From the table in Fig. 6, we see that diffusion of atoms from the $[\bar{1}10]$ to the [001] step edge ($E_{c,x} = 0.38$ eV) is possible already at 150 K. This implies that when the length of an island is reduced, the island's ability to act as a sink for adatoms shrinks; i.e., the smaller the island the faster the decay. In other words, the islands can decay non-linearly by adatoms diffusion from [001] steps to $[\bar{1}10]$ sinks. kMC simulations support our reasoning, showing a net flux of atoms from the [001] island steps to the $[\bar{1}10]$ step of the outer boundary.

Finally, we have determined the exponent $\beta := \beta_X + \beta_Y$ from island decay in kMC calculations for different temperatures from an average of ten runs (Fig. 7b). Although the islands in the kMC simulations are relatively small compared to the experimental ones, they reproduce a transition in exponents around 200 K. Similar kMC simulations by Ferrando *et al.* [21] showed a transition from the growth of monatomic stripes at 200 K to 2D islands at 250 K.

## 5. Summary

Two qualitatively different decay regimes for the island decay on an anisotropic surface have been revealed. At low temperatures, a novel quasi-1D decay mode exists involving both the $[\bar{1}10]$ and the [001] step edges. A transition to a quasi-isotropic decay mode sets in at a surprisingly low temperature $T_c$, and above $T_c$ the islands equilibrate to their equilibrium shape. We have been able to account for these results by identifying the following rate limiting processes: Below $T_c$, atoms detach only from the [001] step but diffuse two-dimensionally. Above $T_c$, atoms that detach from kink sites on the $[\bar{1}10]$ steps are responsible for the 2D decay. Atoms can diffuse along the $[\bar{1}10]$ step and around the corners to the $[\bar{1}10]$ step, enabling shape equilibration of the islands.

In conclusion, we presented an atomistic picture for the decay of islands on an anisotropic surface that can be employed to estimate the thermal stability of metastable structures built of atoms on both metal and semiconductor surfaces. The model presented is of general validity and expected not to be limited to the special case of Ag(110), but rather to be generally applicable to anisotropic surfaces at appropriate temperatures.

We acknowledge financial support from the Danish National Research Foundation through the Center for Atomic-scale Materials Physics (CAMP).

## References

1. D.R. Peale and B.H. Cooper, *J. Vac. Sci. Technol.* **A10**, 2210 (1992).
2. W. Theis, N.C. Bartelt, and R.M. Tromp, *Phys. Rev. Lett.* **75**, 3323 (1995).
3. N.C. Bartelt, W. Theis, and R.M. Tromp, *Phys. Rev.* **B54**, 11741 (1996).
4. K. Morgenstern, G. Rosenfeld, and G. Comsa, *Phys. Rev. Lett.* **76**, 2113 (1996).
5. J.-M. Wen, J.W. Evans, M.C. Bartelt, J.W. Burnett, and P.A. Thiel, *Phys. Rev. Lett.* **76**, 652 (1996).
6. J.B. Hannon, C. Klünker, M. Giesen, H. Ibach, N.C. Bartelt, and J.C. Hamilton, *Phys. Rev. Lett.* **79**, 2506 (1997).
7. A. Ichimiya, Y. Tanaka, and K. Ishiyama, *Surf. Sci.* **386**, 182 (1997).
8. M. Giesen, G. Schulze Icking-Konert, and H. Ibach, *Phys. Rev. Lett.* **80**, 552 (1998).
9. K. Morgenstern, G. Rosenfeld, E. Lægsgaard, F. Besenbacher, and G. Comsa, *Phys. Rev. Lett.* **80**, 556 (1998).
10. K. Morgenstern, G. Rosenfeld, and G. Comsa, *Surf. Sci.* **441**, 289 (1999).
11. M. Poensgen, J.F. Wolf, J. Frohn, M. Giesen, and H. Ibach, *Surf. Sci.* **274**, 430 (1992); L. Kuipers, M.S. Hoogeman, and J.W.M. Frenken, *Phys. Rev. Lett.* **71**, 3517 (1993); K. Morgenstern, G. Rosenfeld, B. Poelsema, and G. Comsa, *Phys. Rev. Lett.* **74**, 2058 (1995); W.W. Pai, N.C. Bartelt, and J.E. Reutt-Robey, *Phys. Rev.* **B53**, 15991 (1996).
12. W.W. Pai, A.K. Swan, Z. Zhang, and J.F. Wendelken, *Phys. Rev. Lett.* **79**, 3210 (1997).
13. K. Morgenstern, E. Lægsgaard, I. Stensgaard, and F. Besenbacher, *Phys. Rev. Lett.* **83**, 1613 (1999); K. Morgenstern, E. Lægsgaard, I. Stensgaard, F. Besenbacher, M. Böhringer, W.-D. Schneider, R. Berndt, F. Mauri, A. De Vita, and R. Car, *Appl. Phys.* **A69** , 559 (1999).
14. L.M. Lifshitz and V.V. Slyozov, *J. Phys. Chem. Solids* **19**, 35 (1961); C. Wagner, *Z. Electrochem.* **65**, 581 (1961).
15. N.C. Bartelt and J. Evans, *Europhys. Lett.* **21**, 99 (1993).
16. L.D. Landau and E.M. Lifshhitz, *Course of Theoretical Physics, Vol. 5: Statistical Physics*, Pergamon Press, 1959, p. 482.
17. F. Hontinfinde, R. Ferrando, and A.C. Levi, *Surf. Sci.* **366**, 306 (1996).
18. F. Besenbacher, *Rep. Prog. Phys.* **59**, 1737 (1996).
19. The lowest temperature reported in literature is (725±25) K: G.A. Held, J.L. Jordan - Sweet, P.M. Horn, A. Mak, and R.J. Birgeneau, *Phys. Rev. Lett.* **59**, 2075 (1987).
20. Y.-W. Mo, B.S. Swartzentruber, R. Kariotis, M.B. Webb, and M.G. Lagally, *Phys. Rev. Lett.* **63**, 2393 (1989); Y.-W. Mo, J. Kleiner, M.B. Webb, and M.G. Lagally, *Phys. Rev. Lett.* **66**, 1998 (1991).
21. R. Ferrando, F. Hontefinde, and A.C. Levi, *Phys. Rev.* **B56**, R4406 (1997).
22. K. Morgenstern, G. Rosenfeld, B. Poelsema, and G. Comsa, *Surf. Sci.* **352 - 354**, 956 (1996); J. Li, R. Berndt, and W.-D. Schneider, *Phys. Rev. Lett.* **76**, 1888 (1996).
23. M. Zinke-Allmang, L.C. Feldman, and M.H.Garbow, *Surf. Sci. Rep.* **16**, 337 (1992); J.G. McLean, B. Krishnamachari, D.R. Peale, E. Chason, J.P. Sethna, and B.H. Cooper, *Phys. Rev.* **B55**, 1811 (1997) *and references therein.*
24. G. Rosenfeld, K. Morgenstern, and G. Comsa, in: *Surface Diffusion: Atomistic and Colletive Processes*, M.C. Tringides (Ed.), Plenum Press, New York 1997.
25. P. Stoltze, *J. Phys. Condens. Matter* **6**, 9495 (1994); J. K. Nørskov, K. W. Jacobsen, P. Stoltze, and L.B. Hansen, *Surf. Sci.* 283, 277 (1993)
26. Y. Li, M.C. Bartelt, J.W. Evans, N. Waelchi, E. Kampshoff, and K. Kern, *Phys. Rev.* **B56**, 12539 (1997).

# SURFACE MORPHOLOGY AND DYNAMICS: USING *AB-INITIO* TOTAL ENERGIES TO MAKE THE MOST OF STM DATA

PETER J. FEIBELMAN
*Sandia National Laboratories*
*Albuquerque, NM 87185-1413*

**ABSTRACT.** *Ab-initio* total energies test typically model-dependent results of interpreting STM morphologies, and can explain what mechanisms underlie morphology change with time. *Ab-initio* step- and kink-formation energies for Pb(111) and Pt(111), e.g., show that the experimental results must be revised. Calculations of S-decorated Cu adclusters point to ad-$Cu_3S_3$ as an agent of S-enhanced Cu-island decay on Cu(111), in a novel "skyhook effect."

## 1. Introduction

Scanning Tunneling Microscopy (STM) has revolutionized surface science by providing atomic-resolution images of a broad range of surfaces. Still, STM data are often as challenging to interpret as any from the host of techniques used to understand surface structure and dynamics. I offer examples here that reveal the power of modern first-principles total-energy calculations to help derive maximum value from STM observations of surface morphology and its time-evolution, whether in the form of reliable, quantitative surface energetic information or a plausible mechanistic scenario.

In the first instance,[1,2] I consider step- and kink-formation energies, fundamental parameters in the quasi-continuum description of surface morphology, and key to predicting how surfaces evolve in time. Because step wandering is governed by kink-formation energetics and the orientation-dependence of step-formation cost determines equilibrium island shapes, "experimental" values of these energies are generally obtained as best-fit parameters in comparisons between continuum surface models and STM data.[3-7]

As in the case of any fit, it is important to question the systematic reliability of the results. In this case we are fortunate that surface geometry imposes testable relationships between kink- and step-formation energies.[1,2] Supported by evidence from large-scale, *ab-initio*, total-energy calculations,[1,2] these tests necessitate revisions of the kink energies published for both Pt(111)[3] and Pb(111).[6]

The other example discussed in this review bears on how surface impurities, whether adventitious or deposited purposely, affect growth morphology. Surface scien-



*M.C. Tringides and Z. Chvoj(eds.),*
*Collective Diffusion on Surfaces: Correlation Effects and Adatom Interactions,* 213–224.

tists have devoted a decade's work to "surfactant-directed" self-assembly of ultra-thin films, with the idea that judicious addition of impurities might make it possible to build desired nanostructures more easily -- or in some cases, at all.[8] However, recent difficulties in interpreting experimental data have focused concern on the equally important effects of impurities *not deposited purposely.*[9] The evidence is that they need to be considered even when the impurity concentration is low. The reason is that impurities are attracted to sites where valence is relatively unsaturated, namely, defects. If surface defects, such as steps or vacancies, control the development of surface morphology, then a small impurity concentration may saturate the small number of defects and thus have a large effect.

In the case discussed here, time-resolved STM shows that S adatoms cause monolayer-high Cu islands on thick Cu(111) films to decay ~3-4 orders of magnitude faster than when impurities are absent.[10] To understand how this can be, I ask what $Cu_nS_m$ clusters form more readily on Cu(111) than a Cu adatom, and diffuse easily.[11] A systematic *ab-initio* search implies that the smallest such cluster is ad-$Cu_3S_3$. Its formation energy is ~0.5 eV lower than a Cu adatom's, and, corresponding to tight internal bonding, its diffusion barrier is ≤0.35 eV. This means that S can speed island decay on Cu(111) by acting as a "skyhook,"[12] weakening the bonds of a Cu adatom cluster to the substrate and promoting its transport.

## 2. Step- and Kink-Formation Energies on *fcc*(111) Metal Surfaces

Several methods have been used to extract formation energies of steps and kinks on fcc(111) surfaces from the statistics of step- and island morphologies observed in STM. To understand them one must first appreciate that a hexagonal substrate supports two types of unkinked steps, (100)-microfacets, called A-type steps, and (111)-microfacets, known as "B-type" steps (See Fig. 1). The two types, in principle, have different formation energies, reflecting their distinct geometries. The energies needed to form kinks in each may also be different for the same reason.

### 2. 1. EXPERIMENTAL STEP-FORMATION ENERGY RATIOS

Michely and Comsa (MC) were the first to measure a ratio of A- and B-step formation energies. They did so, on Pt(111),[13] starting from the observation that aspect ratios of vacancy and adatom islands of various sizes are all the same after the sample is raised to >700K and then cooled rapidly. Since this implies that the ratio of A- to B-step lengths $L_A/L_B$ is thermodynamically determined, they used the thermodynamic formula for the step formation energy ratio,

$$E_A/E_B = (2 + L_A/L_B) / (2L_A/L_B + 1) \quad , \qquad (1)$$

which applies when $L_A$ and $L_B$ are measurably greater than zero. For Pt(111), MC found that $L_A/L_B$=0.66, corresponding to $E_A/E_B$=1.15. Thus, to form an A-type step requires 15% more energy, per unit length, than to form a B-type step. Application of essentially the same strategy subsequently produced values of $E_A/E_B$ = 1.011±0.0007, 0.996±0.0025 and 1.11 for step formation on Cu(111),[14] Ag(111)[14] and Pb(111).[8]

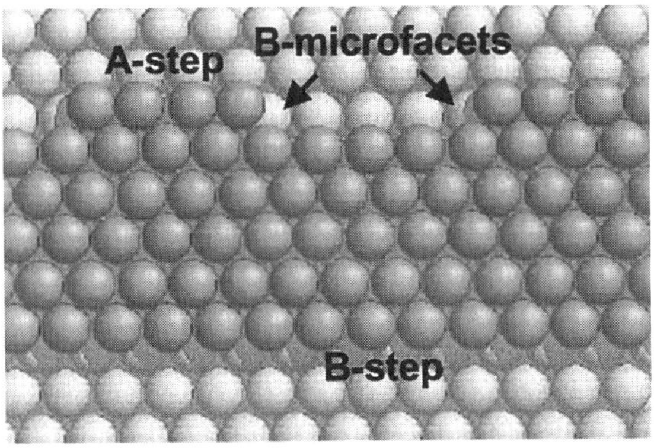


*Fig. 1.* A kinked, stripe island, 6 atom-rows wide, on an fcc(111) surface. The "A-type" edges of the stripe are (100)- and (111)-microfaces. The kinks in the A-type edge are B-type microfacets one atomic spacing long.

## 2.2. EXPERIMENTAL ABSOLUTE STEP-FORMATION ENERGIES

Measuring an *absolute* step-formation energy demands observing a quantity with dimensions of energy. In a method recently invented by Icking-Konert, Giesen and Ibach (IGI),[4] the sample temperature, T, is the needed energy reference. In an Ising model framework, IGI formally compute the T-derivative of the ratio of the free energies of a "100% kinked step," e.g., a $<11\bar{2}>$-oriented step on a (111) surface, and a straight, $<01\bar{1}>$-oriented step. To within known constants, they show that at low T this derivative equals the inverse of the $<01\bar{1}>$ step's formation energy.

The required free-energy ratio can be measured, according to thermodynamics, by comparing the distances from an equilibrium island's center to its sides and corners. Thus, IGI determine step formation energies by observing the change with T of an appropriate distance ratio. This group's most recent results are by Giesen, Steiner and Ibach (GSI),[14] who report average straight-step formation energies of 0.27±0.03 and 0.25±0.03 eV/atom on Cu and Ag(111).

An alternate way to determine absolute step formation energies, proposed by Schlößer et al. is to measure island shape fluctuations as a function of temperature.[5] From the results one extracts the step-formation energy as the line tension of the island perimeter. The method is formulated for circular rather than hexagonal islands and is based on a continuum model. Nonetheless, the result for steps on Cu(111), a formation energy of 0.22 eV/atom, is within 25% of GSI's value.[14]

Yet a third way to extract step formation energies from data is used by Arenhold et al., in which the energy reference is an average surface energy.[6] Since the surface energies of solids are rather poorly known, the accuracy of this approach is inherently limited. Notwithstanding, Arenhold et al. use the fact that the ratio of the average radius of a (111) facet on a crystallite to the crystallite radius equals the ratio of the average step and surface energies. Based on extrapolated measurements on Pb liquid droplets, which yield a Pb surface energy ~38meV/Å$^2$, Arenhold et al. determine an average step formation energy on Pb(111) of ~87meV/atom at 440K.[6]

### 2.3. RELATION BETWEEN STEP- AND KINK-FORMATION ENERGIES

As Swartzentruber et al.[15] first pointed out and exploited, kink- and step-formation energies are related because creating a kink amounts to creating additional step length. On a (111)-surface, where A- and B-steps intersect at 120° angles (see Fig. 1), creating a kink in an A-type step: *a)* adds a B-type microfacet one inter-atomic spacing long, *b)* eliminates one-half that length of the original A-type step, and *c)* adds two 120° corners, one convex, the other concave. To form such a kink thus requires energy,

$$E_K(A;1) = E_S(B) - E_S(A)/2 + E_C(A) \quad , \tag{2}$$

where $E_S(A)$ and $E_S(B)$ are the formation energies per atom of A- and B-type steps and the last term is, by definition, a "corner-formation" energy for an A-step.[15] Extending the new B-step from one to $n_k$ inter-atom spacings, the formation-energy increases to

$$E_K(A;n_k) = n_k[E_S(B) - E_S(A)/2] + E_C(A;n_k) \quad , \tag{3}$$

where $E_C(A;n_k)$ approaches a constant as $n_k$ becomes sufficiently large that the convex and concave corners no longer interact. The same applies for a B-step yielding formation-energy,

$$E_K(B;n_k) = n_k[E_S(A) - E_S(B)/2] + E_C(B;n_k) \quad . \tag{4}$$

For asymptotically-large $n_k$, Eqs. 3 and 4 combine to allow evaluation of the step-formation energy difference. The reason is that if the concave and convex corners that delimit a kink are too far apart to interact, then geometry dictates that $E_C(A;n_k)$ and $E_C(B;n_k)$ must be equal, i.e.,

$$E_C(A;n_k) = E_C(B;n_k) \equiv E_C \quad , n_k \to \infty \, . \tag{5}$$

Subtracting Eq. 4 from Eq. 3 in this limit, one finds,

$$E_K(A;n_k)-E_K(B;n_k)=3n_k[E_S(B)-E_S(A)]/2,\ n_k\to\infty, \tag{6}$$

which yields the step-formation energy difference in terms of calculated kink-formation energies, $E_K(A;n_k)$ and $E_B(B;n_k)$. From the numerical standpoint, an important question is whether one is already in the asymptotic region when $n_k$ is small enough that calculations of kinked edges are practical. In fact, reasonable results are obtained even assuming that Eq. 5 is valid for $n_k=1$.

## 2.4. EXPERIMENTAL KINK ENERGIES VS. EQ. 6

The bending of a surface step, whether at the rounded corner of an island, or as it wanders diffusively, is governed in continuum models by its "stiffness" and in microscopic models by the kink-formation energy.[7] Thus, kink energies are measured by observing island shapes or step-wandering and parameterizing the results via a lattice gas model of the stiffness.

Though the parameterization, in principle, includes both an energy per unit kink length and a corner energy, the latter has typically been assumed to be small enough that it can be neglected. What has only recently been recognized[1,2] is that if the corner energies are negligible, then Eq. 5 is automatically true, and Eq. 6 follows. This being the case, Eq. 6 should be used to check the consistency of the results derived from experiment.

Applying this test immediately points to a problem for Pt(111), namely that Giesen et al.'s measured ratio of 1 for kink-formation energies on A- and B-type steps[3] is inconsistent with Michely and Comsa's ratio of 1.15 for the step-formation energies.[13] According to Eq. 6, equality of the kink-formation energies on A- and B-type steps implies equal step-formation energies.

This argument is reinforced by Density Functional Theory (DFT) calculations in the Local Density Approximation (LDA)[16] (see Ref. 1 for a detailed exposition of the method), according to which,

$$E_S(A) + E_S(B) = 0.75\ \text{eV/atom} \tag{7}$$

and the difference of kink-formation energies is

$$E_K(B;1) - E_K(A;1) = 0.07\ \text{eV}. \tag{8}$$

Combining Eqs. 6 and 8, one obtains

$$E_S(A) - E_S(B) = 0.05\ \text{eV/atom}. \tag{9}$$

Solving Eqs. 7 and 9 then yields $E_S(A) = 0.40$ eV/atom, $E_S(B) = 0.35$ eV/atom, and $E_S(A)/E_S(B) = 1.14$. The first two of these results agree reasonably well with Boisvert et al.'s vicinal surface calculations.[17] The last agrees closely with Michely and Comsa's ratio, 1.15.

Thus, the relationship between step and kink energies based on Eq. 5 appears to hold, theoretically, and revisiting the interpretation of the experimental results is necessary. The authors of Ref. 3 suggest that difficulties in determining the crystal orientation of observed steps may be the source of the problem.[18]

Reconsideration of Arenhold et al.'s kink energies for Pb(111) also reveals a problem, but its source is quite different.[6] In Ref. 6, kink-formation energies $E_K(A;n_k)=n_k\times 61$meV and $E_K(B;n_k)=n_k\times 87$meV are extracted from measurements of island corner rounding, once again under the assumption that the corner energies, $E_C = 0$. Substituting these results into Eqs. 3 and 4 along with the measured A- and B-step formation energies, 90.7 meV/atom and 81.9 meV/atom, immediately indicates inconsistencies, namely the corner energies for A- and B-type steps cannot be independent of $n_k$, they cannot be equal and certainly are not zero. Assuming that the corner energies vanish thus leads to a contradiction.

In this case, the trouble is that the formula used in Ref. 6 to obtain kink energies from step-stiffness measurements was one derived for steps on a square lattice, which turns out not to be a good approximation to Akutsu and Akutsu's (AA's) result for the hexagonal case.[19]

Reanalyzing the experimental data with AA's formula, one obtains revised experimental kink-formation energies, namely, $E_K(A) = 42$ meV and $E_K(B) = 69$ meV. As in the case of step-formation energies, these results are known only as well as the Pb(111) surface energy, to which they are scaled. If the surface energy were 35 instead of 38 meV, the *ab initio* result obtained using the Generalized Gradient Approximation (GGA),[20] the formation of kinks on A- and B-steps would instead be found to cost 39 and 64 meV. These values are in good agreement with the GGA kink formation energies, 41 and 60 meV.

## 2.5. LESSONS LEARNED

Important lessons can be drawn from this work:

a) Because steps and kinks are geometrically related, it is a mistake to view step- and kink-formation energies in isolation. The two types of formation energy are related, and their inter-comparison can signal error in the analysis of experimental information.

b) Care is essential in assuming that formulae valid for one surface symmetry extend to another.

c) GGA provides a reasonably accurate description of the defect energetics of Pb. This result, together with a similar conclusion[21] for Cu, suggests that GGA studies of Pb wetting of Cu, i.e., in solder joint formation, are likely to produce information of value.

### 3. S-Enhanced Decay of Cu Islands on Cu(111)

Confirming the interpretation of data from well-characterized surfaces is only one way that theory may help in STM studies related to collective diffusive behaviors. In this section I review an example of another[11] - uncovering the mechanism by which surface impurities modify the time evolution of a surface morphology.

The problem is to understand why monolayer-high Cu islands on thick Cu(111) films decay ~3-4 orders of magnitude faster when S is adsorbed than when the surface is clean.[10] To solve it, one must first decide what process is rate-limiting in the clean situation, then learn how S accelerates it.

There are two general possibilities. The first is that the Cu current across a (111) terrace is dominated by the barrier that impedes dissociation from kinks. The other is that island decay is "diffusion limited," or in other words, that transport across terraces by the quasi-equilibrium sea of Cu adspecies is rate-limiting.[22]

In principle, the power law that describes the time-dependence of island decay reveals which possibility is operative. However, in the present case, the decay proceeds very slowly when no S is present and very quickly once S is deposited. Accordingly, there is insufficient data as yet to permit experimental determination of the rate-limiting barriers.

In the absence of experimental information, I argue on geometric grounds that the barrier to splitting an adatom off a kink on a Cu island should be close to the sum of the adatom formation energy and the adatom diffusion barrier. This is the same energy that controls the diffusion current on a terrace, which suggests focusing exclusively on this energy sum.

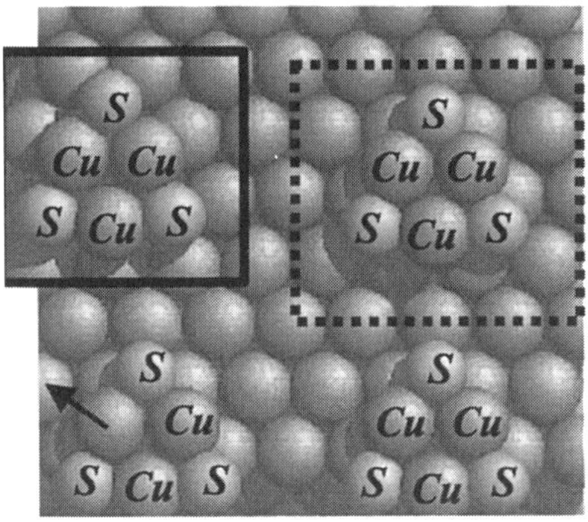


*Fig. 2.* $Cu_3S_3$ ad-clusters on Cu(111), in the $4 \times 2\sqrt{3}$ supercell indicated by the dotted rectangle. The inset cluster is displaced by the distance between *fcc* and *hcp* hollows, in the direction indicated by the arrow. As a result, the S-atoms that cover it are on (111)-microfacets, rather than the (100)-microfacets of the undisplaced clusters. This change of S-adsorption geometry is the main source of the 0.35 eV $Cu_3S_3$ diffusion barrier.

The big contribution to this sum is the energy needed to form a Cu adatom or a cluster of them. The reason is that closer-packing means better coordinated first-layer atoms, and therefore that self-adsorption is maximally costly on close-packed surfaces. The GGA implies, e.g., that 0.79 eV is needed to form a Cu adatom on Cu(111). This means that at 300K (=26 meV), the equilibrium concentration of self-adsorbed Cu atoms is low and might be significantly larger when S adatoms are added to the surface. (See Ref. 11, for details of the *ab-initio* calculations.)

The same logic that predicts a big adatom formation energy suggests that the self-diffusion barrier on clean Cu(111), $E_{diff}$(ad-Cu), should be small. Again the GGA agrees, predicting $E_{diff}$(ad-Cu) ≈ 57 meV. Since this energy is small with no S on the surface, only a small contribution to island decay rate can come from reducing it further by depositing S.

Accordingly, I postulate that S impurities promote island decay by forming tightly-bound, and thus plentiful and mobile $Cu_nS_m$ ad-species,[23] and, via *ab-initio* total-energy calculations, I systematically search for those that are favored compared to an unadorned Cu adatom. The search points to *ad-*$Cu_3S_3$ (See Fig. 2) as the best candidate.

### 3.1. COMPARISON OF CLUSTER FORMATION ENERGIES

The abundance of ad-$Cu_nS_m$ is determined by its chemical potential. It equals the chemical potential of ad-$Cu_n$ less that gained by *m* ad-S's attaching to form a decorated cluster. I compute this gain by reference to the isolated ad-S chemical potential, $\mu$(ad-S)=$E_B$(ad-S)-$k_BT\ln\theta_S$, where $E_B$(ad-S) is the ad-S binding energy and $k_BT\ln\theta_S$ accounts for the entropy loss per S, at ad-S concentration $\theta_S$. Entropy-loss drives up the free energy cost of clusters according to the number of S atoms they abstract from the terrace lattice-gas. This inhibits formation of clusters containing many S adatoms.

### 3.2. Cu ADATOMS

Values of $E_{form}$(ad-Cu) and $E_{diff}$(ad-Cu) are derived from total energies of adatom-containing supercells large enough that inter-adatom interactions are small. I find that Cu adatoms prefer *fcc* to *hcp* 3-fold sites on the 6-layer slab by only 7 meV. Correspondingly, the ad-Cu diffusion barrier lies almost equidistant from the 3-fold hollows at a twofold bridge. To an excellent approximation, $E_{diff}$(ad-Cu) is thus the difference in energies for an ad-Cu in an *fcc* hollow and at the symmetric bridge. The computed sum, $E_{form}$(ad-Cu) + $E_{diff}$(ad-Cu) = 0.85 eV, compares well with the value, 0.78 ± 0.04 eV, obtained from STM observations of Cu island decay on clean Cu(111).[4] Effective Medium Theory calculations by Stoltze yield 0.71 eV and 53 meV for $E_{form}$(ad-Cu) and $E_{diff}$(ad-Cu),[24] in relatively good agreement with the best *ab-initio* results, 0.79 eV and 57 meV. (Calculated results are summarized in Table I.)

Table I: Cluster formation energies, $E_{form}$, on a 6-layer, Cu(111) slab. $N_S$ and $N_{Cu}$ are the numbers of S and Cu adatoms in each cluster. For Cu dimers and trimers, I indicate the face that the S atoms decorate. The "B" cases correspond to Cu's in hcp hollows. In all other cases the Cu's occupy fcc sites. For the Cu monomer plus one S adatom, the three $E_{form}$ values are for S down, Cu down and S-beside-Cu configurations.

| $N_S$ | $N_{Cu}$ | face | $E_{form}$(eV) |
|---|---|---|---|
| 0 | 1 | | 0.79 |
| 1 | 1 | | 1.26, 1.87, 1.87 |
| 0 | 2 | | 1.33 |
| 1 | 2 | A | 1.29 |
| 2 | 2 | | 1.06 |
| 0 | 3 | A | 1.69 |
| 3 | 3 | A | 0.28 |
| 3 | 3 | B | 0.62 |
| 4 | 4 | | 0.49 |
| 5 | 4 | | 1.09 |

## 3.3. Cu-S AD-DIMERS

If ad-CuS is the plentiful species that accounts for S-enhanced Cu transport,[11] at a minimum ad-Cu and ad-S must attract. But they do not. A S adatom *loses* 1.08 eV binding energy in approaching an ad-Cu closely. Thus, forming a Cu-S dimer with the S beside the Cu *costs* 1.87 eV.[25] Presumably because the S cannot conveniently form two bonds, repulsion of the same magnitude also blocks formation of an ad-CuS with the Cu-end down. On the other hand, even with the S-end down, allowing S and Cu valence requirements to be satisfied, the dimer formation energy (with the CuS in an fcc hollow) is 1.26 eV, compared to $E_{form}$(ad-Cu)=0.79 eV. The Cu-S adcluster responsible for enhanced Cu-island decay thus contains more than one Cu atom.

## 3.4. CLEAN AND S-DECORATED Cu DIMERS

One expects Cu's adsorbed on Cu(111) to attract, and indeed Cu-Cu attraction lowers $E_{form}$(ad-$Cu_2$) by 0.27 eV. But does attaching S atoms to such a dimer reduce its $E_{form}$ by *another* 0.53 eV, to make S-decorated Cu-dimers more plentiful than Cu monomers? At least for the most obvious S-decorations of Cu addimers, the answer is no. Placing a single S on the side of the dimer, where it is 4-coordinated, lowers the formation energy by 0.04 eV. Adding another, on the other side of the dimer lowers it 0.19 eV more, not enough to compensate the cost of the second ad-Cu. Other geometries, e.g., Cu's decorating a S-addimer, or S's and Cu's alternating to form a flat tetramer, seem unfavorable. The former requires S's to be nearest neighbors even while $S_2$ dissociates on Cu(111). The latter is unlikely because an ad-S repels an adjacent ad-Cu.

### 3.5. CLEAN AND S-DECORATED Cu TRIMERS

Per adatom, forming Cu ad-trimers should cost less than dimers, because each ad-Cu has *two* ad-Cu neighbors, not one. Counting bonds, with a Cu-Cu bond strength of 0.27eV (see above), one expects $E_{form}(\text{ad-}Cu_3) \approx 1.59$ eV, i.e., about triple the monomer formation energy, 2.4 eV, minus $3 \times 0.27$ eV.

Direct calculations confirm this logic. The energy needed to form a trimer (see Fig. 2), with the three Cu adatoms in neighboring fcc hollows bounded by (100)- or "A-type" microfacets, is ~1.7 eV. If the trimer is rotated 60°, so that its sides are (111)- or "B-type" microfacets, its $E_{form}$ is 0.02 meV less.[26]

Though bond-counting predicts $E_{form}(\text{ad-}Cu_3)$ rather well, a similar approach *greatly* underestimates how much S-decoration reduces it. Rather than by 0.1 to 0.2 eV per added S-atom, as S+Cu-dimer results would suggest, decorating a Cu trimer with three S atoms reduces the formation energy of the complex by 0.47 eV/S-atom, for a trimer bounded by A-type microfacets. *Forming S-decorated Cu trimers thus costs only 0.28 eV, much less than Cu monomers!*

### 3.6. DIFFUSION OF S-DECORATED Cu TRIMERS

Given that creating a $Cu_3S_3$ ad-complex costs just 0.28 eV, and the related fact that S-decoration lifts the Cu adatoms ~0.14 Å higher above the nearest surface Cu's, the cluster diffusion barrier should be low. A plausible diffusion path involves moving each Cu from its initial hollow, say an *fcc* site,[27] (along the arrow in Fig. 3) over a neighboring bridge to an adjacent *hcp* hollow, the three S atoms following more or less rigidly. A lower bound for the barrier along this path is the energy difference between the initial and final configurations of the decorated trimer, or (see Fig. 2) between S-decorated trimers with A- vs. B-type sides.

This bound is significant, because S atoms have an affinity for Cu's arranged in a square.[28] In the present case, the affinity amounts to an energetic preference of ~0.33 eV for an A-sided trimer. It remains to learn if the B-trimer represents a transition geometry or a metastable state, and if the latter, whether the minimum barrier is much bigger than 0.33 eV.

Applying Jónsson's Nudged Elastic Band (NEB) method[29] with two replicas of the S-decorated trimer along the path between A-trimer in *fcc*- and B-trimer in *hcp*-hollows, I find a transition state close to the B-trimer geometry and a barrier of 0.35 eV. Thus $E_{form}(Cu_3S_3)+E_{diff}(Cu_3S_3) \approx 0.63$ eV, *which is 0.22eV lower than the similar sum for a Cu adatom.*

Assuming "diffusion-limited" Cu-island decay,[22] the decay-*rate* scales with the concentration of Cu-carrying adspecies times their diffusion constant. This product is proportional to $D_0$(ad-Cu)exp{−[$E_{form}$(ad-Cu)+$E_{diff}$(ad-Cu)]/$k_BT$}, for Cu-adatom transport, and to $D_0(Cu_3S_3)\theta_S{}^3$exp{−[$E_{form}(Cu_3S_3)$+$E_{diff}(Cu_3S_3)$]/$k_BT$}, for $Cu_3S_3$ clusters, where the $D_0$'s are diffusion prefactors.[30] The S-induced speedup is proportional to three times the latter divided by the former.[31] So the predicted $Cu_3S_3$-mediated decay rate ≈ $15000\ \theta_S{}^3 \times D_0(Cu_3S_3)/D_0$(ad-Cu) that for clean Cu(111). This result makes it plausible that $Cu_3S_3$ clusters account for the speedup seen in Ref. 2. Whether they *really* do depends largely on the $D_0$'s.

### 3.7. CLEAN AND S-DECORATED Cu TETRAMERS

The advantages of additional Cu-Cu bonds and S-decoration persist beyond Cu-trimers. To form ad-$Cu_4S_4$, e.g., requires only ~0.49 eV.[32] However, since barriers to concerted diffusion of $Cu_nS_n$ clusters likely rise with $n$, and entropy loss reduces the concentration of clusters containing $p$ S-atoms by $\theta_S{}^p$, Cu-transport via S-decorated tetramers, pentamers, etc. should be less important than via $Cu_3S_3$.

### 3.8. OBSERVATION OF Cu-S AD-CLUSTERS

Quench experiments that directly reveal $Cu_3S_3$ on terraces would obviously be desirable. A suggestive result in this direction is that STM of a low-T "honeycomb phase" of S/Cu(111), exhibits one surface protrusion per deposited S, separated by ~4Å.[33] I find an S-S separation of 4.4Å for ad-$Cu_3S_3$.

## 4. Conclusion

Because *ab-initio* methods can now confidently be applied to problems involving rather complicated surface morphologies and virtually any elemental composition, we should check assumptions made in the course of STM data analysis with reliable, parameter-free calculations. Moreover, we can and should try to fill in key mechanistic details that underlie the observed time-evolution of STM images. In short, *ab-initio* total energy calculations should now be viewed as an indispensable tool for the analysis of scanning probe data.

## ACKNOWLEDGMENTS

I thank N. C. Bartelt and B. S. Swartzentruber for many helpful discussions. This work was supported by the U. S. Department of Energy under Contract No. DE-AC04-94AL85000. Sandia is a multi-program laboratory operated by Sandia Corporation, a Lockheed Martin Company, for the U. S. Department of Energy.

## REFERENCES


1. P. J. Feibelman, *Surf. Sci.* **463**, L661(2000).
2. P. J. Feibelman, *Phys. Rev.* **B62,** xxxx (2000).
3. M. Giesen, G. S. Icking-Konert, D. Stapel and H. Ibach, *Surf. Sci.* **366**, 229 (1996).
4. G. S. Icking-Konert, M. Giesen and H. Ibach, Surf. Sci. **398**, 37(1998), *Phys. Rev. Lett.* **83**, 3880(1999).
5. D. C. Schlößer, L. K. Verheij, G. Rosenfeld and G. Comsa, *Phys. Rev. Lett.* **82**, 3843(1999).
6. K. Arenhold, S. Surnev. H. P. Bonzel and P. Wynblatt, *Surf. Sci.* **424**, 271(1999); *ibid.*, **441**, 223(1999).
7. H-C. Jeong and E. D. Williams, *Surf. Sci. Rep.* **34**, 171(1999).
8. For an impressive bibliography, see W. F. Egelhoff, P. J. Chen, C. J. Powell, M. D. Stiles and R. D. McMichael, *J. App. Phys.* **79**, 2491(1996).
9. E.g., P. J. Feibelman, *Phys. Rev. Lett.* **81**, 168(1998) and M. Kalff, G. Comsa, T. Michely, *ibid.* **81**, 1255 (1998).
10. K. Pohl, J. de la Figuera, R. Q. Hwang and P. J. Feibelman, unpublished.
11. P. J. Feibelman, *Phys. Rev. Lett.* **85**, 606(2000)..
12. R. Stumpf, Phys. Rev. **B53**, R4253 (1996); S. Horch, H. T. Lorensen, S. Helveg, E. Laegsgaard, I. Stensgaard, K. W. Jacobsen, J. K. Nørskov, F. Besenbacher, *Nature* **398**, 134(1999).
13. Th. Michely and G. Comsa, *Surf. Sci.* **256**, 217(1991).
14. M. Giesen, C. Steiner and H. Ibach, preprint.
15. B. S. Swartzentruber, Y.-W. Mo, R. Kariotis, M. G. Lagally, and M. B. Webb, *Phys. Rev. Lett.* **65**, 1913(1990).
16. P. Hohenberg and W. Kohn, *Phys. Rev.* **136**, B864 (1964); W. Kohn and L. J. Sham, *Phys. Rev.* **140**, A1133 (1965).
17. G. Boisvert, L. J. Lewis and M. Scheffler, *Phys. Rev.* **B57**, 1881(1998).
18. M. Giesen and H. Ibach, private communication.
19. N. Akutsu and Y. Akutsu, *J. Phys. Cond. Mat.* **11**, 6635(1999), Eq. 4.33 for step stiffness and Eq. 4.31 for step free energy. Limiting applicability to the Pb(111) surface, somewhat, results in this article are derived for a six-fold symmetric hexagonal Ising model with equal A- and B-step energies and vanishing corner energies.
20. J. P. Perdew, in *Electronic Structure of Solids '91*, edited by P. Ziesche and H. Eschrig (Akademie Verlag, Berlin, 1991); J. P. Perdew and Y. Wang, unpublished.
21. P. J. Feibelman, *Phys. Rev.* **B60**, 11118 (1999).
22. J. G. McLean, B. Krishnamachari, D. R. Peale, E. Chason, J. P. Sethna, B. H. Cooper., *Phys. Rev.* **B55**, 1811(1997).
23. Ad-$Cu_nS_m$ may also underlie the S/Cu(111) phases reported by E. Wahlström, I. Ekvall, H. Olin, S.-Å. Lindgren and L. Walldén *Phys. Rev.* **B60**, 10699(1999).
24. P. Stoltze, *J. Phys. Cond. Mat.* **6**, 9495(1994).
25. This is not an unusual effect. For the case of Al and S on Al(100), see P. J. Feibelman, *Phys. Rev.* **B38**, 12133 (1988).
26. Consistent with a lower B- than A-step formation energy, as in Ref. 21.
27. Starting the A-trimer with Cu's in *hcp* hollows should make little difference.
28. M. Foss, R. Feidenhans'l, M. Nielsen, E. Findeisen, T. Buslaps, R. L. Johnson and F. Besenbacher, *Surf. Sci.* **388**, 5(1997).
29. H. Jónsson, G. Mills, K. W. Jacobsen, in *Classical and Quantum Dynamics in Condensed Phase Simulations*, ed. B. J. Berne, G. Ciccotti and D. F. Coker (World Scientific, 1998).
30. $\theta_S^3$ represents entropy loss when three lattice-gas S adatoms attach to a trimer.
31. The factor 3 accounts for the *three* Cu atoms that each trimer contains.
32. The Cu's of the ad-$Cu_4S_4$ form a parallelogram with two Cu's roughly in 3-fold hollows and two in bridge-sites. A fifth S atop ad-$Cu_4S_4$ is unhelpful, with $E_{form}$(ad-$Cu_4S_5$) $\approx$ 1.09eV.
33. E. Wahlström, *et al.*, *op. cit.*

# SUBSTRATE-MEDIATED INTERACTION ON AG(111) SURFACES FROM FIRST PRINCIPLES

KRISTEN A. FICHTHORN
*The Pennsylvania State University*
*University Park, PA 16802 USA*

AND

MATTHIAS SCHEFFLER
*Fritz-Haber-Institut der Max-Planck-Gesellschaft*
*D-14 195 Berlin-Dahlem, Germany*

**Abstract.** When two or more atoms bind to a solid surface, the substrate can mediate an interaction between them. In this paper, we use density-functional theory to quantify the substrate-mediated pair interaction between two adatoms on a compressively strained Ag(111) surface and on unstrained Ag(111). On the strained surface, the elastic interaction is significant over the short range and leads to a net attraction between two adatoms. However, at the longest distances probed, the interaction is primarily electronic and repulsive. The repulsion can be as strong at 50 meV, and it forms a ring-like structure around an atom. On unstrained Ag(111), the interaction is primarily electronic in origin and it is weak relative to the interaction found on the strained surface. We calculate the energy barrier for an isolated atom to diffuse in each of these systems. For the strained surface, the magnitude of the diffusion barrier is comparable to that of the adsorbate interaction. We discuss the implications of our findings for growth at surfaces.

## 1. Introduction

Adsorbate-adsorbate interactions at surfaces play a key role in controlling the structure and properties of interfacial materials. Such interactions can be classified as through-space (direct) or through-substrate (indirect), although a clear distinction is not possible [1]. Quantification of the indirect,



*M.C. Tringides and Z. Chvoj (eds.),*
*Collective Diffusion on Surfaces: Correlation Effects and Adatom Interactions*, 225–236.

substrate-mediated interaction between two adsorbates is the primary subject of this paper.

Substrate-mediated interactions can have an elastic and an electronic component. Elastic interactions arise when an adsorbed atom perturbs the positions of surrounding surface atoms [2]. Additionally, an adsorbate may perturb the charge density near the surface and alter the binding energies of neighboring adsorbates - giving rise to an electronic interaction [1,3-6]. The electronic interaction is oscillatory in nature and, in the limit that the separation $d$ between two adsorbates is much greater than the Fermi wavelength $\lambda_F = 2\pi/\mathbf{k}_F$, the interaction energy $\Delta E$ is predicted [5] to have the form

$$\Delta E \sim \frac{cos(2\mathbf{k}_F d)}{d^x} \quad , \tag{1}$$

where $x = 2$ if the substrate has a surface state and $x = 5$ in the absence of a surface state. Oscillatory, charge-density waves associated with the electronic interaction have been imaged in scanning-tunneling microscopy (STM) studies of noble-metal surfaces [7, 8], which possess a surface state. For example, on Ag(111), $\mathbf{k}_F = 0.08\mathring{A}^{-1}$ [8] and charge density waves can be imaged at distances greater than 70 Å [8] in STM studies.

Electronic, substrate-mediated interactions should also be manifested at "intermediate" distances between the short separations required for chemical bonds and the long separations for which Eq. (1) is applicable. For example, Einstein and Schrieffer showed in tight-binding calculations that the intermediate-range interaction is present and is, in fact, more significant than the long-range interaction [3]. Recent, low-temperature STM studies also indicate this is the case [9] for Cu adatoms on Cu(111). The intermediate range is accessible to modern, electronic-structure calculations. In this study, we use density-functional-theory (DFT) calculations to quantify the substrate-mediated interaction between two adatoms on a (compressively) strained and on an unstrained Ag(111) surface. We discuss the ramifications of these interactions for adsorbate diffusion and growth at surfaces.

## 2. Model and Methods

Total energies and forces are calculated using the DFT code described by Bockstedte *et al.* [10], employing fully separable, norm-conserving pseudopotentials of the Kleinman-Bylander form [11]. These were generated using the code developed by Fuchs and Scheffler [12]. For our calculations with Ag, the $s$ pseudopotential component is taken as local and we use $p$ and $d$ projectors. The $4d$-electrons are taken as valence states and their wave functions are expanded in plane waves, which truncate at a cut-off energy of $E_{cut} = 50$ Ry. We use the generalized gradient approximation (GGA) of Perdew, Burke, and Ernzerhof (PBE) [13] for the exchange-correlation functional.

The supercell approach is used to describe the surface. For most results presented here, the surface is modeled as a repeating $(4 \times 2)$ slab with 16 atoms per unit cell and a thickness of four layers. The vacuum spacing is five interlayer distances, or $\sim$ 12Å. Adsorbed atoms are placed on one side of the slab and field effects due to the asymmetry are correctly treated in the code by a compensating dipole field. For some of the adsorbate interactions probed, it is expedient to utilize smaller slabs. Thus, we also use four-layer $(2 \times 2)$, $(4 \times 1)$, and $(2 \times 1)$ slabs for some of the calculations. For all of the main results presented here, the **k**-space integration is performed with the equivalent of 4 special **k** points in the full, surface Brillouin zone of the $(4 \times 2)$ slab. To improve **k**-space integration, the electronic states are occupied according to a Fermi distribution with $k_BT$=0.1 eV and the total energies are extrapolated to zero temperature.

To verify the accuracy of the pseudo-potential, we calculated the lattice constant of bulk Ag in its ground state. For these calculations, we use $E_{cut} = 50$ Ry and 182 **k** points in the irreducible portion of the Brillouin zone. We find a lattice constant of $a_0 = 4.20$Å, which is within 1% of the values found by Yu and Scheffler [14] and by Ratsch *et al.* [15], who both used DFT with the GGA functional of Perdew *et al.* [16]. As an additional test of the accuracy, we calculated the interlayer relaxation $d_{12}$ of the first layer of Ag(111). For these calculations, the first layer atoms are relaxed until the forces on them are less than 0.005 eV/Å. We find that the first layer relaxes inward by 2.0% relative to the bulk, interlayer spacing $d_b$, in good agreement with DFT local-density approximation (LDA) results by Narasimhan and Scheffler [17], who found an inward relaxation of 1.7%, and by Xie *et al.* [18], who found an inward relaxation of 1.0%. Medium-energy ion scattering experiments indicate that $d_{12}$ contracts by 2.5% relative to $d_b$ at temperatures below 670 K [19]. A more recent low-energy electron diffraction investigation indicates no contraction of $d_{12}$ occurs on this surface [20]. In previous theoretical studies of Ag(111) [15, 17, 18], it was found that differences between $d_{23}$ and $d_b$ are negligible. Thus, in the calculations presented here, $d_{23}$ is fixed to $d_b$.

We study adsorbate interactions on both Ag(111) and on a strained Ag(111) surface. For the strained Ag(111) surface, we employ a bulk lattice constant of 4.01 Å. This value is taken from the results of DFT-GGA calculations of the lattice constant of Pt by Ratsch *et al.* [15]. We choose this substrate to address experimental issues related to Ag epitaxy on a Pt(111) substrate covered by a monolayer (ML) of Ag [15,21-23]. To describe the bare, strained Ag(111) substrate, the top layer of atoms is relaxed and the rest are fixed to the strained, bulk positions. This results in an outward relaxation of the top layer by 9.0% relative to $d_b$ for the strained substrate. The outward relaxation allows for relief of the strain induced by the bulk lattice, which is compressed by 4.61% relative to bulk Ag, as described by our DFT calculations with the GGA.

The total interaction energy $\Delta E$ for a periodic slab containing $N$ adatoms, of which $M$ are at binding site $a$ and $(N-M)$ are at binding site $b$, is given by

$$\Delta E = E_{S+N}^{a,b} - ME_{S+1}^{a} - (N-M)E_{S+1}^{b} + (N-1)E_S \quad . \qquad (2)$$

Here, $E_{S+N}^{a,b}$ is the total energy of a slab with $N$ adatoms, $E_{S+1}^{a}$ and $E_{S+1}^{b}$ are the total energies of slabs containing one adatom, and $E_S$ is the total energy of a bare slab. We obtain $E_S$ from the calculations described above. In an attempt to isolate the electronic component of $\Delta E$, we obtain $E_{S+N}^{a,b}$, $E_{S+1}^{a}$, and $E_{S+1}^{b}$ as follows: A single atom is placed in a binding site (an fcc or hcp three-fold hollow site) on a relaxed (as described above) bare slab and its height is optimized with respect to the fixed substrate. From this calculation, we obtain $E_{S+1}^{a}$ and $E_{S+1}^{b}$. To calculate $E_{S+N}^{a,b}$, adatoms are placed in binding sites on the relaxed (and fixed) substrate with heights fixed to values from the single-adatom calculations. We use this procedure to obtain most of the results described below. Additionally, to resolve the elastic contribution to $\Delta E$, we obtain $E_{S+N}^{a,b}$, $E_{S+1}^{a}$, and $E_{S+1}^{b}$ by simultaneously relaxing the adatoms and the first-layer substrate atoms for a few trial configurations.

In the DFT supercell approach, the total interaction energy $\Delta E$ is comprised of interactions between different adatoms in the slab and interactions between adatoms in the slab and the periodic-image adatoms. We can express $\Delta E$ in terms of these interactions using the lattice-gas Hamiltonian approach (see, *e.g.*, [24]), which yields

$$\Delta E = \frac{1}{2}\sum_{i,j} V^{(2)}(d_{i,j})\sigma_i\sigma_j + \frac{1}{3}\sum_{i,j,k} V^{(3)}(d_{i,j}, d_{i,k}, d_{j,k})\sigma_i\sigma_j\sigma_k + \ldots \quad . \qquad (3)$$

Here, the summations run over all sites $i$ in the slab and all sites $j$ and $k$ in the supercell (which includes both the slab and its periodic images), $\sigma_m$ is unity if site $m$ ($m = i, j, k$) is occupied and zero, otherwise, $d_{m,n}$ is the distance between sites $m$ and $n$, $V^{(2)}(d_{i,j})$ is the pair interaction between two adatoms on sites $i$ and $j$, and $V^{(3)}(d_{i,j}, d_{i,k}, d_{j,k})$ is the trio interaction between three adatoms on sites $i$, $j$, and $k$. We neglect higher-order interactions. Another assumption implicit in Eq. (3) is that the interaction between adatoms at a fixed distance is independent of whether these atoms occupy fcc or hcp sites. Actually, in the results presented below, interactions between two adatoms occupying the same type of binding site are

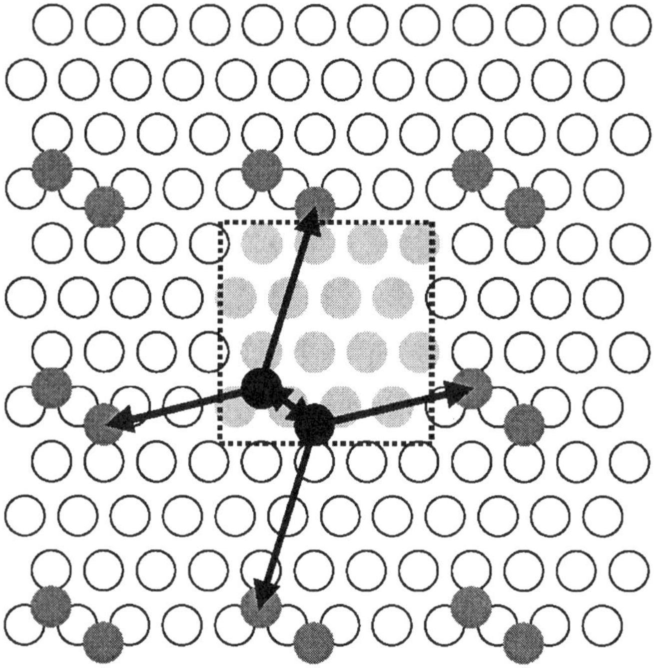

*Figure 1.* Sample lattice-gas Hamiltonian configuration. Atoms in the slab are enclosed in the dashed box. The lines with arrows indicated relevant distances between adatoms in the slab and periodic-image adatoms.

always obtained for adatoms on fcc sites. We performed one test of this assumption for two adatoms on hcp sites. This calculation showed that the interaction was identical to that found for atoms on fcc sites. Finally, in calculations designed to resolve the electronic interaction (which comprise the bulk of the results presented here), the adatom binding energies on fcc and hcp sites are virtually equal: The fcc site is favored by less than 3 meV.

To illustrate our approach, we show one of the structures probed in our calculations in Fig. 1. In this structure, two adatoms are in the $(4 \times 2)$ surface unit cell at a separation of 3.43 Å (distance 2, as can be inferred from Fig. 4 below). Each of these adatoms has two, neighboring, periodic-image adatoms at a distance of 9.08 Å (distance 12 shown on Fig. 4 below). From Eq. (3), $\Delta E$ for this configuration is comprised of one pair interaction at distance 2 and two at distance 12. Thus, for a given adatom configuration, we express $\Delta E$ as a sum of pair and trio interactions with unknown coefficients. From 13 different configurations, we obtain a system of linear equations and solve these for pair-interaction coefficients up to the $13^{th}$-

neighbor. We assume that all other interaction coefficients are zero.

## 3. Results and Discussion

Before proceeding to our results, we discuss our tests for their accuracy. We test the convergence of our results with respect to $E_{cut}$, the number of **k** points, the slab thickness, and the exchange-correlation functional for different adsorbate configurations on Ag(111) and strained Ag(111). The adsorbate configurations in these tests are chosen for two major reasons. First, for each configuration, the total interaction energy is composed of interactions at a single, adsorbate-pair distance. Thus, it is possible to extract pair interactions without solution of the full system of equations for the lattice-gas Hamiltonian. A second reason for choosing these configurations is that they possess a single mirror plane - the highest symmetry that could be achieved here. The (relatively) high symmetry makes it possible to retain up to 16 **k** points in the full surface Brillouin zone of a $(4 \times 2)$ unit cell, so that we can assess the convergence of the **k**-point summations for these configurations. For configurations of lower symmetry, 4 **k**-points in the $(4 \times 2)$ unit cell is the maximum that we could use with the computational resources available to us. In Table 1 we show the influence of the cut-off energy, the slab thickness, and the number of **k** points on three, pair-interaction coefficients for Ag adatoms on Ag(111). The distances for the pair interactions can be seen below in Fig. 4. For each result in Table 1, each of the three energies necessary to obtain the pair-interaction coefficient (cf., Eq. 2) is obtained with the indicated cut-off energy, slab thickness, and number of **k** points.

TABLE 1. Convergence tests for $V^{(2)}(d)$.

| Distance, $d$ | $N_{\mathbf{k}}$ | $E_{cut}$, Ry | Layers | $V^{(2)}(d)$, eV |
|---|---|---|---|---|
| 5 | 4 | 50 | 4 | -0.012 |
| 5 | 4 | 60 | 4 | -0.015 |
| 5 | 4 | 50 | 5 | -0.005 |
| 5 | 16 | 50 | 4 | 0.012 |
| 4 | 4 | 50 | 4 | 0.004 |
| 4 | 16 | 50 | 4 | 0.001 |
| 9 | 4 | 50 | 4 | 0.005 |
| 9 | 16 | 50 | 4 | -0.001 |

From Table 1, we see that the tests for distance 5 indicate satisfactory convergence with respect to the cut-off energy and the slab thickness. The results of our **k**-point test for distance 5 are less satisfying and there is a (relatively) large difference between the pair-interaction coefficients obtained with 4 and 16 **k** points. To further probe the convergence of our results with respect to the **k**-point summation, we investigated two additional adsorbate configurations containing distances 4 and 9. For these configurations, differences in the pair-interaction coefficients are small and indicate a satisfactory **k**-point convergence. It appears that distance 5 is somehow unique. We will discuss this configuration in more detail below.

To assess the dependence of our results on the exchange-correlation functional, we obtain pair-interaction coefficients for three, different adatom separations on strained Ag(111) using the LDA [25]. The pseudopotentials for these calculations are generated as described above for the GGA calculations, except that the LDA is used for the exchange-correlation functional. We test this pseudopotential by calculating the bulk lattice constant for Ag using the same conditions as for the GGA calculations. We find $a_0 = 4.06$ Å, which is in exact agreement with the value found by Narasimhan and Scheffler [17] and within 1% of the value found by Ratsch *et al.* [15]. For strained Ag(111), we employ a lattice constant with a value of 3.92 Å, as found by Ratsch *et al.* [15] in DFT-LDA calculations of the lattice constant for bulk Pt.

In Fig. 2, we show a comparison of the LDA and GGA results. Here, we see a small difference between the pair-interaction coefficients for distances 4 and 9 and a (relatively) large difference between the LDA and the GGA at distance 5. Some of the discrepancy between the LDA and GGA results can likely be attributed to differences in strain. The lattice in the GGA calculations is compressively strained by 4.61% relative to the bulk, while the strain for the LDA system is 3.21% (for a reference, the experimental strain is 4.2% [22]). As we will discuss in more detail below, we find that pair interactions on Ag(111) follow the same trend as those on compressively strained Ag(111), but they are consistently lower. A comparison of the LDA and GGA results on Fig. 2 also indicates this trend. Thus, it is likely that strain plays some role in the differences between the LDA and the GGA results.

We now return to our discussion of $V^{(2)}(5)$, for which we observe large discrepancies with different **k**-point summations (Table 1) and with the LDA vs. the GGA (Fig. 2). Although the exact reason for this large error bar is unclear, one possible explanation is that longer-range pair or trio interactions contribute to the total interaction energy of the configuration that we used to obtain $V^{(2)}(5)$. As we see in Fig. 3, this configuration consists of a row of atoms occupying alternating fcc sites along the $\langle 110 \rangle$ direction. A trio in this configuration would most likely consist of three adatoms in alternating fcc sites along the $\langle 110 \rangle$ direction. Although we did

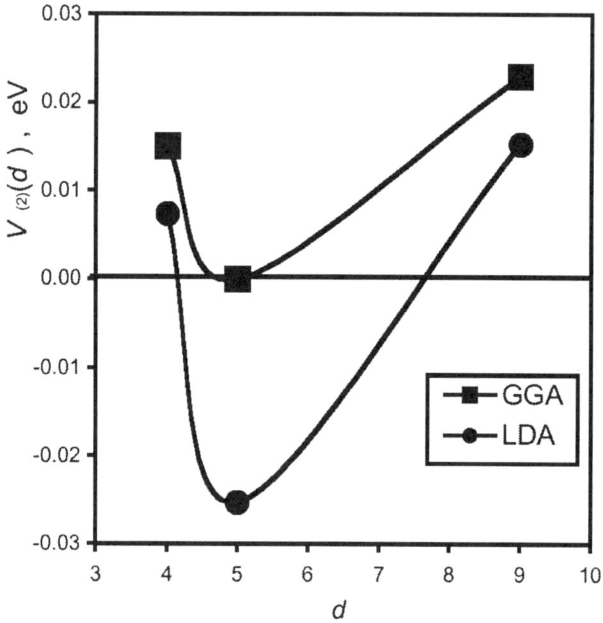


*Figure 2.* Comparison of LDA and GGA pair-interaction energies $V^{(2)}(d)$ for three different adsorbate separations $d$.

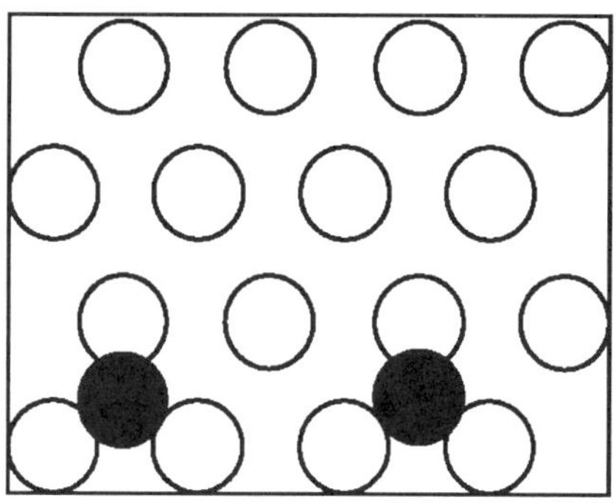

*Figure 3.* Configuration of adatoms used to obtain $V^{(2)}(5)$.

not quantify this interaction, we found that the trio interaction $V^{(3)}(4, 5, 9)$ is attractive by 12 meV [26]. Einstein found in tight-binding calculations that the magnitude of the trio interaction is determined primarily by the two, shortest adatom separations in the trio [27]. In light of these results, a non-zero trio interaction seems possible. If this is the case, then one trio interaction is incorporated into our pair parameterization for $V^{(2)}(5)$ and the combined uncertainties in both interactions could lead to the relatively large error bar for this interaction coefficient. Similarly, longer-range pair interactions (which we assume to be zero) could be incorporated into the values of $V^{(2)}(5)$ and cause large uncertainty.

Figure 4 shows the pair interaction for the various distances probed. To facilitate resolution of the long-range interactions, the short-range interactions at distances 1 (the nearest-neighbor fcc site to the central adatom) and 2 (the second-closest hcp site to the central adatom) are not shown. For both surfaces, the interaction is relatively weak at moderate distances and it is the strongest and the most repulsive at the longest distances probed. For strained Ag(111) [cf., Fig. 4(a)], the interaction can be as repulsive as $\sim$ 50 meV at the longest distances and the repulsion forms a ring-like structure around the central adatom. The interactions are weaker on Ag(111) [Fig. 4(b)] and do not form a complete ring over the distances probed here. Generally, we find that the adatom pair interaction is lower for Ag(111) than it is for strained Ag(111) - at all distances for which the pair interaction is appreciable, the attractions on Ag(111) are stronger and the repulsion is weaker. We note that Bogicevic and co-workers [28] have recently obtained interactions similar to ours in DFT-LDA calculations of Al on Al(111) and Cu on Cu(111).

Since the surface atoms have the same configuration in all of the results discussed above, the elastic interaction due to adsorbate-induced rearrangement of the surface atoms is suppressed. To assess the role of elastic interactions, we recompute the pair-interaction coefficients in a few trial structures for which both the adsorbate and first-layer surface atoms are fully relaxed until the forces are typically less than 0.025 eV/Å.

In Table 2, we show the elastic contribution to the pair-interaction coefficients at several adsorbate separations. This is given as $\Delta V^{(2)}(d)$, where $\Delta V^{(2)}(d)$ is the difference between the pair-interaction coefficient obtained for a fully relaxed structure and that for the electronic interaction. From Table 2, we see that the elastic interaction plays a very small role in the total interaction energy for Ag(111). The largest relaxation is for a dimer, an attraction of 14 meV. The elastic interaction is stronger for strained Ag(111) and is seen to be appreciable up to distance 5 on Fig. 4. At distances 4 and 5, the reduction of the pair interaction upon relaxation is greater than the electronic interaction [cf., Fig. 4(a)], indicating the elastic interaction is dominant at these distances. At longer distances, the elastic interaction is weak and we conclude that the repulsive ring seen on strained Ag(111) is electronic in origin.

Adsorbate interactions at surfaces can influence both thermodynamically favored structures and adsorbate diffusion. To assess the impact of

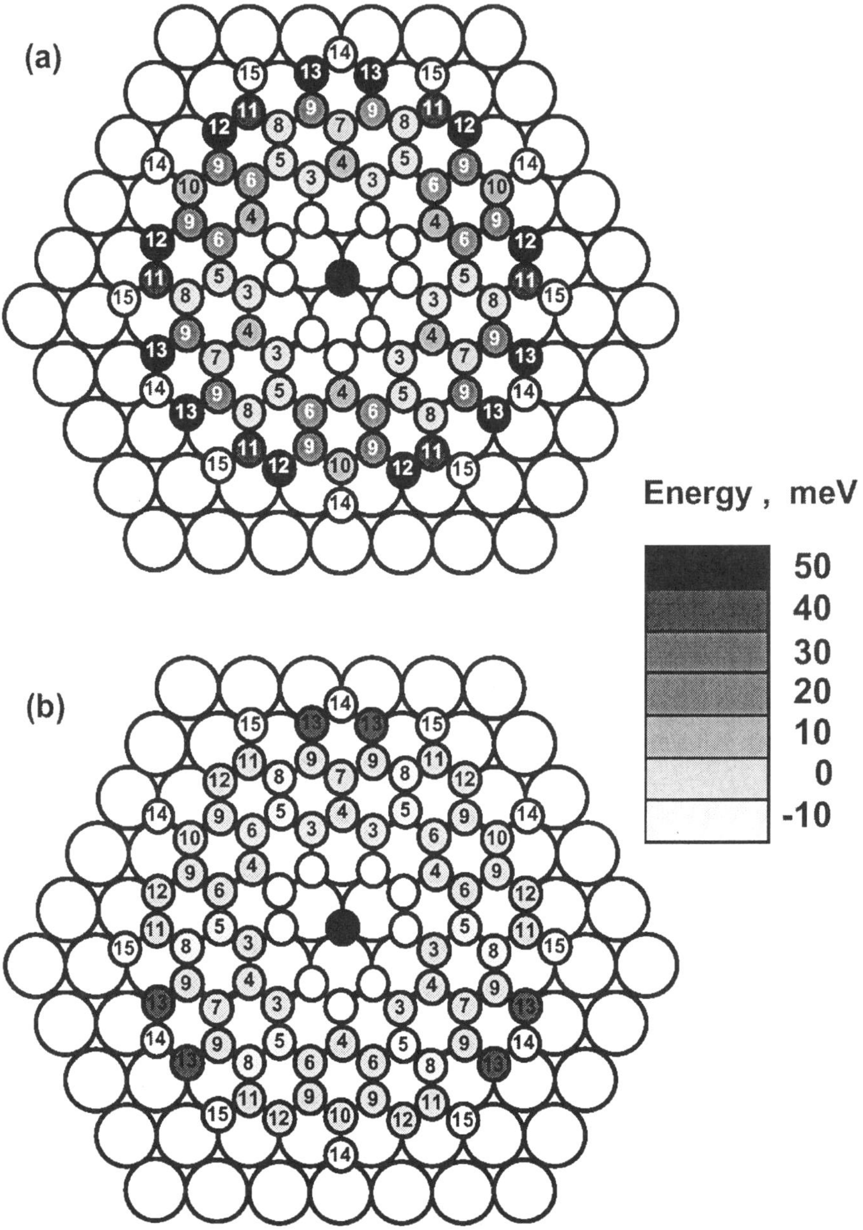


*Figure 4.* Pair interaction as a function of distance from a central adatom (shown in black) for strained Ag(111) (a) and unstrained Ag(111) (b).

TABLE 2. The elastic contribution $\Delta V^{(2)}(d)$ (units of eV) to the pair interaction between two adsorbates separated by a distance $d$.

| Distance, $d$ | $\Delta V^{(2)}(d)$– Ag(111) | $\Delta V^{(2)}(d)$– Strained Ag(111) |
|---|---|---|
| 1 | -0.014 | -0.051 |
| 4 | 0.003 | -0.016 |
| 5 | 0.003 | -0.023 |
| 9 | 0.003 | -0.007 |
| 11 | 0.003 | -0.003 |

adsorbate interactions on surface diffusion, we obtain the diffusion barrier for an isolated adatom to hop between adjacent three-fold hollow sites on both surfaces. In these calculations, both the adatom and the surface atoms are relaxed. The diffusion barrier is given by the difference between the total energy when the adatom is at a bridge site and when the atom is at a fcc three-fold hollow site. We find for Ag(111) that the diffusion barrier is 87 meV. For strained Ag(111), we find a barrier of 52 meV. Our barriers are in good agreement with those of Ratsch *et al.* [15], who found 81 meV for Ag(111) and 63 meV for strained Ag(111) in DFT-LDA calculations. Differences between the barriers that we find and those found by Ratsch can be attributed to differences in the exchange-correlation functional (LDA vs. the GGA used here). Additionally, there are differences between the lattice constant predicted for the strained Ag(111) surface in the two studies. As discussed above, the lattice in our GGA calculations is compressively strained by 4.61% relative to the bulk, while the strain for the LDA system is 3.21%. Ratsch *et al.* found that the value of the diffusion barrier on a strained Ag(111) surface is very sensitive to the degree to which the surface is strained and that smaller diffusion barriers arise on surfaces under greater compression [15]. The difference between our diffusion barrier and that found by Ratsch *et al.* [15] is adherent to this trend. Our diffusion barriers are also in good agreement with experimental values [22] for Ag on 1-ML-Ag/Pt(111) (60 meV) and on Ag(111) (97 meV).

Comparing the diffusion barrier for an adatom on strained Ag(111) to the magnitude of the pair interactions on this surface, we see that the two are comparable. Thus, one conclusion arising from this work is that substrate-mediated interactions can influence adsorbate diffusion and morphologies formed during growth at surfaces. Our recent kinetic Monte Carlo simulation studies of island nucleation and growth show that this is the case [23, 26]. Recent experimental STM studies have begun to probe the role of adsorbate lateral interactions in determining the structure of adsorbate islands at surfaces [9, 29]. Theoretical-experimental interplay in this area would be fruitful.

## 4. Acknowledgments

This research is supported by the Alexander von Humboldt Foundation and NSF grant No. DMR-9617122.

## References

1. Bradshaw, A. and Scheffler, M. (1979) *J. Vac. Sci. Technol.* **16**, 447.
2. Lau, K. H. and Kohn, W. (1977) *Surf. Sci.* **65**, 607.
3. Einstein, T. L. and Schrieffer, J. R. (1973) *Phys. Rev. B* **7**, 3629.
4. Einstein, T. L. (1996) *Handbook of Surface Science*, W. N. Unertl, Ed. (Elsevier, Amsterdam), Vol. 1, p. 577.
5. Lau, K. H. and Kohn, W. (1978) *Surf. Sci.* **75**, 69.
6. Hyldgaard, P. and Persson, M. (2000) *J. Phys.: Condens. Matter* **12**, L13.
7. Briner, B. G., Hoffman, Ph., Doering, M., Rust, H.-P., Bradshaw, A. M., Petersen, L., Sprunger, P. T., Laegsgaard, E., Besenbacher, F., and Plummer, E. W. (1997) *Europhysics News* **28**, 148.
8. Jeandupeux, O., Burgi, L., Hirstein, A., Brune, H., and Kern, K. (1999) *Phys. Rev. B* **59**, 15926.
9. Repp, J., Moresco, F., Meyer, G. and Rieder, K.-H. (2000) *Phys. Rev. Lett.* **85**, 2981.
10. Bockstedte, M., Kley, A., Neugebauer, J. and Scheffler, M (1997) *Comput. Phys. Comm.* **107**, 187.
11. Kleinmann, L and Bylander, D. M. (1982) *Phys. Rev. Lett.* **48**, 1425.
12. Fuchs, M. and Scheffler, M. (1999) *Comput. Phys. Commun.* **119**, 67.
13. Perdew, J. P., Burke, K., and Ernzerhof, M. (1996) *Phys. Rev. Lett.* **77**, 3865.
14. Yu, B.-D. and Scheffler, M. (1997) *Phys. Rev. B* **55** 13916.
15. Ratsch, C., Seitsonen, A. P., and Scheffler, M. (1997) *Phys. Rev. B* **55**, 6750.
16. Perdew, J. P., Chevary, J. A., Vosko, S. H., Jackson, K. A., Pederson, M. R., Singh, D. J., and Fiolhais (1992) *Phys. Rev. B* **46**, 6671.
17. Narasimhan, S. and Scheffler, M. (1997) *Z. Phys. Chem.* **202**, 253.
18. Xie, J., de Gironcoli, S., Baroni, S., and Scheffler, M. (1999) *Phys. Rev. B* **59**, 965.
19. Statiris, P., Lu, H. C., and Gustafsson, T. (1994) *Phys. Rev. Lett.* **72**, 3574.
20. Soares, E. A., de Carvalho, V. E., de Castillo, C. M. C., de Carvalho, A. V., Toomes, R., and Woodruff, D. P. (1999) *Surf. Sci.* **419**, 89.
21. Ratsch, C. and Scheffler, M. (1998) *Phys. Rev. B* **58**, 13163.
22. Brune, H., Bromann, K., Röder, H., Kern, K., Jacobsen, J., Stoltze, P., Jacobsen, K., and Norskov, J. (1995) *Phys. Rev. B* **52**, R14380.
23. Fichthorn, K. A. and Scheffler, M. (2000) *Phys. Rev. Lett.* **84**, 5371.
24. Stampfl, C., Kreuzer, H. J., Payne, S. H., Pfnür, H., and Scheffler, M. (1999) *Phys. Rev. Lett.* **83**, 2993.
25. Ceperly, D. M. and Alder, B. (1980) *Phys. Rev. Lett.* **45**, 567.
26. Fichthorn, K. A. and Scheffler, M. (to be published).
27. Einstein, T. L. (1979) *Surf. Sci.* **84**, L497.
28. Bogicevic, A., Ovesson, S., Hyldgaard, P., Lundqvist, B. I., Brune, H., and Jennison, D. R. (2000) *Phys. Rev. Lett.* **85**, 1910.
29. Österlund, L., Pederson, M. O., Stensgaard, I., Laegsgaard, E., and Besenbacher, F. (1999) *Phys. Rev. Lett.* **83**, 4812.

# DIFFUSION OF ADATOMS AND SMALL CLUSTERS ON MISSING-ROW-RECONSTRUCTED SURFACES

F. MONTALENTI AND R. FERRANDO
*INFM and CFSBT/CNR, Dipartimento di Fisica dell'Università di Genova*
*via Dodecaneso 33, 16146 Genova, Italy*

**Abstract.** The mobility of adatom, dimers, and atomic chains on missing-row reconstructed surfaces is investigated by molecular-dynamics simulations and analytical calculations. Results are reported for $Au(110)$. In the case of adatoms, diffusion can take place either by in-channel jumps or by walks on the channel facets (metastable walks). In the case of dimers and chains, the dominant mechanism is the leapfrog, in which one of the side adatoms of the chain is promoted above the chain itself, diffuses there until it is retrapped at the opposite end of the chain.

## 1. Introduction

The study of the diffusion of adatoms and small clusters on surfaces is a fundamental step for the understanding of the growth properties of materials [1].

On isotropic metal surfaces, isolated adatoms can move by different mechanisms, such as jumps or exchanges (which may involve two or more atoms [2-8]). On the other hand, dimers can perform concerted translational, and rotational moves [9] on the fcc(111) surfaces, and jumps and exchanges on fcc(100) surfaces. Larger clusters can displace by several different mechanisms, involving concerted motion of chains, diffusion along the periphery, diffusion of vacancies in the inner cluster, and adatom detachment and reattachment [10-12].

On anisotropic surfaces, one could think that the situation is simpler. In fact, if the anisotropy is strong, adatoms and clusters move essentially along one direction (the easy-diffusion direction). From this point of view, good candidates are the (110) surfaces of fcc transition metals and noble metals.



*M.C. Tringides and Z. Chvoj (eds.),*
*Collective Diffusion on Surfaces: Correlation Effects and Adatom Interactions,* 237–245.

These surfaces are strongly anisotropic, characterized by channels along the easy $[1\bar{1}0]$ in-channel direction. However, in the unreconstructed $(1\times1)$ geometry, these surfaces show also a significant diffusion along the cross-channel [001] direction, due to the occurrence of the exchange mechanism [13, 14]. On the contrary, in the reconstructed $(1\times2)$ geometry (missing-row geometry), which is the most favorable for *Au* and *Pt*, the channels are much deeper, and cross-channel diffusion is practically absent for isolated adatoms. Therefore these surfaces are in principle the most natural candidates to study quasi-one-dimensional diffusion processes. However, as we shall show in the following, even if diffusion occurs only along a single direction, it is not correct to treat it within simple one-dimensional models, because novel mechanisms play an important role on these surfaces, both for adatoms [15, 16] and clusters [17, 18] (the latter are referred to as chains in the following when the number of constituent atoms $n$ is at least 3).

In this paper, we study the diffusion of adatoms, dimers and chains on missing-row $Au(110)$, and review the results obtained by classical molecular-dynamics simulations and analytical methods. The molecular dynamics (MD) simulations are employed to calculate the energy barriers for the most important microscopic diffusion mechanisms; the analytical calculations are employed to extract the diffusion coefficients for clusters from the microscopic barriers. In the MD simulations, the metals are modeled by many-body potentials (as developed in the framework of the tight-binding (TB) model, see Ref. [19]); the form and the parameters of the potentials can be found in [20].

## 2. Adatom diffusion

As said in the introduction, the diffusion of isolated adatoms on the $(110)(1\times2)$ surfaces takes place essentially along the $[1\bar{1}0]$ direction, cross-channel diffusion [21] being characterized by much higher barriers (see the energy barriers in Table I). A possible mechanism is the jump along the channel bottom (the sequence $1a \rightarrow 1f \rightarrow 1e$ in Fig. 1). This is the most trivial diffusion path, and according to both TB calculations, it also the most favorable one. However, another mechanism is possible, and we shall refer to it as the metastable walk (MW) [15] mechanism. In a MW, the adatom climbs up on one of the (111) facets of the channel, diffuses there passing through a network of metastable minima (see $1b$, $1c$ and $1d$ in Fig. 1), and finally falls down again to the channel bottom. The barriers between the metastable minima are low, being very close to the barriers for diffusion on a (111) flat surface. Our calculations show that the MW mechanism has a larger barrier than the simple jump on *Au*; however we remark that first-principle calculations indicate that the MW trajectory should be the most energetically favorable in the case of $Pt$.

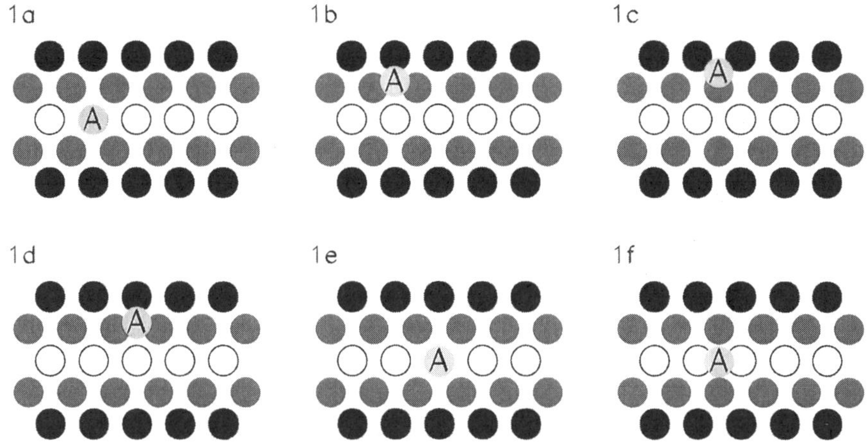


*Figure 1.* Schematic representation of the metastable-walk sequence ($1a \rightarrow 1b \rightarrow 1c \rightarrow 1d \rightarrow 1e$) and of the jump sequence ($1a \rightarrow 1f \rightarrow 1e$). Positions $1b$, $1c$ and $1d$ are metastable minima on the facet.

In any case, it is relatively easy to make effective long jumps by the MW mechanism, because once the adatom is on the (111) facet, it can diffuse rather easily there for several lattice spacings before being retrapped at the channel bottom; at sufficiently low $T$ the probability of an effective jump by MW of a length $l$ of lattice spacings decays as $\exp[-l\Delta E/(k_B T)]$ [15], where $\Delta E$ is the difference between the barriers for going from $1b$ to $1c$ and from $1b$ to $1a$ in Fig. 1, provided this difference is positive (as we have found for $Au$, see Table I, and as it has been found by Lorensen et al. for $Pt$ [16]). $\Delta E$ can be however rather small, and this leads to a considerable percentage of effective long jumps induced by the MW mechanism. This may explain why [16] a considerable percentage of long jumps is observed in adatom diffusion on missing-row $Pt/Pt(110)$ [22]; in fact the in-channel trajectories would justify a much lower percentage [23]. In our high-temperature molecular dynamics simulations on $Au$ [15] we find that the most part of the single moves (i.e. of the moves to nearest-neighbor sites) is due to in-channel jumps; on the other hand, more than 25% of the double moves is due to metastable walks, and the majority of the triple moves is due again to the MW mechanism.

## 3. Dimer diffusion

In the case of dimer diffusion, climbing on channel facets is even more important. In fact, our calculations predict that the leapfrog mechanism should be the dominant one.

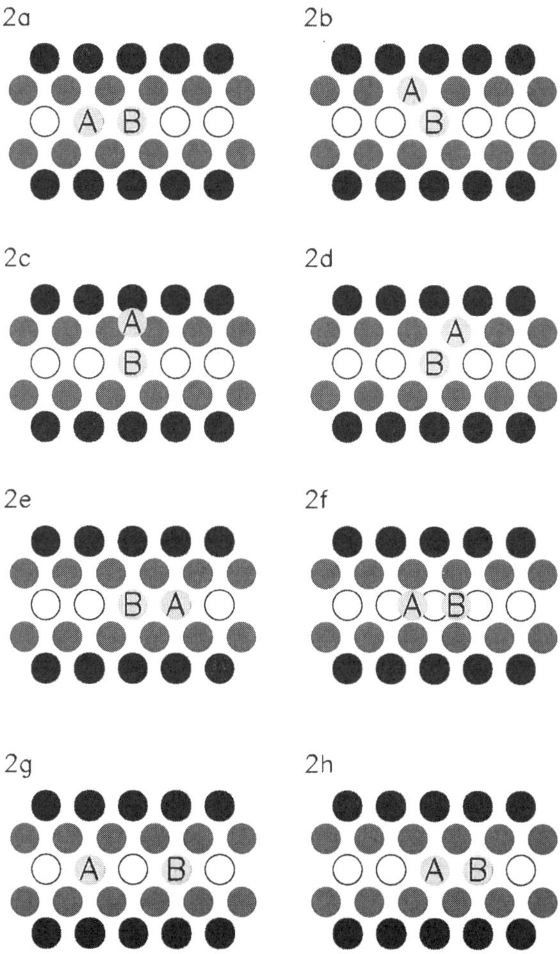


*Figure 2.* Schematic representation of the dimer leapfrog sequence ($2a \rightarrow 2b \rightarrow 2c \rightarrow 2d \rightarrow 2e$), of the concerted-jump sequence ($2a \rightarrow 2f \rightarrow 2h$) and of the dissociation-reassociation sequence ($2a \rightarrow 2g \rightarrow 2h$). In the leapfrog sequence, positions $2b$ and $2d$ are metastable minima, whereas $2c$ is a saddle point.

In the dimer leapfrog, one of the atoms of the dimer is promoted on the (111) facet, where it finds two equivalent metastable minima, see $2b$

and $2d$ in Fig. 2, connected through a saddle point ($2c$). Again, the barrier between these metastable minima is low for pure $Au/Au$; the situation is different in the case of heterodimers, such as $Cu_2$ on $Au$ or $AuCu$ on $Au$ [24]. On the other hand, for these heterodimers the leapfrog is not a favorable diffusion mechanism. Once the adatom is promoted, it jumps back and forth between the two minima several times. At the end, the adatom has almost an equal probability of coming back to the site of departure and jumping down at the opposite side of the dimer, thus completing a diffusion move. The leapfrog mechanism resembles a dimer rotation on fcc(111) surfaces, the latter being quite an important mechanism on such surfaces [9]. Our calculations predict that leapfrogging is the lowest-energy dimer-diffusion process for both metals. Even if dimer-leapfrog has not yet been experimentally observed (and probably it would be extremely difficult to observe due to its very fast time scale [17]), its importance is supported also by the results of first-principle calculations on $Pt$ [25], even if there is some experimental indication in favor of the dissociation-reassociation mechanism. In the latter mechanism (see $2a$, $2g$, and $2h$ in Fig. 2), the dimer breaks up, then the two adatoms diffuse independently, and finally meet again in a position which can be quite far apart from the original position of departure. If the diffusion of isolated adatoms is much faster than dissociation itself, only the initial and the final position of the dimer are observed. Dissociation-reassociation can thus induce effective long jumps of the dimer, whose probability distribution falls off as $1/l^2$, where $l$ is the effective-jump length [26]. Finally, we remark that the concerted jump of the dimer (sequence $2a \to 2f \to 2h$ in Fig. 2), in which both atoms move together at the channel bottom, seems to be unfavorable.

## 4. Chain diffusion

When chains with $n \geq 3$ atoms are considered, the concerted jump can be easily ruled out: its activation barrier is roughly proportional to $n$ (see Table I), because $n$ atoms have to be displaced together from their hollow position to a close saddle point. Also the moves caused by dissociation-reassociation have increasing effective barriers with $n$. Let us consider, for example, a trimer. In order to displace the trimer by dissociation-reassociation, two bonds must be broken, instead of the single bond of the dimer case. Analytical calculations [27] show that, at sufficiently low temperatures, the effective-jump rate follows an Arrhenius law with effective barrier $E_{eff}(3)$

$$E_{eff}(3) = E_{d3} + \frac{(E_{d2} - E_j)}{2}, \tag{1}$$

where $E_{d3}$ is the barrier needed to detach a side atom from the trimer, $E_{d2}$ is the dissociation barrier for the dimer and $E_j$ is the energy barrier for jumping of an isolated adatom. The addendum in parenthesis is the contribution to the effective barrier from the breaking of the second bond. For a tetramer, the barrier $E_{eff}(4)$ is even larger. Moreover, one should consider that the probability that dissociated adatoms, during their walk, meet other adatoms or chains and stick to them increases with $n$, so there is a considerable probability of observing a final chain with a different length from the initial chain.

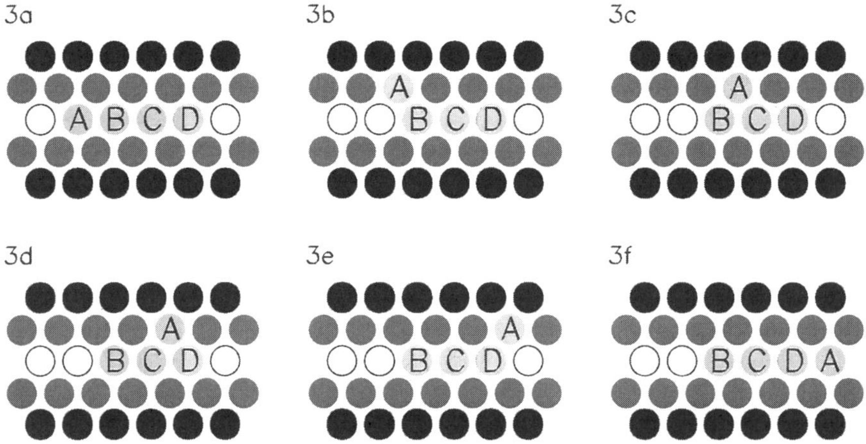


*Figure 3.* Schematic representation of the chain leapfrog process.

Due to the above facts, chains should either change their length or be practically immobile. However, this is not the case for chains on missing-row surfaces, where, from the STM experiments performed by Linderoth et al. [18], there is evidence of chain diffusion. Therefore chain diffusion can take place only by another mechanism, different from concerted jump and dissociation-reassociation, and the mechanism is again the leapfrog (see Fig. 3). In fact, it is easily understood that the leapfrog diffusion barrier should depend weakly on the chain length $n$. Essentially, the rate-limiting step of the leapfrog sequence is the upward promotion (in Fig. 3, from the initial configuration *3a* to the quite deep metastable minimum in *3c* passing through *3b*), whose barrier $E_{up}$ depends on the local environment of the promoted adatom, which is initially at the end of the chain. This local environment is practically independent of the chain length. In fact, as can be seen in Table I, upward promotion energies are indeed nearly the same for dimers, trimers and so on. However, a further point must be taken into account. Once one of the side adatoms is promoted above the chain, there is a probability $P_{LF}$ that it will jump down at the opposite end of the chain, thus completing the leapfrog sequence. Before jumping down, the adatom must diffuse above the chain (process $3c \to 3d$ in Fig. 3) with a rate $r_{above}$ (this is indeed the rate for diffusion on a locally unreconstructed $(1 \times 1)$ surface) and, once it has reached the border, it can jump down with a rate $r_{down}$ (in Fig. 3, the sequence is $3d \to 3e \to 3f$; usually the descent is by jump, the exchange barrier being much larger). The probability $P_{LF}$ does depend on $n$ according to the following expression [27]:

$$P_{LF} = \frac{b}{n - 3 - b(n - 5)}, \tag{2}$$

where

$$b = \frac{r_{above}}{r_{above} + r_{down}}. \tag{3}$$

In $Au$ we find that $r_{above} \gg r_{down}$ at the temperatures of experimental interest, so that $b \simeq 1$. This implies that $P_{LF}$ is constant for a wide range of $n$. Asymptotically, $P_{LF} \propto n^{-1}$ for $n \to \infty$, but this limit is not physically important because multiple promotions become likely for very long chains. Multiple promotions stop the leapfrog mobility because they lead to the formation of new chains above the original chain. These new chains are in a locally unreconstructed geometry, so they are practically immobile.

## 5. Conclusions

The mobility of adatoms, dimers, and linear clusters on missing row $Au(110)$ has been investigated by molecular dynamics simulations and analytical calculations. Even if cross-channel diffusion takes place on a much slower time scale than in-channel diffusion, the atomic motion on this surface (and on the analogous $Pt(110)$ surface) cannot be treated at all by simple one-dimensional models. In fact, for adatoms, dimers, and chains, there are diffusion mechanisms involving the motion on the channel facets. According to our results [15], in the case of adatoms, these mechanisms are not dominant (but first-principle calculations on $Pt$ indicate that they should be dominant [16, 25]), but they give a large proportion of effective long jumps. On the contrary, in the case of dimers, both our calculations [17] and first-principle results show that the leapfrog mechanism is dominant. This mechanism is even more important for chains, where it causes a chain mobility which is independent on the chain length to a large extent. The latter length-independent mobility is typical of chains on missing-row surfaces: on unreconstructed surfaces, such as silver or copper (110) we expect chains either to change their length or to be practically immobile, already for lengths of the order of 4 atoms [27].

TABLE 1. Energy barriers (in $eV$) of the elementary diffusion mechanisms for adatoms, dimers and linear chains on $Au(110)\,(1\times 2)$. Many of the mechanism are represented in the figures.

| Mechanism | Barrier |
|---|---|
| adatom jump diffusion ($E_j$, $1a \to 1f \to 1e$) | 0.31 |
| adatom promotion on the facet ($1a \to 1b$) | 0.40 |
| adatom walk through metastable minima ($1b \to 1c$) | 0.10 |
| adatom jump to the channel bottom ($1d \to 1e$) | 0.06 |
| adatom cross-channel jump barrier | 0.70 |
| adatom cross-channel exchange barrier | 0.66 |
| dimer upward promotion ($2a \to 2b$) | 0.45 |
| dimer switch ($2b \to 2d$ through the saddle $2c$) | 0.08 |
| dimer jump descent ($2d \to 2e$) | 0.14 |
| dimer dissociation ($E_{d2}$) | 0.51 |
| dimer concerted jump | 0.52 |
| trimer upward promotion | 0.45 |
| trimer jump descent | 0.37 |
| trimer dissociation in adatom plus dimer ($E_{d3}$) | 0.52 |
| trimer concerted jump | 0.72 |
| trimer exchange descent | 0.56 |
| tetramer concerted jump | 0.91 |
| pentamer concerted jump | 1.07 |
| hexamer concerted jump | 1.24 |
| descent from the chain ($E_{down}$, $3d \to 3e \to 3f$) | 0.37 |
| diffusion above the chain ($E_{above}$,$3c \to 3d$) | 0.28 |

We acknowledge financial support form the Italian MURST under the project N. 9902112831.

## References


1. A. Pimpinelli and J. Villain, *Physics of Crystal Growth*, Cambridge University Press, Cambridge, 1998.
2. G.L. Kellogg and P.J. Feibelman, *Phys. Rev. Lett.* **64**, 3143 (1990).
3. P.J. Feibelman, *Phys. Rev. Lett.* **65**, 729 (1990).
4. J.E. Black and Z.J. Tian, *Phys. Rev. Lett.* **71**, 2445 (1993).
5. J.M. Cohen, *Surf. Sci.* **306**, L545 (1994).
6. G.A. Evangelakis, N.I. Papanicolaou, *Surf. Sci.* **347**, 376 (1996); G.C. Kallinteris, G.A. Evangelakis, N.I. Papanicolaou, *Surf. Sci.* **369**, 185 (1996).
7. G. Boisvert and L.J. Lewis, *Phys. Rev.* **B56**, 7643 (1997).
8. O.S. Trushin, M. Kotrla and F. Maca, *Surf. Sci.* **389**, 55 (1997).
9. G. Boisvert and L.J. Lewis, *Phys. Rev.* **B59**, 9846 (1999).
10. A.F. Voter, *Phys. Rev.* **B34**, 6819 (1986).

11. J. Heinonen, J. Merikoski and T. Ala-Nissila, *Phys. Rev. Lett.* **82**, 2733 (1999).
12. O.S. Trushin, P. Salo and T. Ala-Nissila, *Phys. Rev.* **B62**, 1611 (2000).
13. C.L. Liu, J.M. Cohen, J.B. Adams and A.F. Voter, *Surf. Sci.* **253**, 334 (1991)
14. R. Ferrando, *Phys. Rev. Lett.* **82**, 4195 (1996).
15. F. Montalenti and R. Ferrando, *Phys. Rev.* **B58**, 3617 (1998).
16. H.T. Lorensen, J.K. Nørskov and K.W. Jacobsen, *Phys. Rev.* **B60**, R5149 (1999).
17. F. Montalenti and R. Ferrando, *Phys. Rev. Lett.* **82**, 1498 (1999).
18. T.R. Linderoth, S. Horch, L. Pedersen, S. Helveg, E. Laegsgaard, I. Stensgaard and F. Besenbacher, *Phys. Rev. Lett.* **82**, 1494 (1999).
19. M. Guillopé and B. Legrand, *Surf. Sci.* **215**, 577 (1989).
20. F. Montalenti and R. Ferrando, *Phys. Rev.* **B59**, 5881 (1999).
21. F. Montalenti and R. Ferrando, *Surf. Sci.* **433-435**, 445 (1999).
22. T.R. Linderoth, S. Horch, S. Helveg, E. Læsgaard, I. Stensgaard and F. Besenbacher, *Phys. Rev. Lett.* **78**, 4978 (1997).
23. J. Jacobsen, K.W. Jacobsen and J.P. Sethna, *Phys. Rev. Lett.* **79**, 2843 (1997).
24. F. Montalenti and R. Ferrando, *Surf. Sci.* **432**, 27 (1999).
25. P.J. Feibelman, *Phys. Rev.* **B61**, R2452 (2000).
26. F. Montalenti and R. Ferrando, *Phys. Rev.* **B60**, 11102 (1999).
27. F. Montalenti and R. Ferrando, *J. Chem. Phys.* **113**, 349 (2000).

# EFFECTS OF ADSORBATES ON SUBMONOLAYER GROWTH

MIROSLAV KOTRLA
*Institute of Physics Academy of Sciences of the Czech Republic*
*Na Slovance 2, 182 21 Praha 8, Czech Republic*

JOACHIM KRUG
*Fachbereich Physik, Universität GH Essen*
*D-45117 Essen, Germany*

AND

PAVEL ŠMILAUER
*Institute of Physics, Academy of Sciences of the Czech Republic*
*Cukrovarnická 10, 162 53 Praha 6, Czech Republic*

**Abstract.** The effects of adsorbates on nucleation and growth of two-dimensional islands is investigated by kinetic Monte Carlo simulations and rate equation theory. The variation of island morphology with adsorbate parameters is discussed and the temperature-dependence of island density in the case of immobile adsorbates is studied in detail. A set of rate equations for the description of nucleation in the presence of predeposited mobile and immobile adsorbates is developed.

## 1. Introduction

While a satisfactory understanding of nucleation and growth in homoepitaxial systems is beginning to emerge [1], the dramatic effects of tiny traces of adsorbates continue to surprise researchers. Among a wealth of recent examples, we may mention the influence of CO on island shapes and interlayer transport on Pt(111) [2], the modification of attachment kinetics by H preadsorbed on Si(001) [3] and the H-induced enhancement of Pt self-diffusion on Pt(110) [4]. In this situation it seems useful to explore different generic scenarios for adsorbate effects on submonolayer growth within a reasonably simple, yet flexible, theoretical model. The present paper reports on an ongoing effort aimed in this direction.



*M.C. Tringides and Z. Chvoj (eds.),*
*Collective Diffusion on Surfaces: Correlation Effects and Adatom Interactions*, 247–257.

In Section 2 we briefly review our earlier findings and present new results concerning the dependence of the island density and morphology on the temperature and the characteristics of the adsorbate (mobility and the strength of interaction with adatoms). In the Sec. 3 we sketch a rate equation theory which provides a unified description of two-dimensional nucleation in the presence of predeposited mobile and immobile adsorbates, and derive analytic solutions for some special cases.

## 2. Kinetic Monte Carlo Simulations

### 2.1. MODEL

We employ a recently introduced solid-on-solid growth model with two surface species $A$ and $B$, representing the adatoms of the growing material and the adsorbates, respectively [5, 6]. The simulation starts on a flat substrate composed only of $A$-atoms. The basic microscopic processes are deposition and migration; desorption is not allowed. Two deposition modes can be studied: (i) simultaneous deposition (*codeposition*) of both species, with generally different fluxes, $F_A$ and $F_B$, and (ii) *predeposition* of a certain adsorbate coverage $\theta_B$ prior to growth.

Adatoms and adsorbates migrate with a nearest-neighbor hopping rate $R_D = k_0 e^{-E_D/k_BT}$, where $k_0 = 10^{13}$ Hz and the energy barrier for a particle of type $X =$ A or $B$ is given by

$$E_D^X = \sum_{Y=A,B} \left( n_0^Y E_{\text{sub}}^{XY} + n_1^{XY} E_{\text{n}}^{XY} \right). \tag{1}$$

Here $E_{\text{sub}}^{XY}$ is the hopping barrier for a free $X$ atom on a substrate atom $Y$, $n_0^Y$ is equal to one if the substrate atom is of type $Y$ and zero otherwise, $n_1^{XY}$ is the number of nearest-neighbor $X$-$Y$ pairs, and $E_{\text{n}}^{XY}$ is the corresponding contribution to the barrier (symmetric in $X$ and $Y$). There are no lateral interactions between adsorbate atoms ($E_{\text{n}}^{BB} = 0$). The diffusion rate of a free adatom is $D = k_0 e^{-E_{\text{sub}}^{AA}/k_BT}$.

In previous work [5], we have shown that in addition a process of place exchange between an adsorbate and an adatom with at most one lateral bond is necessary to achieve decorated island edges. This occurs at a rate $k_0 e^{-E_{\text{ex}}/k_BT}$. When the barrier $E_{\text{ex}}$ is sufficiently low, then adsorbates are floating on the edges of growing islands.

Adsorbates can be mobile or immobile and strongly or weakly interacting with the adatoms. Their mobility is controlled by the strength of interaction of adsorbates with the substrate which is described by the energy barrier $E_{\text{sub}}^{BA}$. The interaction with adatoms is determined by the energy barrier $E_{\text{n}}^{AB}$.

## 2.2. SUMMARY OF PREVIOUS RESULTS

We have investigated previously the case of floating mobile adsorbates in the situation when the concentration of adsorbates is comparable with that of the growing material [6]. We studied the dependence of the island density $N$ on the flux $F_A$ and the coverage $\theta_A$ for both codeposition and predeposition. We found that the adsorbates strongly increase the island density without appreciably changing its power-law dependence on flux, the increase being stronger for larger $E_{\mathrm{n}}^{AB}$. The increase was only slightly higher for predeposition than for codeposition.

We also observed a stronger coverage dependence of the island density in comparison with homoepitaxy. This was interpreted as a delay of the saturation regime, where island density becomes independent of coverage. The coverage dependence is again more pronounced for strongly interacting adsorbates, but it is weaker in the case of predeposition than for codeposition. A further noteworthy feature is the much higher density of free adatoms than in homoepitaxy, which shows an intriguing, oscillatory coverage dependence in the case of codeposition [6].

Many of these simulation results could be qualitatively explained in terms of a simple rate equation theory, which rests on the assumption that the adsorbates affect the growth process only by slowing down the diffusion of adatoms. If, in addition, the adsorbates are treated as immobile traps (which may be justified at least for sufficiently high adsorbate coverage $\theta_B$, see Section 3.4), the adatom migration is described by an effective diffusion coefficient

$$\overline{D}(\theta_B) = \frac{D}{1 - \theta_B + \theta_B e^{E_{\mathrm{n}}^{AB}/k_B T}}. \tag{2}$$

Recently, we have considered a situation when only a low concentration of predeposited adsorbates is present on the surface [7]. We investigated the dependence of island density on flux and on the concentration of adsorbates $\theta_B$. Even concentrations as low as $\theta_B = 0.002$ ML can lead to a severalfold increase of the island density. However, the increase significantly depends on the mobility of the adsorbates. In the case of essentially immobile adsorbates ($E_{\mathrm{sub}}^{BA} = 5$ eV), a new feature in the flux dependence of the island density appears. Instead of a single power law relationship, there is a plateau where $N \approx \theta_B$, reflecting the dominance of heterogeneous nucleation in a flux interval $F_1 > F > F_2$. Scaling arguments yield the estimates $F_1 \approx D\theta_B^2$, $F_2 \approx \theta_B D e^{-E_{\mathrm{n}}^{AB}/k_B T}$ for the plateau boundaries in terms of impurity coverage and strength. For sufficiently small $\theta_B$ these were confirmed by scaling plots of $N/\theta_B$ vs. $F/F_1$ and $F/F_2$ [7].

In this work we further extend the above results. In the following subsection, we discuss effects of adsorbates on the island morphology, and in the subsequent subsection we present results on the temperature dependence of the island density. The energy barriers $E_{\text{sub}}^{AA} = 0.8$ eV, $E_{\text{sub}}^{AB} = 0.1$ eV, $E_{\text{sub}}^{BB} = 0.1$ eV, and $E_{\text{n}}^{AA} = 0.3$ eV were fixed in the following. Other energy barriers as well as the ratio $D/F$ were varied.

## 2.3. ISLAND MORPHOLOGY

The island shape in homoepitaxy depends on the rates of the various kinetic processes involved. The size of islands varies with the ratio $D/F$. To be specific, we set in this subsection the temperature to $T = 500$ K. Then, the island shapes in homoepitaxy with the parameters mentioned above in Section 2.2 are regular (approximately rectangular, close to square shape, nondendritic) for fluxes in the interval $F = 0.00025 - 0.64$ ML/s. The shapes become irregular for higher fluxes. In order to evaluate the effect of adsorbates, we performed several sets of simulation with fixed flux $F = 0.001$ ML/s and varying adsorbate properties.

Let us consider first the case of floating adsorbates. It is useful to distinguish two situations: (i) well decorated island and (ii) poorly decorated islands. In the former case, the majority of edge sites is decorated by an adsorbate. We found that the shape is very regular, only the islands are smaller; in fact it can be shown that the decoration *enhances* edge diffusion and thus makes the island edges smoother [6]. The shape is not very sensitive to the mobility of adsorbates. It only has to be sufficiently high so that adsorbates can reach the island edges. The exchange of adatoms approaching the step edge and the adsorbates decorating the step edge is the dominant mechanism.

If the islands are poorly decorated (the number of adsorbates attached to the island edges is considerably less than the number of perimeter sites), then, the shape changes to irregular and regularity is restored again only for very small $\theta_B$. Rather than promoting edge diffusion, in this regime the adsorbates bound on the island edges act as sinks for adatoms provided that the adatom-adsorbate interaction (given by $E_{\text{n}}^{AB}$) is sufficiently strong. Another factor affecting the island shape in this situation is the change of the mobility of adsorbates (given by $E_{\text{sub}}^{BA}$). Examples of the different island mophologies for various values of parameters $E_{\text{n}}^{AB}$ and $E_{\text{sub}}^{BA}$ are shown in Fig. 1. We can see that the increase of both energy barriers leads to more irregular island shapes and also to an increase of island density.

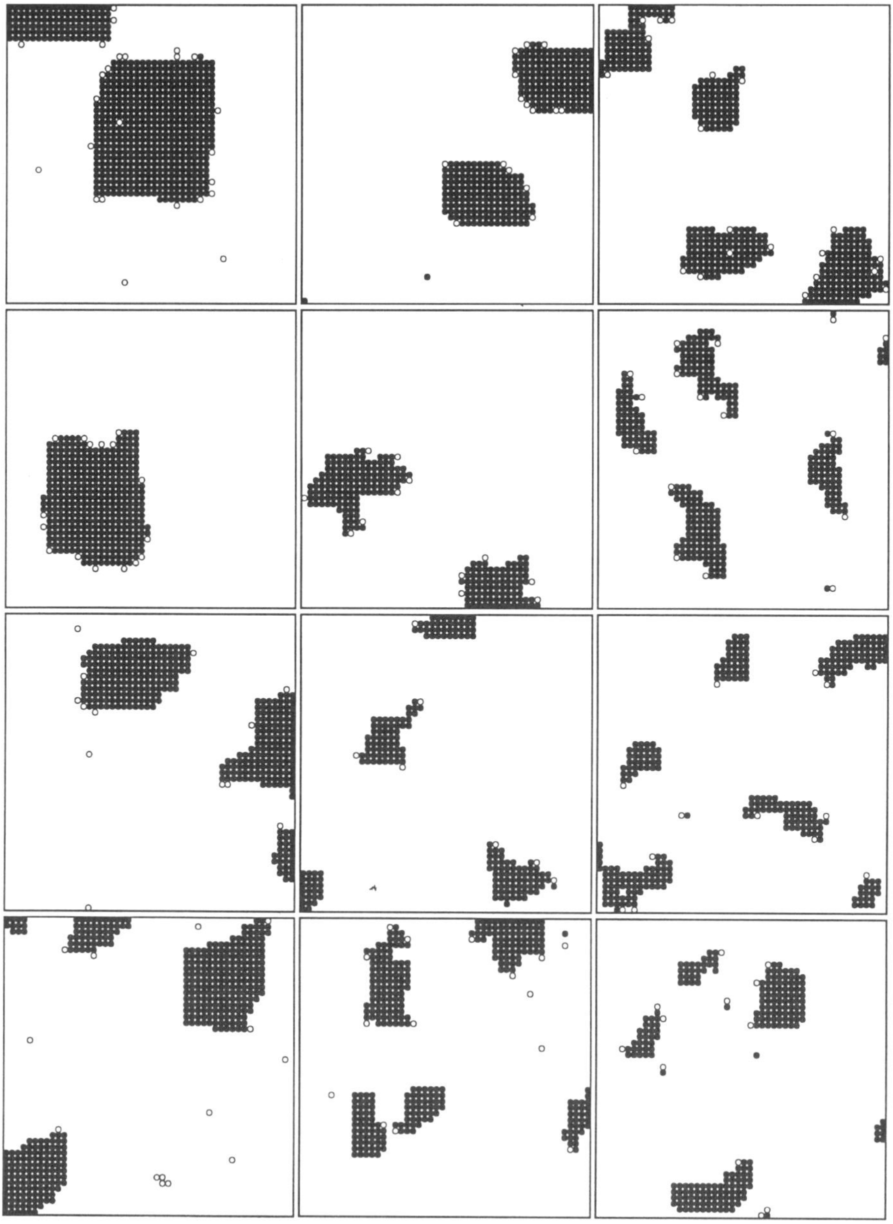

*Figure 1.* Examples of configurations obtained with flux $F_A = 0.001$ ML/s at a coverage $\theta_A = 0.1$ ML after predeposition $\theta_B = 0.005$ ML of adsorbate coverage, $E_n^{AA} = 0.3$ eV, $E_{ex} = 1$ eV and different energy barriers $E_n^{AB} = 0.2$ eV (left column), $E_n^{AB} = 0.4$ eV (middle column), $E_n^{AB} = 0.6$ eV (right column), $E_{sub}^{BA}$: from top to bottom $E_{sub}^{BA} = 0.7$ eV, $E_{sub}^{BA} = 1.0$ eV, $E_{sub}^{BA} = 1.2$ eV, $E_{sub}^{BA} = 1.5$ eV. We show only 50 × 50 sections of larger simulation boxes.

The irregularity is not so apparent for weak adsorbate-adatom interaction (left column) or high mobility of adsorbates (top row). The immobile strongly interacting adsorbate catches more diffusing adatoms and at the same time it is floating on the growing island edge. Hence, the island grows faster in the neighborhood of such adsorbates. This kind of adsorbates also block the motion of adatoms along the edge and thus suppresses the smoothening of the island edges.

Let us discuss briefly the island morphology for non-floating adsorbates. The mechanism of floating becomes less relevant for adsorbates with low mobility and does not act at all in the limit of immobile adsorbates. Non-floating adsorbates are incorporated inside a growing island and beyond a certain island size have no effect on growth. In this context, we want to note that surface defects can be treated in our approach as immobile adsorbates.

### 2.4. TEMPERATURE DEPENDENT ISLAND DENSITY

While in our simulations we have so far focused on the flux dependence of the island density [5-7] in experiments it is most often measured as a function of temperature at fixed flux. Therefore, we complement our previous results by the calculation of the temperature dependence. We restrict ourselves here to the case of predeposition of immobile (non-floating) adsorbates. In Fig.2 we show data for several adsorbate concentrations. We find, similar to the flux dependence [7], a plateau interval $T_1 < T < T_2$, where $N$ is almost independent of $T$. For low temperatures, $T < T_1$, homogeneous nucleation is more important than heterogeneous nucleation by adsorbates, because the adatoms are not sufficiently mobile to reach the adsorbates. In the plateau region nucleation is predominantly heterogeneous. For high temperatures, $T > T_2$, adatoms start to detach from adsorbates and homogeneous nucleation is again important. This kind of behavior is observed experimentally in defect nucleation on oxide and halide surfaces [8] and in growth on surfaces with strain-relief patterns, which act as an ordered array of defects [9].

For a quantitative estimate of the transition temperatures $T_1$ and $T_2$, we recall the three relevant time scales in the problem [7]: (i) The *diffusion time* $\tau_D \approx 1/D\theta_B$, which is the time required for an atom to explore the "capture zone" of area $1/\theta_B$ associated with a single impurity, and hence to get trapped at the impurity. (ii) The *deposition time* $\tau_F \approx \theta_B/F$, which is the time between subsequent arrivals of adatoms within the capture zone. (iii) The trapping time $\tau \approx (1/D)e^{E_{\rm n}^{AB}/k_BT}$ at the impurity. The plateau regime is characterized by $\tau > \tau_F > \tau_D$, which yields the expressions

$$T_1 = \frac{E_{\rm sub}^{AA}}{k_B \ln(k_0\theta_B^2/F)}, \quad T_2 = \frac{E_{\rm sub}^{AA} + E_{\rm n}^{AB}}{k_B \ln(k_0\theta_B/F)}. \tag{3}$$

Using $E_{\text{sub}}^{AA} = 0.8$ eV, $F = 0.001$ ML/s and $\theta_B = 0.001$ ML, we get $T_1 = 403$ K, while $T_2 = 465$ K for $E_{\text{n}}^{AB} = 0.4$ eV and $T_2 = 542$ K for $E_{\text{n}}^{AB} = 0.6$ eV, in good agreement with the simulations. From the expression for $T_2$, we directly see how the width of the plateau grows with increasing $E_{\text{n}}^{AB}$.

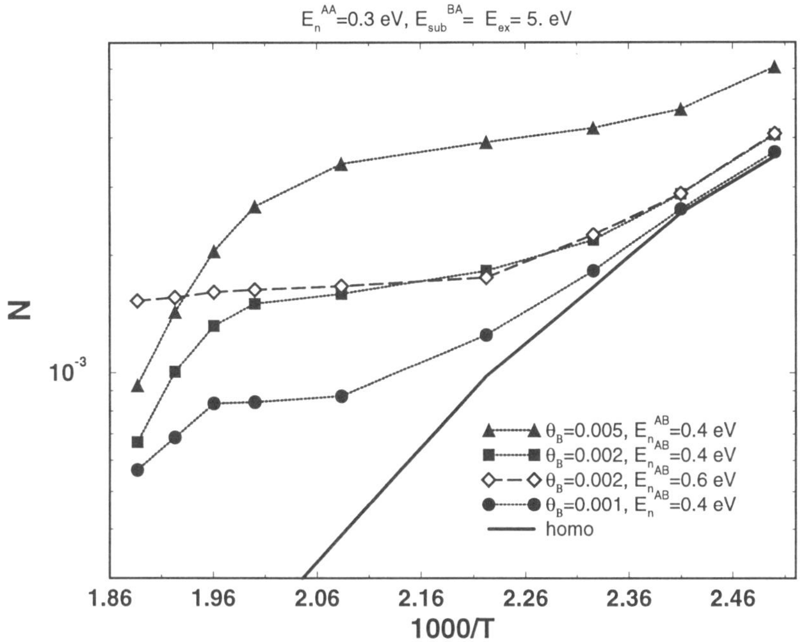


*Figure 2.* Averaged island density as a function of the inverse temperature for different concentration $\theta_B$ of predeposited impurities and different energy of interaction between adsorbates and adatoms: $E_{\text{n}}^{AB} = 0.4$ eV - filled symbols, $E_{\text{n}}^{AB} = 0.6$ eV - open symbols. The adatom interaction energy $E_{\text{n}}^{AA} = 0.3$ eV and the energy barriers $E_{\text{sub}}^{BA} = E_{\text{ex}} = 5$ eV are fixed. Deposition flux is $F_A = 0.001$ ML/s, the coverage of adatoms is $\theta_A = 0.1$ ML. The behavior in the absence of impurities (homoepitaxy, $\theta_B = 0$ ML) is shown for comparison.

## 3. Rate Equations

In this section we develop a set of rate equations for the description of two-dimensional nucleation in the presence of mobile and immobile adsorbates. Throughout, the impurities will be assumed to have been predeposited at a coverage $\theta_B$, though codeposition is easily incorporated along the lines of our earlier work [6]. To keep matters transparent, we also restrict ourselves to the case of irreversible aggregation (critical nucleus size $i^* = 1$). Detachment of impurities from adatom islands will, however, be included, since this process plays a crucial role at low deposition fluxes.

In addition to the adatom density $n$ and the island density $N$, a minimal description of nucleation in the presence of adsorbates turns out to require the monitoring of two additional quantities [8, 10]: The density of free impurities $n_0$, and the density $n_c$ of critical clusters bound at impurities; in the present case $i^* = 1$ the latter is simply the density of bound adatom-impurity pairs. Taking into account all relevant processes, one arrives at the following set of evolution equations:

$$dn/dt = F - Dn(2n + N) - Dn(n_0 + n_c) + n_c/\tau, \tag{4}$$

$$dN/dt = Dn(n + n_c), \tag{5}$$

$$dn_0/dt = -Dnn_0 - \tilde{D}n_0 N + n_c/\tau + N/\tau', \tag{6}$$

$$dn_c/dt = Dn(n_0 - n_c) - n_c/\tau. \tag{7}$$

Here $1/\tau \approx k_0 e^{-(E_{\rm sub}^{AA} + E_{\rm n}^{AB})/k_B T}$ denotes the rate of detachment of an adatom from an impurity, $\tilde{D} = k_0 e^{-E_{\rm sub}^{BA}/k_B T}$ is the impurity diffusion constant, and $1/\tau' \approx k_0 e^{-(E_{\rm sub}^{BA} + E_{\rm n}^{AB})/k_B T}$ is the detachment rate of impurities from islands. In the expression for $\tau$, we have assumed that the impurities are (much) less mobile than the adatoms, so that the relative mobility is dominated by the adatom mobility.

Most of the impurity terms in Eqs.(4-7) should be self-explanatory. We merely point out the term $n_c/\tau$ in (4,6,7) which describes the dissociation of adatom-impurity pairs, and the term $N/\tau'$ in (7) which describes detachment of impurities from the islands. Strictly speaking, all reaction terms should be adorned with dimensionless capture coefficients. However, since our objective here is a qualitative, rather than a quantitative description, all these coefficients have been set to unity.

### 3.1. DIMENSIONLESS FORMULATION

Following Tang [11] we now rescale the adatom and island densities by $\sqrt{F/D}$ and time by $1/\sqrt{DF}$. In addition, the impurity densities $n_0$ and $n_c$ will be rescaled by the initial impurity coverage $\theta_B$. Marking dimensionless quantities with a hat, we obtain the rescaled evolution equations

$$d\hat{n}/d\hat{t} = 1 - \hat{n}(2\hat{n} + \hat{N}) - a\hat{n}(\hat{n}_0 + \hat{n}_c) + b\hat{n}_c, \tag{8}$$

$$d\hat{N}/d\hat{t} = \hat{n}(\hat{n} + a\hat{n}_c), \tag{9}$$

$$d\hat{n}_0/d\hat{t} = -\hat{n}\hat{n}_0 - c\hat{n}_0\hat{N} + (b/a)\hat{n}_c + (bc/a^2)\hat{N}, \tag{10}$$

$$d\hat{n}_c/d\hat{t} = \hat{n}(\hat{n}_0 - \hat{n}_c) - (b/a)\hat{n}_c, \tag{11}$$

where $a = \theta_B\sqrt{D/F}$, $b = \theta_B/F\tau$ and $c = \tilde{D}/D$. The initial conditions are $\hat{n}(0) = \hat{N}(0) = \hat{n}_c(0) = 0$, $\hat{n}_0(0) = 1$.

The parameters $a$ and $b$ may be written in terms of the characteristic fluxes, $F_1$ and $F_2$, which limit the plateau regime in the case of immobile impurities, as $a = \sqrt{F_1/F}$ and $b = F_2/F$. For impurities to be at all relevant, it is necessary that $F_1/F_2 = D\theta_B\tau = a^2/b \gg 1$. By retaining only the most important terms, we can analytically extract the behavior predicted by Eqs.(8-11) in simple cases. So far this has been achieved only for immobile impurities ($c = 0$).

### 3.2. IMMOBILE IMPURITIES WITHOUT DETACHMENT

Here we consider the plateau regime, where $a \gg 1$ and $b \ll 1$. Numerical integration of (8-11) with $b = c = 0$ and $a \gg 1$ shows that the adatom density is time independent, taking the value $n = F/D\theta_B$ to high accuracy. For early times, this reflects the balance between deposition and capture at impurities, while at late times the capture at islands dominates; since $N = \theta_B$ at late times, the resulting adatom density is the same in both regimes. Setting $n \equiv F/D\theta_B$, the remaining three rate equations become linear and are readily integrated, with the result

$$N = \theta_B(1 - (1 + \theta/\theta_B)e^{-\theta/\theta_B}), \tag{12}$$

$$n_0 = \theta_B e^{-\theta/\theta_B}, \tag{13}$$

$$n_c = \theta e^{-\theta/\theta_B}. \tag{14}$$

Both impurity species decay exponentially at late times.

### 3.3. IMMOBILE IMPURITIES WITH DETACHMENT

The regime $F \ll F_2$ is detachment-dominated, in the sense that an adatom typically visits many impurities before being incorporated in an island. The primary effect of the impurities is then to slow down the diffusion by temporarily trapping the adatoms [6]. To see how this emerges from the rate equations, we infer from numerical integration that, in the relevant late time regime, an equilibrium is established between attachment and detachment of adatoms at free impurities. This implies that $Dnn_0 \approx n_c/\tau$. Equation (6) then yields $dn_0/dt \approx 0$, so that the impurity concentration essentially retains its initial value $n_0 = \theta_B$, and $n_c \approx (D\theta_B\tau)n$. In this way, the impurity rate equations (6,7) have effectively been eliminated, and we are left with a pair of modified equations for $n$ and $N$. Taking into account that $D\theta_B\tau = F_1/F_2 \gg 1$, and therefore $n_c \gg n$, they read

$$dn/dt = F - Dn[(D\theta_B\tau)n + N], \tag{15}$$

$$dN/dt = (D\theta_B\tau)Dn^2. \tag{16}$$

Standard analysis [11] shows that the asymptotic island density is of the order $N \sim (F\tau\theta_B)^{1/3}$, which can be brought into the familiar form $N \sim (F/\bar{D})^{1/3}$ by identifying the effective diffusion constant as $\bar{D} = 1/(\tau\theta_B)$. This is precisely the strong impurity limit of (2).

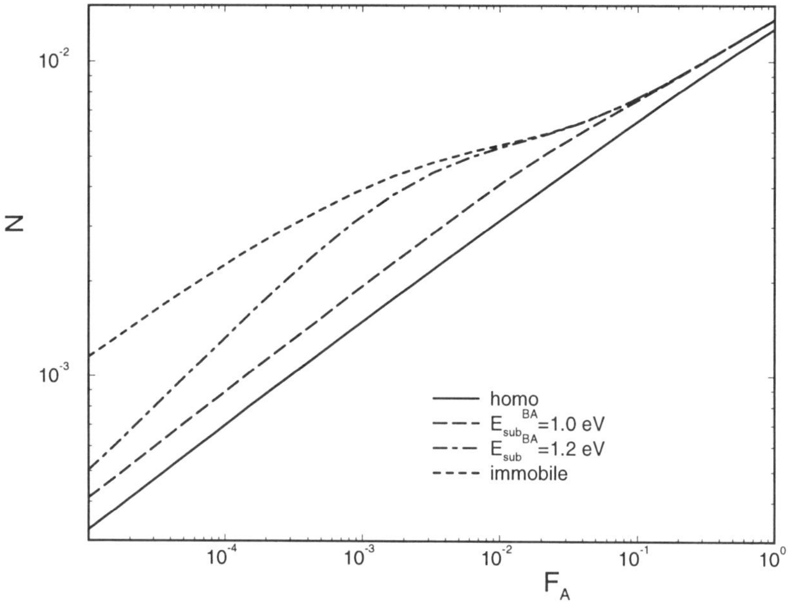


*Figure 3.* Island density as a function of deposition flux, obtained by numerical solution of the rate equations (4-7) for different choices of the impurity diffusion barrier. The adatom coverage is $\theta = 0.1$ ML and the impurity coverage $\theta_B = 0.005$ ML.

## 3.4. THE EFFECT OF IMPURITY MOBILITY

The second term on the right-hand side of Equation (6) describes the depletion of free impurities due to capture at stable islands. This is partly compensated by the detachment of impurities from islands described by the last term, but nevertheless, the net effect is expected to be a decrease of the island density, because the depletion reduces the ability of mobile impurities to act as nucleation centers.

In the absence of an analytic study of the full set of rate equations, in Fig. 3 we provide some sample results obtained by numerical integration. The adatom diffusion constant $D$ and the detachment time $\tau$ were chosen in accordance with the parameters employed in the KMC simulations. Data for three different choices of the impurity diffusion barrier $E_{\text{sub}}^{BA}$ are shown, and the behavior in the absence of impurities is included for reference. The case of low impurity mobility ($E_{\text{sub}}^{BA} = 1.2$ eV, corresponding to $\tilde{D}/D \approx 10^{-4}$) is particularly noteworthy, as it indicates an intermediate scaling regime where the island density scaling exponent, defined by $N \sim F^{\chi}$, is larger than the homoepitaxial value $\chi = 1/3$ (a fit yields $\chi \approx 0.42$). Further exploration of this phenomenon will be presented elsewhere.

*Acknowledgments.* J.K. would like to thank CAMP and the Department of Physics at DTU for their kind hospitality while this paper was prepared. Support by the COST project P3.130 and by Volkswagenstiftung are gratefully acknowledged.

## References

1. Brune, H. (1998) Microscopic view of epitaxial metal growth: nucleation and aggregation, *Surface Science Reports*, **31**, p. 121–229
2. Kalff, M., Comsa, G. and Michely, T. (1998) How sensitive is epitaxial growth to adsorbates?, *Physical Review Letters*, **81**, pp. 1255–1258
3. Šmilauer, P., Mizushima, K. and Vvedensky D.D. (1998) Activated Si-H exchange at Si-island edges on Si(001), *Physical Review Letters*, **81**, pp. 5600–5603
4. Horch, S., Lorensen, H.T., Helveg, S., Lægsgaard, E., Stensgaard, I., Jacobsen, K.W., Nørskov, J.K. and Besenbacher, F. (1999) Enhancement of surface self-diffusion of platinum atoms by adsorbed hydrogen, *Nature*, **398**, pp. 134–136
5. Kotrla, M., Krug, J. and Šmilauer, P. (2000) Submonolayer growth with decorated island edges, *Surface Science*, **454 – 456**, pp. 681–685
6. Kotrla, M., Krug, J. and Šmilauer, P. (2000) Submonolayer epitaxy with impurities: Kinetic Monte Carlo simulations and rate-equation analysis, *Physical Review B*, **62**, pp. 2889–2898
7. Kotrla, M., Krug, J. and Šmilauer, P. (2000) Effects of mobile and immobile impurities on two-dimensional nucleation, submitted to *Surface Science*
8. Venables, J.A. and Harding, J.H. (2000) Nucleation and growth of supported metal clusters at defect sites on oxide and halide (001) surfaces, *Journal of Crystal Growth*, **211**, pp. 27–33
9. Brune, H., Giovannini, M., Bromann, K. and Kern, K (1998) Self-organized growth of nanostructure arrays on strain-relief patterns, *Nature*, **394**, pp. 451–453
10. Venables, J.A. (1997) Nucleation growth and pattern formation in heteroepitaxy, *Physica A*, **239**, pp. 35–46
11. Tang, L.-H. (1993), Island formation in submonolayer epitaxy, *Journal de Physique I*, **3**, pp. 935–950

# SURFACTANTS IN SEMICONDUCTOR HETEROEPITAXY: THERMODYNAMICS AND/OR KINETICS?

IVAN MARKOV
*Institute of Physical Chemistry*
*Bulgarian Academy of Sciences*
*1113 Sofia, Bulgaria*

**Abstract.** The effect of surfactants on the thermodynamics and kinetics of semiconductor heteroepitaxy is briefly discussed. It is argued that the way the surfactants suppress the thermodynamic driving force for 3D islanding depends on the mechanism of exchange of overlayer and surfactant atoms. If the overlayer atoms occupy bulk-like positions provided by the outgoing surfactant dimers, as is the case of Ge/Si(001), large atomic displacements are forcibly inhibited, and the wetting of the overlayer by the substrate becomes nearly complete. This complete wetting of the overlayer by the substrate means a zero thermodynamic driving force for 3D islanding. Once the thermodynamics do not require 3D islanding a change of the growth mode with the temperature should not be observed. Thus, the temperature change of the growth mode appears as an indication for the primary role of the kinetics, as in the case of Ge/Si(111).

## 1. Introduction

The presence of surface active species (surfactants) on the surface of growing crystals changes dramatically the mode of growth, the rate of nucleation, the transition from step flow to two-dimensional (2D) nucleation, the structure of steps, the distribution of stress, surface alloying, the concentration of defects, etc.[1] The major effect is the suppression of the three-dimensional (3D) islanding so that smooth films suitable for fabrication of microelectronic devices could be produced. Although the initial attempts to understand the effect of the surfactants on the mode of growth were based on thermodynamic grounds[2], there is still a debate in the literature between the adherents of the thermodynamic and the kinetic views. This is why we will make an attempt to consider both the thermodynamic and the kinetic aspects of the effect of surfactants on the thin film growth mode. We will do that in the most general way without accounting for the particular properties of the materials. We note only that in the case of semiconductor growth a complete monolayer of the surfactant is needed[3].



*M.C. Tringides and Z. Chvoj (eds.),*
*Collective Diffusion on Surfaces: Correlation Effects and Adatom Interactions*, 259–271.

## 2. Thermodynamics

If we want to change the direction of a process, we have to change the sign of the thermodynamic force that drives it. If we want to prevent a particular process from taking place we must suppress the corresponding thermodynamic driving force making it as close to zero as possible. The thermodynamic driving force which determines the occurrence of one or another mechanism of epitaxial growth is the difference $\Delta\mu = \mu(n) - \mu_{3D}^0$ of the chemical potential, $\mu(n)$, of the overlayer which depends on the film thickness measured in number $n$ of monolayers counted from the interface, and the chemical potential, $\mu_{3D}^0$, of the bulk 3D crystal[4, 5]. The thickness dependence of the film chemical potential $\mu(n)$ originates mostly from the thickness distribution of the strain due to the lattice misfit, but the interaction between the deposit and the substrate, which rapidly decreases with the distance from the interface and could be neglected beyond several monolayers, should be also accounted for[4, 5]. If we deposit a crystal $A$ on the surface of a crystal $B$ in the absence of a surfactant, the thermodynamic driving force can be written in terms of surface energies $\sigma_A$ and $\sigma_B$[6],

$$\Delta\mu = a^2[\sigma_A + \sigma_{AB}(n) - \sigma_B], \tag{1}$$

where $a^2$ is the area occupied by an atom, and $\sigma_{AB}(n)$ is the interfacial energy which includes in itself the misfit strain energy, and the attenuation of the energetic influence of the substrate[6]. In fact, this is the familiar 3-$\sigma$ criterion of Bauer[7].

We can write the above expression in terms of interatomic energies[8]

$$\mu(n) = \mu_{3D}^0 + (E_{AA} - E_{AB}) = \mu_{3D}^0 + E_{AA}\Phi, \tag{2}$$

where $E_{AA}$ and $E_{AB}$ are the energies per atom to disjoin a half-crystal $A$ from a like half-crystal $A$ and an unlike half-crystal $B$. It is in fact the adhesion energy $E_{AB}$ which includes in itself the thickness distribution of the strain energy. In the above equation, $\Phi = 1 - E_{AB}/E_{AA}$ is the so-called adhesion parameter which accounts for the wetting of the overgrowth by the substrate. This is the same parameter which enters the work of formation of nuclei on a foreign substrate in the classical nucleation theory.

Equations (1) and (2) are identical. Equation (1) can be readily obtained from (2) by using the definition of surface free energy and the relation of Dupré[6]. The essential physics, however, is in Eq. (2) as the surface energies are derivatives of the interatomic forces. It shows that *the thermodynamic driving force for occurrence of one or another mode of growth is the relative adhesion/cohesion difference, or in other words, the wetting.* The incomplete wetting, $0 < \Phi < 1$, is the driving force for 3D nucleation and growth, whereas the complete wetting, $\Phi \leq 0$, is the driving force for planar growth by nucleation and growth of 2D monolayer height nuclei[8].

When $0 < \Phi < 1$, $\mu(n) > \mu_{3D}^0$, 3D islands are formed on top of the substrate surface from the very beginning of deposition. This is the well-known Volmer-Weber (VW) growth and $\mu(n)$ goes asymptotically to $\mu_{3D}^0$ from above (the curve denoted by VW in Fig. 1)[4-6].

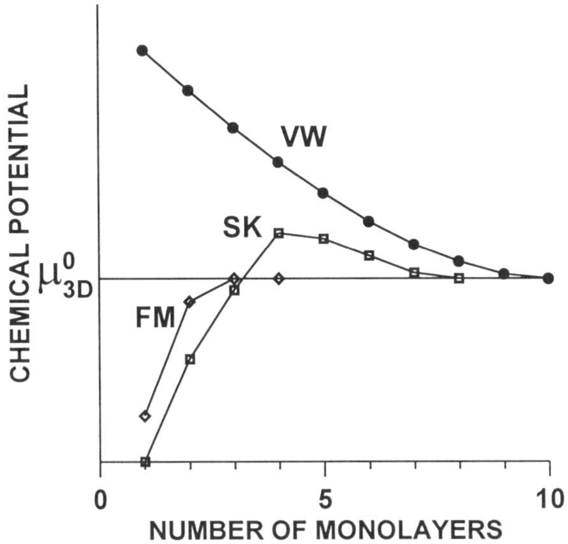


*Figure 1.* Schematic dependence of the film chemical potential on the film thickness in number of monolayers for the three modes of growth denoted at each curve: VW - Volmer-Weber, SK - Stranski-Krastanov, and FM - Frank-van der Merwe.

In the other extreme of complete wetting ($\Phi \leq 0$) either layer-by-layer (Frank-van der Merwe or FM) growth at negligible lattice misfit, or a Stranski-Krastanov (SK) growth (layer-by-layer growth followed by 3D islands) at a perceptible misfit, are thermodynamically favored[9, 10]. In the case of FM growth $\mu(n)$ goes asymptotically to $\mu^0_{3D}$ from below (the curve denoted by FM in Fig. 1). The first monolayer has the lowest chemical potential owing to the strongest interaction with the substrate, and a second monolayer can form only after the completion of the first one.

The SK growth mode appears as a result of the interplay of the film - substrate bonding, strain, and surface energies. A wetting layer consisting of an integer number of monolayers is first formed by a FM mode driven by a negative thermodynamic driving force $\Delta\mu$ (complete wetting). Further FM growth becomes unfavorable as the strain energy accumulates linearly with film thickness. In addition, the stronger attraction from the substrate, that overcompensates the strain energy, disappears beyond several atomic diameters. On top of the wetting layer 3D islands form and grow under the influence of a positive thermodynamic driving force (incomplete wetting). If misfit dislocations (MDs) are introduced immediately after the completion of the wetting layer, unstrained 3D islands are formed and grow on top of it. This is the classical Stranski-Krastanov growth in which the work needed to create new surfaces is overcompensated by a complete strain relaxation. It was found recently that initially coherent (dislocation-free) 3D islands are formed (coherent Stranski-Krastanov growth)[11-13], in which the formation of a new surface is overcompensated by a gradual strain relaxation. In both cases, the 3D islands and the wetting layer represent necessarily different phases separated by an interphase boundary. This boundary is determined by the displacements of the atoms belonging to the first atomic plane of the islands from the bottoms of the corresponding potential troughs of the uniformly strained wetting layer. In the classical SK growth, the displacements are described in terms of MDs[14], whereas in the coherent SK growth the atoms closer to the islands edges are displaced owing to the islands finite size[15]. These displacements give rise to weaker adhesion, or, in other words, to incomplete wetting. It is namely this incomplete wetting (or weaker adhesion) which appears as the thermodynamic driving force for the 3D islanding in the SK mode of growth. In other words, we can treat the SK mode as a FM mode driven by complete wetting ($\Delta\mu < 0$), followed by VW mode driven by incomplete wetting ($\Delta\mu > 0$) (the curve denoted by SK in Fig. 1).

When considering the effect of the surfactant on the mode of growth, we note that the chemical potential $\mu_{3D}^0$ of the infinitely large crystal does not depend on whether there are impurities adsorbed on its surface or not[16, 17]. This, however, is not true for sufficiently thin films. If a monolayer thick film of $A$ is covered by a monolayer of a surfactant (S) atoms, the growth mode criterion (1) turns into[18],

$$\Delta\mu_S = a^2(\sigma_{AB} + \sigma_{SA} - \sigma_{SB}), \tag{3}$$

where the $\sigma$'s are the corresponding interfacial energies. This expression is equivalent to

$$\mu_S(n) = \mu_{3D}^0 + (E_{AA} - E_{AB}) - (E_{SA} - E_{SB}), \tag{4}$$

where $\mu_S(n)$ is the chemical potential of the film covered by S atoms. It is immediately seen that if 3D islanding is thermodynamically favored in absence of a surfactant, i.e., $E_{AA} > E_{AB}$, the inequality (4) may have the opposite sign if the adhesion of the surfactant is stronger to the overlayer rather than to the substrate and thus the third term in (4) overcompensates the second one. Then the surfactant is "good" from a thermodynamic point of view if it adheres more strongly to the overlayer, thus changing the sign of the thermodynamic driving force.

This is valid, however, only for the transition from VW to FM growth as is the case of the As mediated growth of Si on Ge(001)[2, 19]. Even in this case it is valid only at zero misfit. After several monolayers of $A$, the energetic contact between the surfactant and the substrate $B$ is completely lost ($E_{SB} \rightarrow E_{SA}$), and if the misfit is large, 3D islands should appear resulting in SK mode. In systems that follow the SK growth mode where, beyond the wetting layer, $A$ is deposited on strained $A$[20], the difference $E_{SA} - E_{SB}$ will be nearly equal to zero. Many papers both theoretical[21-24] and experimental[25-30] have been devoted to the problem considering mainly the surfactant growth of Ge on Si(001). However, in most of the papers quoted above the mechanism of the exchange of the surfactant (As or Sb) atoms with Ge atoms belonging to the first monolayer on the Si(001) surface have been studied. It is obvious, that such studies cannot explain the effect of the surfactant to suppress the 3D islanding as the first monolayer belongs to the wetting layer and it will grow in a FM mode pseudomorphous with the substrate irrespective of whether a surfactant on top exists or not.

Obviously, in the SK mode the surfactant should make the thermodynamic driving force for 3D islanding on top of the wetting layer equal to zero for considerable film thickness. As mentioned above in the SK mode the atoms in contact with the wetting layer should be displaced from their "bulk-like" positions either to make MDs or near the islands' edges. These displacements make the wetting incomplete, or the same, make the adhesion weaker than the cohesion ($E_{AB} < E_{AA}$), which is the thermodynamic driving force for 3D islanding. As found in a series of papers the surfactant group-V atoms form dimers on Si(Ge)(001) which reside exactly where the Si(Ge) dimers should be located[25,27,31-33]. After the surfactant and overlayer dimers exchange places, the latter occupy the epitaxial sites provided by the outgoing surfactant. The surfactant dimers on top do not permit large displacements of the overlayer atoms underneath, and, in turn, incomplete wetting. In other words, the surfactant atoms press the overlayer atoms into bulk-like positions and do not allow the formation of misfit dislocations. Thus, *the surfactants suppress the thermodynamic driving force for 3D islanding by not allowing large atomic displacements.* Once the thermodynamics do not require 3D islanding, a change of the growth mode with the temperature should not be observed.

## 3. Kinetics

In the absence of a surfactant, the atoms arriving at the crystal surface diffuse on it, join pre-existing steps or islands, or give rise to new islands. At sufficiently high temperatures the diffusivity of the atoms on the terraces is high and they reach the pre-existing steps before meeting with each other. This results in a step-flow growth. At low temperatures and, in turn, low diffusivity the atoms meet with each other before reaching the steps and give rise to 2D nuclei on the terraces. In the presence of a complete monolayer of S atoms, two new phenomena take place. First, the surfactant changes the energetics of incorporation of atoms into ascending and descending steps, and second, the overlayer atoms must exchange places with the S atoms on the terraces or at the steps in order to join the crystal lattice. Both phenomena lead to a change of the diffusivity of the adatoms across the steps and on terraces and, in turn, exert an effect on the direction of transport (upward or downward) of atoms. Moreover, the change of diffusivity on terraces affects strongly the kinetics of nucleation and growth of the islands, thus stimulating either the step flow or 2D nucleation. We consider in more detail the effect of the surfactant on the kinetics of attachment and detachment of atoms to and from the steps, and the exchange-deexchange kinetics on terraces.

Whereas an atom approaching an ascending step joins it upon striking, an atom approaching a descending step must overcome an additional Ehrlich-Schwoebel (ES) barrier[34, 35]. If the ES barrier is low, the overlayer grows more or less in a layer-by-layer mode. Otherwise, instabilities of different kind appear[36]. Zhang and Lagally pointed out the possibility of a reduction of the ES barrier when overlayer atoms exchange sites with S atoms that decorate the steps in such a way that the last barrier before descending is shifted downwards below the level of the surface diffusion barrier[37]. The latter is in good agreement with first principle molecular dynamics calculations of displacement of Sb dimer at the step edge by a Si dimer through a push-out mechanism[38]. In addition, Markov has suggested that atoms that approach ascending steps must overcome an extra energetic barrier owing to the necessity to displace the S atoms already adsorbed at the step edges[39]. These effects, of purely kinetic origin, can suppress the thermodynamic driving force for 3D islanding in heteroepitaxial growth.

When thermodynamics predict 3D islanding, ($\Delta\mu > 0$) the atoms are more strongly bound to the upper layers rather than to the lower layers. In other words, the upper layers have lower chemical potentials (Fig. 1)[4, 5]. This results in a gradient of the chemical potential that drives the atoms upwards and is the driving force for the 2D-3D transformation[4-6]. The appearance of an additional barrier at the ascending steps and the reduction of the ES barrier at the descending steps, owing to the decoration of the steps by S atoms, reverse the asymmetry of the attachment/detachment kinetics and thus leads to a suppression of the thermodynamic driving force for 2D-3D transition. A reverse gradient of adatoms of a *kinetic origin* appears that drives the atoms downward. The latter results in a planar rather than a 3D growth[39, 40]. Thus the attachment/detachment kinetics suppress the thermodynamic driving force for 3D islanding. At high temperatures the thermodynamics prevail and SK growth takes place irrespective of the presence of the surfactant. Surfactant-induced FM growth occurs at lower temperatures. Note this kinetic effect does not require a complete monolayer of the surfactant, but a small amount which is sufficient to decorate the steps.

In the presence of a complete monolayer of the surfactant, a new kinetic effect appears owing to the necessity of the overlayer atoms to join the crystal lattice and of the S atoms to float over the surface. These are the phenomena of exchange and de-exchange of overlayer and S atoms. Zhang and Lagally[37] first assumed that an exchange process between overlayer and S atoms on the terraces should take place on the S precovered surfaces. Kandel and Kaxiras assumed later that a de-exchange process is also possible and can play a significant role in epitaxial growth[3, 47]. They calculated the values of 0.8 and 1.6 eV for the activation energies, $E_{ex}$ and $E_{dex}$, for exchange and de-exchange processes, respectively. Ko, Chang, and Yi computed through first-principles pseudopotential total-energy calculations the values of the barriers for surface diffusion, exchange and de-exchange for Ge on Ga, As and Sb precovered Si(111) surface. They found that in all cases, $E_{dex} > E_{ex}$, but the differences are not as large as the ones calculated by Kandel and Kaxiras[48].

We introduce two time constants which characterize the exchange and the de-exchange processes both being normalized to the time, $t_1 = 1/F$, necessary for deposition of a complete monolayer, $F$ being the the atom arrival frequency[49, 50]. First, this is the mean residence time of the atoms on top of the S layer before exchange $\tau_{ex} = (F/\nu)\exp(E_{ex}/kT)$, where $\nu$ is the attempt frequency. Second, this is the mean residence time before de-exchange of an atom embedded in (or under) the S layer $\tau_{dex} = (F/\nu)\exp(E_{dex}/kT)$. Obviously, if a particular time constant is greater than unity, the corresponding process will not occur. As $\tau_{dex} \gg \tau_{ex}$, three possibilities exist. The first is $\tau_{dex} \ll 1$. This extreme describes the *reversible exchange* as the adatoms have sufficient time to go back on top of the S layer through a de-exchange process. A dynamic *exchange/deexchange equilibrium* is established. With the value of $E_{dex} = 1.6$ eV calculated by Kandel and Kaxiras, we find, at 600K, $\tau_{dex} \approx 0.002$. Then the exchange/deexchange equilibrium is established in the very beginning of the deposition process[49, 50]. The nucleation process takes place in a more or less disordered (because of the different size of the atoms involved) 2D phase – a monolayer consisting of mixed S and overlayer atoms. A considerable fraction of the incoming atoms remains on top of the S layer and diffuse fast to the pre-existing steps. The formation, growth, and decay of nuclei take place through exchange and de-exchange processes. The second extreme is $\tau_{ex} \ll 1 < \tau_{dex}$. This is the case of *irreversible exchange*. The incoming atoms rapidly exchange places with S atoms and remain buried under the S layer. All processes of diffusion, nucleation, and incorporation into islands and steps occur under (or between) the S atoms. The case $1 < \tau_{ex} \ll \tau_{dex}$ should be excluded as it means that the S atoms will remain buried under the arriving overlayer atoms.

We calculate further the nucleus density, $N_S$, in both cases of reversible and irreversible exchange. This is extremely important for two reasons. First, the large number of small islands will promote layer-by-layer growth owing to the existence of a critical island size for second layer nucleation[52-55]. Second, a higher nucleation rate leads to growth by 2D nuclei, whereas in the opposite case step-flow growth will take place. In all cases the expression for $N_S$ can be written in the form

$$N_S = N_0 \exp\left(-\frac{\chi}{i} E_S/kT\right), \tag{5}$$

where $E_S$ combines all energy contributions which depend on the presence of the surfactant[49, 50], and

$$N_0 = \left(\frac{\nu}{F}\right)^{-\chi} \exp\left(\frac{\chi}{i} E_0/kT\right) \tag{6}$$

is the nucleus density in the absence of a surfactant ($E_S = 0$)[49, 50]. Thus, the island density scales with the ratio $\nu/F$, the scaling exponent $\chi$ being a function of the number $i$ of atoms in the critical nucleus[41, 42]. In the case of low barriers for attachment of atoms to 2D nuclei, the latter "feel" each other through the diffusion fields around them, and[41, 42]

$$\chi = \frac{i}{i+2}. \tag{7}$$

Kandel has recently shown, by using a rate equation approach, that in the presence of surfactants which give rise to barriers for incorporation of atoms to the critical nuclei[43],

$$\chi = \frac{2i}{i+3}. \tag{8}$$

The scaling exponent (8) has been independently derived by Markov by using a different approach[44]. He showed that it is characteristic for a kinetic regime of growth where the adatom concentration between the islands is practically constant and the islands "do not feel" the presence of each other.

As has been noted in Ref. [43], the scaling exponent (7) varies with $i$ from 1/3 to 1, whereas (8) has values larger than unity already at $i > 2$. Thus, one could distinguish between diffusion and kinetic regimes of growth if $\chi$ is smaller or greater than unity. Typical examples of the scaling exponent (8) are the homoepitaxy of Si on Sn precovered surface of Si(111)[45], and of Ge on Pb precovered surface of Si(111)[46].

It is immediately seen that the exponential multiplying $N_0$ in Eq. (5) can be smaller or greater than unity depending on the interplay of the energies involved in $E_S$. If $E_S < 0$ the surfactant will stimulate the nucleus formation ($N_S \gg N_0$). The latter means that a greater density of smaller islands will be formed, and the film will grow by 2D nucleation rather than by step-flow. What is more important is that the greater density of smaller nuclei leads to layer-by-layer growth rather than to the mound formation. In the opposite case the surfactant will inhibit the nucleation and will promote the step-flow growth. In the case of reversible exchange $E_S$ contains the term $-i[(E_{dex} - E_{ex}) - \Delta E_{sd}]$ where $\Delta E_{sd} = E^{\circ}_{sd} - E_{sd}$ is the the difference between the barriers for diffusion on the clean substrate, $E^{\circ}_{sd}$, and on the surfactant passivated surface, $E_{sd}$, ($E^{\circ}_{sd} > E_{sd}$). If $\Delta E_{sd}$ overcompensates the difference $E_{dex} - E_{ex}$, we could expect that $E_S > 0$, the nucleus formation will be inhibited, ($N_S \ll N_0$), and surfactant-induced step-flow growth will take place[49, 50].

In the case of irreversible exchange Eq. (5) is still valid but the barriers $E_{dex}$ and $E_{ex}$ do not enter the energy term $E_S$. The largest term in $E_S$ is the increment of the surface diffusion barrier $-i(E_{sd} - E^{\circ}_{sd})$ where now $E_{sd} > E^{\circ}_{sd}$ owing to the fact that the process of diffusion under the S layer is inhibited compared with that on the clean surface. It is unlikely this term to be overcompensated by the smaller positive terms, and hence, we can always expect that the surfactant will stimulate the 2D nucleation rather than the step-flow growth.

Finally, we discuss briefly the reasons that lead to a change of the island density. Usually the latter is attributed to the change of atoms diffusivity. However, the change the island density might be due to other reasons. The nucleation rate depends on the Gibbs free energy for nucleus formation which is in turn a function of the nucleus size. The presence of the surfactant could lead to a change of the number of atoms in the critical nucleus[51]. As a result the values of $E_S$ and $E_0$ in Eqs. (5) and (6), as well as the scaling exponent $\chi$ will change in one and the same temperature interval. In addition, the regime of growth can change from diffusion to kinetic, so that the scaling exponent (7) will be replaced by (8). In this case, the island density will increase by orders of magnitude even if $i$ remains constant.

## 4. Growth of Ge on Si

We discuss in more detail a "model" system such as Ge/Si. As pointed out by Kaxiras[57], one and the same surfactant can have qualitatively different effect depending on the crystallographic orientation of the substrate. On Si or Ge(001) the surfactants form dimers which do not break during the exchange process[21-27]. On (111) surface, three different geometries (substitutional, trimers, and zigzag chains) of the group-V adsorbates (P, As, Sb) are possible[57]. In the first $1 \times 1$ structure, one monolayer of S atoms is incorporated into the upper part of the (111) bilayer, whereas in the last two structures a monolayer of S atoms is bonded on top of a complete (111) bilayer. In addition, the growth of the two surfaces of Si or Ge differs drastically. Whereas the (001) surface grows by single monolayers[58], the (111) surfaces grow by bilayers[59]. Thus, we have completely different mechanisms of exchange of S and overlayer atoms. In the case of (001) surfaces, the growth unit is a dimer so that an exchange of S and overlayer dimers takes place through a single monolayer[21-27]. On (111) surfaces, the mechanism of exchange depends on the surface geometry of the surfactant. For example, if the surfactant geometry is substitutional, we could suppose that single overlayer atoms should displace single S atoms from their positions in the upper monolayer that constitutes the bilayer. After the complete building of the upper monolayer, atoms belonging to the next lower monolayer should be incorporated first in order to serve as a template for the S atoms in the next upper monolayer. Thus, one surfactant monolayer should be displaced by a complete bilayer of the overgrowth, the process taking place by exchange of single atoms. Comparing both surfaces we could speculate that on a (111) surface, the surfactant could not suppress successfully the thermodynamic driving force for 3D islanding by not allowing large atomic displacements (formation of misfit dislocations).

We consider first the growth of Ge on Si(111). Hwang, Chang and Tsong studied this system by using Pb as a surfactant[46]. They established that the Pb atoms make a substitutional $1 \times 1$ structure on top of the Ge surface. It was found that the saturation nucleus density in the submonolayer regime scales with the deposition flux with $\chi = 1.76$. This clearly shows that in the Pb mediated epitaxy of Ge on Si(111), the scaling exponent is given by Eq. (8) and the growth proceeds in a kinetic regime. In other words, significant energy barriers appear for the attachment of atoms to the critical nuclei owing to the necessity of exchange of overlayer and S atoms. The same scaling exponent ($\chi = 1.8$) has been obtained in the homoepitaxial growth of Si on Sn precovered surface of Si(111)[45], after comparing the experimental data for the critical terrace width for step flow growth with the corresponding theory[49, 50]. Under clean conditions in the same system, Voigtländer et al. obtained a scaling exponent smaller than unity ($\chi = 0.85$)[60].

The observations of Hwang et al. show unambiguously that Pb has a prominent kinetic effect on the submonolayer growth of Ge on Si(111)[46]. In a further study, the same authors found that 3D islanding is suppressed below 470K. At temperatures higher than 470K, 3D islanding takes place[61]. This observation could be explained by the suppression of the thermodynamic driving force for 3D islanding by the attachment/detachment kinetics mentioned above. The upward surface flux is hindered, whereas the downward flux is enhanced. We could conclude that the suppression of 3D islanding in Ge growth on Pb precovered Si(111) is due more to kinetic rather than to thermodynamic reasons.

Voigtländer and Zinner studied the growth of Ge on Si(111) mediated by Sb[62]. They established the same substitutional structure of the Sb atoms on the surface of Ge. They found that a transition from surfactant mediated layer-by-layer growth to the equilibrium SK mode takes place above twice the higher temperatures (900K) as compared with the Pb mediated growth. The higher temperatures could be related to stronger chemical bonding but qualitatively the surfactant effect is the same.

The system Ge/Si(001) is the most studied one[3]. Ide has observed that after annealing at about 800K for 10 min, the 2D Ge islands formed in submonolayer regime on As precovered Si(001) did not change, whereas the islands deposited under clean conditions disappeared[30]. This observation shows that 2D islands formed under the surfactant layer are more stable than those without S atoms on top of them. Antimony mediated deposition and annealing experiments carried out by Osten et al. showed that Sb prevents 3D islanding, but also can smooth an already islanded film[29]. Katayama et al.[63] found that as the coverage of the surfactant increases, the intermixing of Ge and Si, as well as the nucleation and the growth of macroscopic Ge islands, are suppressed. Both observations could be explained with the forced occupation of epitaxial sites by the Ge atoms under the influence of the S dimers on top of them. On the other hand, Ide also established that at 800K the Ge atoms reach and join the preexisting steps of Si (step flow growth) on the clean surface, whereas 2D islands are formed in the presence of As[30]. The latter unambiguously demonstrates the kinetic effect of the surfactant on the mode of growth by decreasing the diffusivity of the incoming atoms.

Most of the studies of surfactant mediated growth of Ge on Si(001) surface are carried out at or near to 500°C[2, 25, 26, 28, 30, 64, 65]. To the author's knowledge, higher temperature deposition has been carried out by Tromp and Reuter (630°C)[21], and by Horn von Hoegen et al.[66] at 700°C. No 3D islanding except for mound formation has been noticed on top of the wetting layer at the highest temperature used[66]. A change of the growth mode with temperature as in the case of Ge/Si(111) has not been reported so far[67]. This can be interpreted as follows. The surfactant suppresses successfully the thermodynamic driving force for 3D islanding by not allowing large atomic displacements and the growth continues in a layer-by-layer mode (with the unavoidable roughness to relieve the strain) until MDs are introduced at the interface. Once the thermodynamics do not require 3D islanding, a change of the growth mode with the temperature should not be observed.

Jenkins and Srivastava carried out first principle density functional theory (DFT) calculations of the growth mode of Ge on Sb precovered Si(001)[24]. They studied the behavior of the chemical potential of the Ge film relative to the bulk chemical potential, $\mu_{3D}^0$, in the absence and presence of Sb as a surfactant. They found that in the absence of Sb, the chemical potentials of the first three monolayers of Ge, which belong to the wetting layer, are smaller (more negative) than $\mu_{3D}^0$ (complete wetting). The chemical potential of the forth monolayer becomes more positive than $\mu_{3D}^0$ (incomplete wetting), which is an indication of 3D islanding after the first three monolayers. In the presence of Sb the chemical potential of the first monolayer is slightly more negative than $\mu_{3D}^0$, but the chemical potentials of the remaining three monolayers are equal to $\mu_{3D}^0$ within the accuracy of the calculations. An error is introduced in the calculations by not allowing the Ge overlayers to relax laterally. Thus, after the exchange of Ge and Sb dimers, the Ge atoms occupy the bulk-like positions left by the outgoing Sb dimers and the surfactant successfully suppressed the thermodynamic driving force for 3D islanding. In fact, the surfactant replaced the curve SK by the curve FM in Fig. (1).

Similar first principle DFT calculations have been carried out by González-Méndez and Takeuchi for the growth of Si on Ge(001) by using As as a surfactant[19]. In the absence of a surfactant, Si grows in the VW mode as 3D islands directly on the Ge surface[68]. In the presence of As, the overlayer prefers to grow in a layer-by-layer mode due to thermodynamic reasons, the energy of the 3D islands being always lower than that of the complete monolayers. The deposit atoms are almost in bulk-like positions. Equally important, the authors established that As greatly reduces the Si and Ge intermixing. This is another indication of the thermodynamic influence of the surfactant as the alloying also requires large atomic displacements.

## 5. Conclusions

The presence of a third element in the system unavoidably changes the kinetics of growth. Hence, we must decide whether the kinetics is solely responsible for the suppression of the 3D islanding, or the thermodynamics play a decisive role, with the kinetics exerting an additional effect. We argue that this question could be answered by studying the temperature dependence of the capability of the surfactant to suppress 3D islanding. If the kinetics play a decisive role, we should expect a change of the mode of growth from layer-by-layer at lower temperature to 3D (SK or VW) islands at higher temperature. The physical reason is that at high temperatures the system is closer to equilibrium whereas at lower temperatures the kinetic effects suppress the thermodynamic driving force for 3D islanding. This is most probably the case of the system Ge/Si(111) where one monolayer of the surfactant should be displaced by two monolayers of the overgrowth and the exchange process takes place randomly by single atoms. The S atoms do not press the overlayer atoms into bulk-like positions and the formation of misfit dislocations or displacements of the edge atoms is not inhibited. We have seen that the growth is accompanied by typical kinetic phenomena as the kinetic regime of growth ($\chi > 1$). On the contrary, the system Ge/Si(001) demonstrates the decisive role of the thermodynamics. The kinetics play an auxiliary role in the same direction. The physical reason for suppression of the 3D islanding is that a monolayer of the surfactant is displaced by a monolayer of the film during growth, and the exchange process occurs by dimers rather than by single atoms. As a result the overlayer atoms occupy nearly bulk-like positions provided by the outgoing surfactant dimers. Large atomic displacements are forcibly prohibited, which, in turn, lead to complete wetting and a zero thermodynamic driving force for 3D islanding. It is thus clear that the third element in the system does not act as a surfactant at all.

We conclude that the relative weight of the effect of the third element on the thermodynamics and kinetics in semiconductor heteroepitaxy depend mostly on the mechanism of exchange of overlayer and surfactant atoms. Undoubtedly, the third element changes the surface energies but it is the change of the relative adhesion of the overlayer to the substrate that is crucial from a thermodynamic point of view.

The author is greatly indebted to M. Paunov for the fruitfull discussion and the critical reading of the manuscript.

## References


1. M. Zinke-Allmang, *Thin Solid Films* **346**, 1 (1999).
2. M. Copel, M. C. Reuter, E. Kaxiras, and R. M. Tromp, *Phys. Rev. Lett.* **63**, 632 (1989).
3. D. Kandel and E. Kaxiras, *Solid State Phys.* **54**, 219 (2000).
4. S. Stoyanov, *Surf. Sci.* **172**, 198 (1986).
5. I. Markov and S. Stoyanov, *Contemp. Phys.* **28**, 267 (1987).
6. I. Markov, *Crystal Growth for Beginners*, World Scientific, 1995, p. 389.
7. E. Bauer, *Z. Kristallographie* **110**, 372 (1958).
8. R. Kaischew, *Commun. Bulg. Acad. Sci.* (Phys.) **1**, 100 (1950).
9. R. Kern, G. LeLay, and J. J. Metois, in *Current Topics in Materials Science*, vol. 3, ed. by E. Kaldis, (North-Holland, 1979).
10. M. H. Grabow and G. H. Gilmer, *Surf. Sci.* **194**, 333 (1988).
11. D. J. Eaglesham and M. Cerullo, *Phys. Rev. Lett.* **64**, 1943 (1990).
12. D. Leonard, M. Krishnamurty, C. M. Reaves, S. P. Denbaars, and P. M. Petroff, *Appl. Phys. Lett.* **63**, 3203 (1993).
13. Vinh Le Thanh, P. Boucaud, D. Débarre, Y. Zheng, D. Bouchier, and J.-M. Lourtioz, *Phys. Rev. B* **58**, 13115 (1998).
14. J. W. Matthews, D. C. Jackson, and A. Chambers, *Thin Solid Films* **29**, 129 (1975).
15. E. Korutcheva, A. M. Turiel, and I. Markov, *Phys. Rev. B* **61**, 16890 (2000).
16. G. Bliznakov, *Commun. Bulg. Acad. Sci.* **3**, 23 (1953).
17. N. Pangarov, *Electrochim. Acta* **28**, 763 (1983).
18. M. Paunov, *Cryst. Res. Technol.* **33**, 165 (1998).
19. M. E. González-Méndez and N. Takeuchi, *Surf. Sci.* **441**, L897 (1999).
20. C. W. Snyder and B. G. Orr, *Phys. Rev. Lett.* **70**, 1030 (1993).
21. R. M. Tromp and M. C. Reuter, *Phys. Rev. Lett.* **68**, 954 (1992).
22. B. D. Yu and A. Oshiyama, *Phys. Rev. Lett.* **72**, 3190 (1994).
23. M. Jiang, X. Zhou, B. Li, and P. Cao, *Phys. Rev. B* **60**, 8171 (1999).
24. S. J. Jenkins and G. P. Srivastava, *Surf. Sci.* **398**, L308 (1998).
25. R. Cao, X. Yang, J. Terry, and P. Pianetta, *Appl. Phys. Lett.* **61**, 2347 (1992).
26. B. D. Yu, T. Ide, and A. Oshiyama, *Phys. Rev. B* **50**, 14631 (1994)
27. M. A. Boshart, A. A. Bailes III, and L. E. Seiberling, *Surf. Sci.* **348**, L75 (1996); *Phys. Rev. Lett.* **77**, 1087 (1996).
28. J. M. C. Thornton, A. A. Williams, J. E. Macdonald, R. G. van Silfhout, M. S. Finney, and C. Norris, *Surf. Sci.* **273**, 1 (1992).
29. H. J. Osten, J. Klatt, G. Lippert, E. Bugiel, and S. Hinrich, *Appl. Phys. Lett.* **60**, 2522 (1992); H. J. Osten et al. *Phys. Rev. Lett.* **69**, 450 (1992).
30. T. Ide, *Phys. Rev.* B **51**, 5397 (1995).
31. J. Nogami, A. A. Baski, and C. F. Quate, *Appl. Phys. Lett.* **58**, 475 (1991).
32. R. I. G. Uhrberg, R. D. Bringans, R. Z. Bachrach, and J. E. Northrup, *Phys. Rev. Lett.* **56**, 520 (1986).
33. R. S. Becker, T. Klitsner, and J. S. Vickers, *J. Microscopy* **152**, 157 (1988).
34. G. Ehrlich and F. G. Hudda, *J. Chem. Phys.* **44**, 1039 (1966).
35. R. L. Schwoebel and E. J. Shipsey, *J. Appl. Phys.* **37**, 3682 (1966).
36. P. Politi, G. Grenet, A. Marty, A. Ponchet, and J. Villain, *Phys. Rep.* **324**, 271

(2000).
37. Z. Zhang and M. Lagally, *Phys. Rev. Lett.* **72**, 693 (1994).
38. C. W. Oh, E. Kim, and Y. H. Lee, *Phys. Rev. Lett.* **76**, 776 (1996).
39. I. Markov, *Phys. Rev. B* **50**, 11 271 (1994).
40. I. Markov, *Mater. Chem. Phys.* **49**, 93 (1997).
41. J. A. Venables, G. D. T. Spiller, and M. Hanbücken, *Rep. Progr. Phys.* **47**, 399 (1984).
42. S. Stoyanov and D. Kashchiev, in *Current Topics in Materials Science*, edited by E. Kaldis, (North-Holland, 1981), Vol. 7, p. 69.
43. D. Kandel, *Phys. Rev. Lett.* **78**, 499 (1997).
44. I. Markov, *Phys. Rev. B* **56**, 12 544 (1997).
45. S. Iwanari and K. Takayanagi, *J. Cryst. Growth* **119**, 229 (1992).
46. I.-S. Hwang, T.-C. Chang, and T. T. Tsong, *Phys. Rev. Lett.* **80**, 4229 (1998).
47. D. Kandel and E. Kaxiras, *Phys. Rev. Lett.* **75**, 2742 (1995).
48. Young-Jo Ko, K. J. Chang, and Jae-Yel Yi, *Phys. Rev. B* **60**, 1777 (1999).
49. I. Markov, *Phys. Rev. B* **59**, 1689 (1999).
50. I. Markov, *Surf. Sci.* **429**, 102 (1999).
51. I. Markov, *Phys. Rev. B* **53**, 4148 (1996).
52. S. Stoyanov and I. Markov, *Surf. Sci.* **116**, 313 (1982).
53. J. Tersoff, A. W. Denier van der Gon, and R. M. Tromp, *Phys. Rev. Lett.* **72**, 266 (1994).
54. J. Krug, P. Politi, and T. Michely, *Phys. Rev.* B **61**, 14037 (2000).
55. S. Heinrichs, J. Rottler, and P. Maass, e-print cond/mat 0004031.
56. S. Stoyanov, *J. Cryst. Growth* **94**, 751 (1989).
57. E. Kaxiras, *Europhys. Lett.* **21**, 685 (1993).
58. J. E. Griffith and G. P. Kochanski, *Crit. Rev. Solid State Mater. Sci.* **16**, 255 (1990).
59. U. Köhler, J. E. Demuth, and R. J. Hamers, *J. Vac. Sci. Technol. A* **7**, 2860 (1989).
60. B. Voigtländer, A. Zinner, T. Weber, and H. Bonzel, *Phys. Rev. B* **51**, 7583 (1995).
61. I.-S. Hwang, T.-C. Chang, and T. T. Tsong, *Surf. Sci.* **410**, L741 (1998).
62. B. Voigtländer and A. Zinner, *J. Vac Sci. Technol. A* **12**, 1932 (1994).
63. M. Katayama, T. Nakayama, M. Aono, and C. F. McConville, *Phys. Rev. B* **54**, 8600 (1996).
64. M. Copel, M. C. Reuter, M. Horn von Hoegen, and R. Tromp, *Phys. Rev. B* **42**, 11 682 (1990).
65. F. K. LeGoues, M. Copel, and R. M. Tromp, *Phys. Rev. Lett.* **63**, 1826 (1989).
66. M. Horn-von Hoegen, A. Al-Falou, B. H. Müller, U. Köhler, L. Andersohn, B. Dahlmeier, and M. Henzler, *Phys. Rev. B* **49**, 2637 (1994).
67. M. Horn-von Hoegen, private communication.
68. P. M. Marée, K. Nakagawa, F. M. Mulders, J. F. van der Veen, and K. L. Kavanagh, *Surf. Sci.* **191**, 305 (1987).

# NONLINEAR DIFFUSION AND SLIDING FRICTION

S.C. YING
*Department of Physics, Brown University,*
*Providence, Rhode Island 02912*

E. GRANATO
*Laboratório Associado de Sensores e Materiais, Instituto Nacional de Pesquisas Espaciais,*
*12201-190 São José dos Campos, São Paulo, Brazil*

AND

T. ALA-NISSILA
*Helsinki Institute of Physics and Laboratory of Physics,*
*Helsinki University of Technology, P.O. Box 1100,*
*FIN-02015 HUT, Espoo, Finland*

**Abstract.** In this paper, we discuss the relation between the collective diffusion constant $D_c$ and the sliding friction $\bar{\eta}$, two transport coefficients that are of central importance in different areas of surface science. In the linear regime, they are very simply related. However, in the nonlinear regime, the sliding friction is a more useful entity as it measures the dynamic response of an adsorbed layer under a large external driving force. Analytic considerations and numerical simulations are applied to study the nonlinear sliding friction for two distinct systems: boundary lubrication layers and flux vortex lattice in type II superconductors. The stick-slip behavior in macroscopic sliding motion and the peak effect in type II superconductors are explained in terms of dynamical phases and the transitions between them.

## 1. Relation Between Diffusion and Sliding Friction

The focus of this workshop is the collective diffusion constant $D_c$ which is a transport coefficient introduced through the Fick's law in the current response of a system to a density gradient that takes the form



*M.C. Tringides and Z. Chvoj (eds.),*
*Collective Diffusion on Surfaces: Correlation Effects and Adatom Interactions,* 273–283.

$$\mathbf{J} = -D_c \nabla n \quad . \tag{1}$$

For simplicity of discussion, we have assumed an isotropic system so that $D_c$ is just a scalar. It is important to note that for the Fick's law to be valid, one has to be in the linear regime of vanishingly small density gradient.

In recent years, the dynamical response of a system to an external driving force has been the focus of many studies in connection with the problem of boundary lubrication as well as the conductance of type II superconductors [1, 2]. For these systems, the dynamical response is commonly analyzed in term of the sliding friction $\bar{\eta}$. This is defined through the relation

$$\langle \mathbf{v} \rangle = \frac{1}{m\bar{\eta}} \mathbf{F} \quad . \tag{2}$$

Thus $\frac{1}{m\bar{\eta}}$ is the ratio of the induced drift velocity to a driving force F, and m is the mass. It is important to note that the sliding friction is a macroscopic transport coefficient and is in general different from the microscopic friction that enters the Langevin description of the stochastic dynamics of the system [3]. We can now easily relate the sliding friction $\bar{\eta}$ to the collective diffusion constant $D_c$ defined in Eqs. (1) and (2) in the linear regime. Consider a closed system with the driving force $\mathbf{F}$ arising from an external potential $\phi$. At equilibrium, the current resulting from this external driving force is exactly balanced by the current generated by the density gradient in the opposite direction. Thus from Eq.(1) and Eq.(2) the following relation holds

$$D_c = -\frac{n}{m\bar{\eta}} \frac{\partial \phi}{\partial n} \quad . \tag{3}$$

Here $n$ is the number density. The quantity $\frac{\partial \phi}{\partial n}$ can be related to the isothermal compressibility $\kappa$ through the well know relation [3],

$$\frac{\partial \phi}{\partial n} = -\frac{1}{n^2 \kappa} \quad . \tag{4}$$

The relation between $D_c$ and $\bar{\eta}$ as shown in Eq.(3) has also been derived by Persson [1] in a slightly different form. It follows from general theory of linear response [3] that the sliding friction defined in Eq.(2) can be expressed in terms of equilibrium time dependent correlation function as

$$\frac{1}{m\bar{\eta}} = \frac{1}{dNk_BT} \int_0^\infty dt \langle \mathbf{V}_T(t) \cdot \mathbf{V}_T(0) \rangle \quad . \tag{5}$$

Here $d$ is the dimension,$N$the total number of particles and $\mathbf{V} = \sum_{i=1}^{N} \mathbf{v}_i$ is the total velocity. Combining the results of Eq.(3) and Eq.(5) the leads to the well know expression [4] for the collective diffusion $D_c$ as

$$D_c = \frac{1}{dNk_BTn\kappa} \int_0^\infty dt \langle \mathbf{V}_T(t) \cdot \mathbf{V}_T(0) \rangle \quad . \tag{6}$$

Thus the sliding friction is related to the so-called *mobility* factor of the diffusion constant $D_c$ without the *thermodynamic factor* which depends only on equilibrium properties.

In the nonlinear regime where there exist large density gradients and the system is strongly inhomogeneous, the diffusion constant introduced through the Fick's law is not of much use and at best one can only hope to describe a local region by a density dependent $D_c$. However, in the context of a system subject to a large external driving force $F$, the sliding friction defined through Eq.(2) remains a perfectly useful concept. For an open system or a system with a source and sink, the driven system under large external force $F$ can be in a steady state with a constant drift velocity and uniform density. The corresponding sliding friction $\bar{\eta}$ would of course depend on the value of the force $F$ itself in the nonlinear regime. Beyond a critical force $F_c$, the nature of the diffusion changes from activated form with a finite diffusion barrier to a nonactivated behavior with the drift velocity $\mathbf{v}_d$ rising rapidly in proportion to the driving force. The whole subject of the dynamical response of an adsorbate layer under a driving force has received much attention recently. It is relevant for such diverse areas as boundary lubrication and vortex lattice dynamics in Type II superconductors. The critical force basically determines the static friction between two macroscopic blocks with a boundary layer of lubricant in the microscopic contact areas in the case of boundary lubrication and the magnitude of the critical current for the type II superconductors. In the steady drifting state, the nonlinear nature of the problem gives rise to a rich variety of dynamical phases and the transitions between them lead to macroscopic observable behavior such as stick-slip motion.

## 2. Boundary Lubrication and Sliding Friction

In the boundary lubrication problem for interfaces [1], the macroscopic sliding friction is controlled by the response of the tightly bound lubricant layer between the surfaces to the external driving force. It has been shown that there are many dynamical phases, and the transitions between these phases can lead to such macroscopic behavior as the stick and slip motion [5, 6]. The essence of the physics in the boundary lubrication has been reasonably well understood with the study of a zeroth order model consisting in an overlayer of interacting atoms adsorbed in a periodic potential subject to an external driving force.

Consider a model of interacting adatoms in a periodic potential subject to an external driving force [1, 5]. The dynamics is described by the Langevin equation

$$m\ddot{\mathbf{r}}_i + m\eta\dot{\mathbf{r}}_i = -\frac{\partial U}{\partial \mathbf{r}_i} - \frac{\partial V}{\partial \mathbf{r}_i} + \mathbf{f}_i + \mathbf{F} \quad , \tag{7}$$

where $\mathbf{r_i}$ is the adsorbate position, $U = \sum_i u(r_i)$ is the periodic substrate pinning potential, $V = \sum_{i \neq j} v(|\mathbf{r_i} - \mathbf{r_j})$ is the interaction potential between adatoms, $\mathbf{F}$ is the uniform external force acting on each adatom and $\mathbf{f}_i$ is an uncorrelated stochastic force, with zero average, and variance related to the microscopic friction parameter $\eta$, through the fluctuation dissipation relation $< f_i^\alpha(t)^2 >= 2\eta m k_B T$, where $\alpha$ represents the vector components.

As shown previously for an overlayer with the Lennard-Jones interaction potential [5] and also for a model with pure elastic interaction potential [6], the behavior of the drift velocity $\mathbf{v}_d$ as a function of $\mathbf{F}$ along the high symmetry direction $x$ or $y$ depends strongly on the initial equilibrium phase at $\mathbf{F} = \mathbf{0}$. We first summarize the known behavior for the longitudinal response along the high symmetry $x$-direction. When the overlayer at $F = 0$ is in an initial fluid or incommensurate state, then $v_d$ will be nonzero for arbitrarily small external force leading to a finite sliding friction $1/\bar{\eta}_L = v_d/F$ for arbitrarily small values of $F$. For an initial commensurate state below the melting temperature $T_m$, the drift velocity is essentially zero below a critical value of $F$, neglecting the contribution from creep motion due to thermal activation. The $F$ vs $v_d$ characteristics shows hysteresis behavior with unequal critical forces $F_a$ and $F_b$ for increasing or decreasing external force $F_x$, corresponding to the static and kinetic friction forces, respectively. This behavior is shown in Fig. 1 where the simulation results are obtained for particles moving in a periodic pinning potential with square symmetry

$$u(\mathbf{r}) = U_o[2 - \cos(2\pi x/a) - \cos(2\pi y/a)] \quad , \tag{8}$$

and interacting with Lennard-Jones pair potential

$$v(r_{ij}) = \epsilon[(r_o/r_{ij})^{12} - 2(r_o/r_{ij})^6] \quad . \tag{9}$$

Here $\epsilon$ is the well depth and $r_0$ is the particle separation at the minima in the pair potential. Periodic boundary conditions are used, with $r_o/a = 1.56$ and the adsorbate coverage set to $\theta = 1/2$. With this choice, the ground state of the overlayer is a commensurate pinned $c(2 \times 2)$ structure Dimensionless units are used where $a = 1$, $m = 1$, $U_o = 1$ and the force is normalized by $2\pi U_o$ and velocities by $2\pi U_o/\eta$. A system consisting of $10 \times 10$ to $20 \times 20$

substrate atoms was considered with the time variable discretized in units of $\delta t = 0.002 - 0.01\tau$, where $\tau = (ma^2/U_o)^{1/2}$. Most calculations were performed on the smaller system with typically $1 - 4 \times 10^6$ time steps for each calculation of time-averaged quantities allowing equal time steps for equilibration.

It is now understood that for large sliding velocities when $F > F_c$, the overlayer behaves as a sliding solid [5, 6]. As the force is decreased, the sliding overlayer undergoes a dynamic melting transition at $F_c$, where the effective temperature of the overlayer $T^*$ obtained from the variance of the particle velocity distribution ($kT^* = m(< v^2 > - < v >^2)/2$), different from the substrate temperature $T$ in general, reaches a value above the melting transition temperature $T_m$ and the overlayer melts into a sliding liquid phase. With further decrease of the external force, the overlayer is finally pinned at the critical kinetic frictional force $F_b$ to become the commensurate phase again. It is important to note in Fig. 1 that at $F_b$ there is a velocity gap $v_b$. Smooth sliding is not possible for sliding velocities less than $v_b$. As a consequence, the motion of the overlayer can exhibit the familiar stick and slip behavior. This is demonstrated In Fig.2, where the response of an adlayer driven by a spring moving with a velocity $v_k = 0.08 < v_b$ is shown. Instead of moving with a constant drift velocity $v_k$ , the adlayer motion shows the characteristic stick and slip behavior. Only the average velocity of the adlayer is equal to $v_k$.

Besides the dynamical phases and the transitions between them, the other interesting aspect of nonlinearity is the induced asymmetry. For the potential with four-fold symmetry, the diffusion constant and the sliding friction in the linear regime are isotropic. Thus, in the sliding state along the $x$-direction, an additional force along the $y$-direction would just cause the adlayer to slide along the direction of the resultant force, with a drift velocity determined by the linear sliding friction. However, in the nonlinear regime, the external force defines an unique direction and the symmetry is broken. In a recent study [7], it was found that in the sliding solid phase that exists at $F_x > F_c$ in Fig. 1, the pinning potential in the transverse direction remains essentially static in a reference frame moving with the overlayer [8]. As a result, once the system has reached a steady state for a fixed $F_x > F_c$, the overlayer remains pinned along the transverse direction at temperatures below a critical value $T_p$, which is much smaller than the melting transition temperature $T_m$. Above $T_p$, the sliding friction $\bar{\eta}$ has a finite value. In the temperature range $T_p < T < T_m$, there is no transverse critical force and the overlayer is an incommensurate sliding solid in all directions. The behavior of the nonlinear response in the transverse direction correlates with the structure of the sliding state at $F_y = 0$. In the transverse pinned phase below $T_p$, the particles are essentially moving in one-dimensional channels along

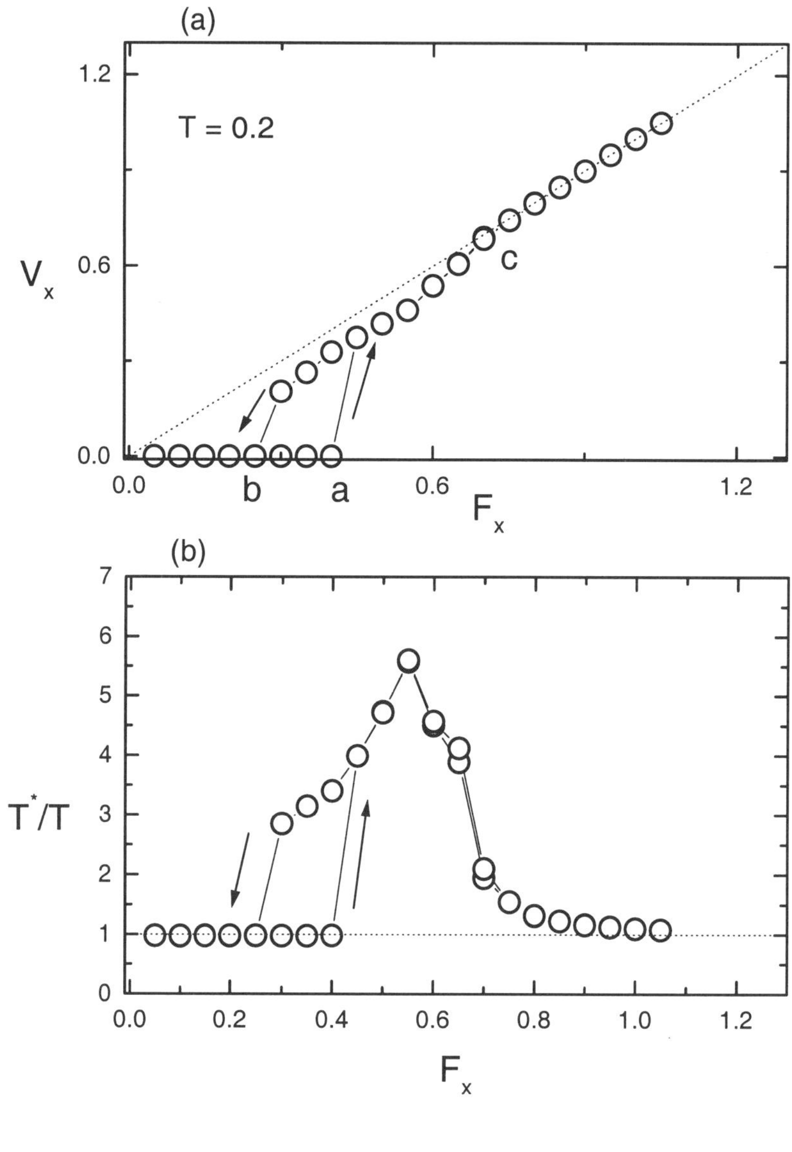


*Figure 1.* Drift velocity $V_x$ and effective temperature of the overlayer $T^*$, for the Lennard-Jones model, as a function of the external force $F_x$ along the $x$-axis for $\epsilon = 1$ and $\eta = 1$. The direction of variation of $F_x$ is indicated by arrows.

the principal $x$-axis of the pinning potential and the particle coordination number is 4, leading to a square lattice structure. For $T > T_p$, the particle

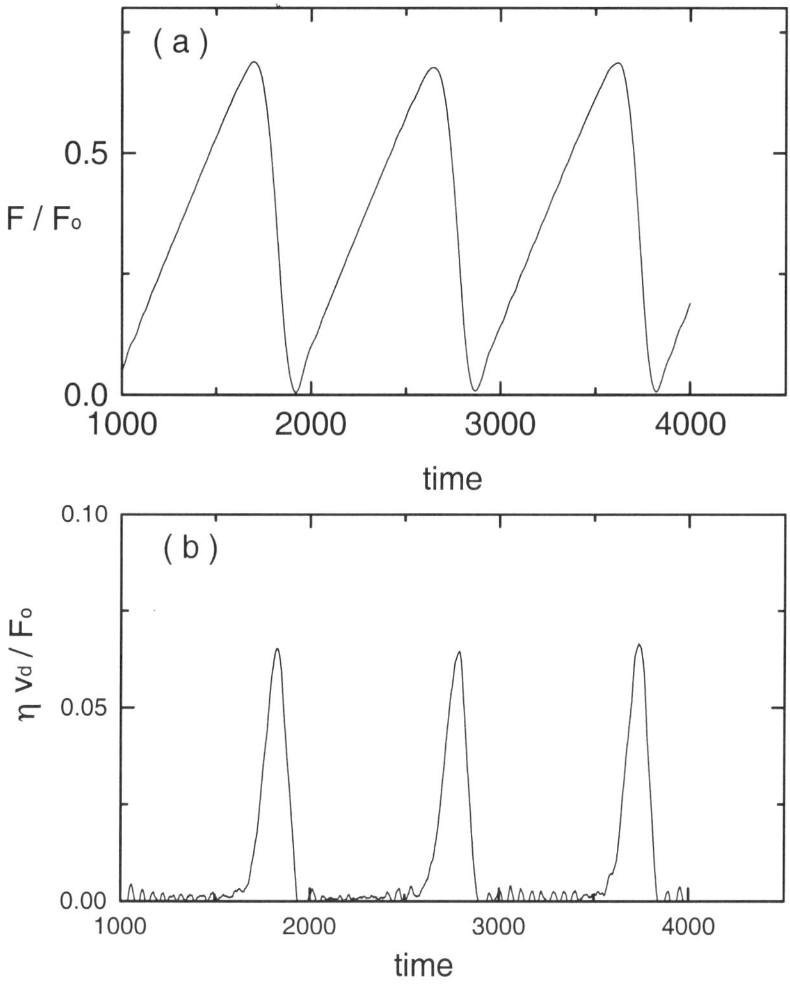


*Figure 2.* Time evolution of the $v_d$ and $F$, for the Frenkel-Kontorova model, when the system is pulled by a spring at constant velocity $v_s = 0.01$, below $v_b$, for $\eta = 0.6$, $T = 0.5$ and $K_s = 0.5$.

trajectories are free to fluctuate in the transverse direction and an isotropic triangular lattice structure occurs. This is exactly the structure that the overlayer would have in the absence of a periodic pinning potential.

## 3. Peak Effect in Type II Superconductors

One of the central unsolved problems in type-II superconductivity concerns the so-called "peak effect". When a current $I$ is passed through the superconductor in the mixed phase, the flux line lattice (FLL) moves in response to the Lorentz force, leading to dissipation and an induced voltage $V$. Naively, the nonlinear conductance $C = I/V$ is expected to decrease monotonically towards the $H_{c2}$ phase boundary because of the diminishing order parameter and hence a reduced pinning strength. However, experimentally it was observed a long time ago that instead of a monotonic behavior, the conductance peaks sharply to a large value before dropping to the normal metal value at the superconducting-normal transition [9]. It has also been established that this peak effect is not just limited to conventional superconductors, but shows up in a similar fashion in the high $T_c$ YBCO superconductors also [10, 11]. To date, however, there has been no satisfactory explanation for this peak effect. Two recent studies have now established conclusively the relation of the peak effect with the underlying phase transition in the FLL [12, 13].

In the FLL, the driving force $F$ is the Lorentz force proportional to the current passing through the superconductor, and the moving FLL produces a changing flux and an induced voltage that is proportional to the average drift velocity $\langle v \rangle$ of the FLL. Thus, in the language of the boundary lubrication problem, the static friction of the adlayer corresponds to the critical current $J_c$ in the superconductor, and the nonlinear sliding friction $\bar{\eta}$ of the adlayer defined as $\bar{\eta} = F/\langle v \rangle$ is essentially the nonlinear conductance $C$ for the superconductor. Note that $\bar{\eta}$ is just the inverse of the usual definition of the mobility for the adlayer. Results from various numerical studies of the boundary layer problem [5, 6] have shown that both the static friction and the nonlinear sliding friction depend in a complicated manner on the interplay of the strength of the pinning potential, interactions among the particles (vortices for the FLL) and the bare frictional damping $\eta$ from the environment. In recent studies [14], it has been argued that for the FLL, it is the variation of the nonadiabatic frictional damping $\eta$ and not the adiabatic pinning strength that develops anomalous temperature and magnetic field dependence near the melting transition. This anomalous behavior of the frictional damping then leads to the peak effect for the nonlinear sliding friction $\bar{\eta}$ for the FLL and hence the conductance $C$ of the superconductor. We present the arguments here, briefly.

Consider the random force acting on the pinning center at the position $\mathbf{r}$ by the flux lattice. In the simple pair interaction model, this can be expressed in terms of the linear displacement $u_{\mathbf{q},\alpha}$ of each vortex from its equilibrium position as

$$f_\alpha(\mathbf{r}) = \sum_{\mathbf{q}} W(\mathbf{r}, \mathbf{q}) u_{\mathbf{q},\alpha} \quad . \tag{10}$$

Here, $\mathbf{q}$ stands for the normal mode index of the FLL, $\alpha$ is the Cartesian component label and $W$ represents the coupling function. In response to this, there is an equal and opposite reaction force on the FLL by the pinning center. When correlations between the different pinning centers are neglected, the frictional damping (in the Markovian limit) on the FLL is given by [15, 16]

$$\eta = \sum_{\mathbf{q},\mathbf{r}} W^2(\mathbf{r}, \mathbf{q}) S(\mathbf{q}, \omega = 0) \quad , \tag{11}$$

where $S(\mathbf{q}, \omega = 0) = \int_0^\infty dt \langle u_{\mathbf{q},\alpha}(t)^* u_{\mathbf{q},\alpha}(0) \rangle$ is the dynamic structure factor. Correlations between the random forces from pinning centers at different positions would lead to higher order terms in the pinning center concentration $n_p$ in Eq.(11), and are negligible in the limit $n_p \to 0$. According to general dynamical scaling arguments [17, 18], $S(\mathbf{q}, \omega)$ should take the scaling form near $T_c$ for a continuous phase transition as

$$N^d S(\mathbf{q}, \omega) = \xi^{z+\gamma/\nu} g_\pm(q\xi, \omega\xi^z) \quad . \tag{12}$$

where $g_\pm$ is a scaling function, $\xi \propto |T/T_c - 1|^{-\nu}$ is the divergent order parameter correlation length, $d$ is the system dimension, $\gamma$ the susceptibility exponent and $z$ is the dynamical critical exponent. Substitution of Eq. (12) back into Eq. (11) then leads to the conclusion that as one approaches $T_c$, the friction $\eta$ has a singular part that goes as $\eta \propto |T/T_c - 1|^{-x}$ with $x = \nu(z-d)+\gamma$. The dimension $d$ enters explicitly through the $q$ integration in Eq. (11) where we have assumed a typical short ranged coupling potential $W(q)$ that is regular at $q = 0$. Thus the friction $\eta$ can either diverge if $x > 0$ or be finite with a cusp only . A similar anomaly has also been predicted for adatom diffusion near the surface reconstruction transition of the W(100) surface [19]. For this case, the exponent $x$ has been explicitly evaluated for a model Hamiltonian and shown to have the value [19, 20] $x \approx 1.8$. Thus the diffusion constant of adatoms on this surface is predicted to vanish at the transition.

The friction $\eta$ calculated in Eq. (11) corresponds to the bare friction acting on the center of mass (CM) degree of freedom of the FLL, playing the same role as the friction $\eta$ introduced earlier in in Eq.(7) for the boundary lubrication problem. Experimentally, the mobility measurements of the FLL have all been performed in the nonlinear regime. The nonlinear response of the flux lattice under a driving current can only be determined by solving the Langevin equation for the moving vortices in the FLL. In general, the nonlinear sliding friction $\bar{\eta}$ of the FLL depends on the details of the vortex interaction, strength of the pinning potentials, and the driving force. However, in various recent studies of the nonlinear sliding friction of an adsorbed overlayer on an substrate [5, 6], it has been shown that the magnitude of $\bar{\eta}$ is determined by the bare friction $\eta$ as given in Eq.(11), with $\bar{\eta}$ approaching the bare friction $\eta$ in the limit of large driving force. Even for a system with a positive exponent $x$ leading to a divergent behavior for $\eta$ and $\bar{\eta}$ near the transition, the conductance peak at the transition in practice will be significantly rounded by nonlinear effects due to the nonzero driving current. It has been argued, in general terms, that the current density $J$ sets an additional length scale in resistance measurements [21] $L_J^{d-1} \sim kT/J$ due to thermal fluctuations. The divergent critical fluctuations at $T_c$ will be then be cut off by this length when $\xi \sim L_J$, giving rise to a nonlinear resistance behavior $R \sim I^{x/\nu(d-1)}$. Experimentally [12], a strong nonlinearity is indeed observed for the conductance maxima which decreases for increasing $I$. Thus, we conclude that for a FLL system with a positive exponent $x$, its nonlinear sliding friction $\bar{\eta}$ has a peak at the melting transition, its origin being the strong critical fluctuation near the melting transition. This then leads to the peak in the conductance $C$. In the case of a weak first order transition or finite size system, the divergence or the cusp singularity of $\eta$ would be rounded off even in the linear regime and thus we expect the peak effect for these systems to be much weaker.

While most previous theoretical considerations of the peak effect focus on the adiabatic pinning strength [2], we have identified the origin of the peak effect through the non-adiabatic coupling of the pinning centers to the strong critical fluctuations near the transition point. This leads to a vanishing linear mobility for the flux lattice at the transition. In the nonlinear regime, the finite driving current provides a cutoff for the divergent critical fluctuations, and this leads to a finite peak in the nonlinear sliding friction for the FLL and hence the conductance for the superconductor, with the strength of the peak dependent on the magnitude of the driving current. The recently observed gigantic peak effect in high $T_c$ superconductors and the strong correlation between the peak effect and the observed melting transition [12, 13] provide strong support for the mechanism proposed here.

ACKNOWLEDGEMENT:

The work described in this paper is partially supported by a NSF-CNpq international collaboration grant(SCY, E.G. and T.A-N.), by FAPESP(E.G.) and by the Academy of Finland through its center of excellence program (T. A-N.).

## References


1. Persson B.N. J., *Sliding Friction: Physical Principles and Applications* (Springer-Verlag Berlin Heidelberg, 1998).
2. Larkin A. I., Marchetti M. C.and Vinokur V. M., Phys. Rev. Lett. **75**, 2992 (1995).
3. Kubo R., Toda M. and Hashitsume N.,*Statiscal Physics II*(Springer-Verlag Berlin Heidelberg, 1985).
4. Gomer R., Rep. Prog. Phys. **53**, 917 (1990).
5. Persson B.N.J., Phys. Rev. Lett. **71**, 1212 (1993); Phys. Rev. B **48**, 18140 (1993); J. Chem. Phys. **103**, 3449 (1995).
6. Granato E. and Ying S.C., Phys. Rev. B **59**, 5154 (1999); Granato E., Baldan M.R., and Ying S.C., in *Physics of Sliding Friction*, edited by Persson B.N.J.and Tosatti E.(Kluwer, Dorbrecht, 1996).
7. Granato E. and Ying S. C. , submitted to Phys. Rev. Lett. (2000)
8. Giamarchi T. and Le Doussal P., Phys. Rev. Lett. **76**, 3408 (1996); Le Doussal P. and Giamarchi T., Phys. Rev. B **57**, 11356 (1998).
9. LeBlanc M. A. and Little W. A.in: *Proceedings of Conference on Low Temperature Physics*, (University of Toronto Press, Toronto), p.198 (1960).
10. Ling X. S.and Budnick J. I., in *Magnetic Susceptibility of Superconductots and Other Spin Systems*, edited by Hein R. A. et al., Plenum Press, New York, N.Y., p.377 (1991).
11. Kwok W. K., Fendrich J. A., Van der Beek C. J. and Crabtree G. W., Phys. Rev. Lett. **73**, 2614 (1994).
12. Shi J., Ling X. S., Liang R., Bonn D. A. and Hardy W. N., Phys. Rev. B **60**, R12593 (1999); Ling X.S., Berger J.E. and Prober D.E., Phys. Rev. B **57**, R3249 (1998).
13. Ling X. S., Park S. R., Choi S. M., Dender D. C. and Lynn J. W. , submitted to Phys. Rev. Lett. (2000).
14. Ala-Nissila T., Granato E. and Ying S. C., J. Phys.: Condensed Matt. **2**, 8537 (1990) and to appear in Phys. Rev. B, (2000).
15. Mori H., Prog. Theor. Phys. **34**, 399 (1965).
16. Ying S. C., Phys. Rev. B **41**, 7068 (1989).
17. Halperin B. I. and Hohenberg P. C., Phys. Rev. **177**, 952 (1969).
18. Hohenberg P. C. and Halperin B. I., Rev. Mod. Phys. **49**, 435 (1977).
19. Ala-Nissila T., Han W. K. and Ying S. C., Phys. Rev. Lett. **68**, 1866 (1992).
20. Baldan M. R., Granato E. and Ying S. C., Phys. Rev. B **(62),2146(2000).**
21. **D. S. Fisher, M. P. A. Fisher and D. Huse, Phys. Rev.B 43, 130 (1991).**

# SELECTIVE ACTIVATION DYNAMICS: A MINIMAL PATH APPROACH

L. Y. CHEN AND P. L. NASH
*Department of Physics, University of Texas at San Antonio, San Antonio, TX 78249 USA*

AND

S. C. YING
*Department of Physics, Brown University, Providence, RI 02912 USA*

## 1. Introduction

Surface diffusion[1] or transition between stable (or metastable) states presents well-known computational difficulties, particularly when the dynamical bottlenecks separating the initial and final (stable or metastable) states are not known. Conventional methods for studying rare event activations are based on transition state theory[2-4].The application of this approach requires prior knowledge of the mechanism or pathway for the activation process. For multidimensional systems, it is a challenging problem to determine the transition state numerically without assuming any particular configurations[5,6].Various innovations have been put forward to improve or go beyond the transition state theory. Examples are the accelerated dynamics approach[7] and the transition path approach[8-10].In reference [11], we have proposed a novel approach to the calculation of the rate for activated processes based on the path integral solution of Langevin equation. In this paper, we pursue further the idea proposed there and fully establish the minimal path approximation. According to the path integral formulation of the Langevin equation, the probability for an arbitrary path connecting the initial and the final configurations is determined by a positive-definite "action" functional. It is shown that the paths having the minimal action dominate over all others for low temperatures. Taking advantage of this fact, the expectation value of a physical observable is computed as the weighted average of its values along the minimal paths. The minimal paths



*M.C. Tringides and Z. Chvoj (eds.),*
*Collective Diffusion on Surfaces: Correlation Effects and Adatom Interactions,* 285–293.

are found through integrating an activation equation and a deactivation equation (both second order differential equations). This minimal path approach does not require prior knowledge of the "transition states" nor that of the stable/metastable states. Instead, through sampling of intermediate states and integrations along the minimal activation/deactivation paths, both can be found by a numerical effort on the level of Monte-Carlo sampling in the phase space. It is important to note that not only is the prior knowledge of the transition state unnecessary, the implementation of the minimal path method also avoids the usual approximations involved in the transition state theory, which becomes increasingly inaccurate in the limit of low frictional damping.

## 2. The Minimal Path Approximation

We start with the standard Langevin equation,

$$m\ddot{x}(t) = -\eta\dot{x}(t) - \nabla V(x(t)) + \xi(t) \tag{1}$$

that has a wide variety of applications [12-14] and has been the basis for study of surface diffusion in references [15-17]. Here, $x(t)$ stands for a multicomponent vector with dimension $ND$, where $N$ is the number of particles and $D$ is the physical dimension of the system. $V(x)$ is a general potential including both the external potential and interparticle interactions. The random force $\xi(t)$ characterizes the coupling of this system to its environment (a heat bath). Its correlation can be written in the form $\langle\xi(t)\xi(t')\rangle = 2k_BT\eta\delta(t-t')$ as we consider in this paper only the Markovian limit. $\eta$ is the frictional (damping) coefficient. In general, potential $V(x)$ has many local minima separated by activation barriers.

Due to the stochastic nature of the Langevin equation, there exists an entire distribution of solutions (paths) to Eq. (1) for a given set of boundary conditions, and physical observables are obtained as averages over this distribution of paths. It has been shown that, with regards to the solution of Eq. (1), the probability functional $P[x(t)]$ for the system to take a particular path $x(t)$ starting at $(x_1, v_1, t_1)$ and ending at $(x_2, v_2, t_2)$ is given by the following expression [18]

$$P[x(t)] = \frac{1}{\mathcal{N}} \exp\left\{-\int_{t_1}^{t_2} \frac{dt}{4\eta k_BT} \left[m\ddot{x} + \eta\dot{x} + \nabla V(x)\right]^2\right\}\Bigg|_{\{x(t_1)=x_1,\dot{x}(t_1)=v_1\}}^{\{x(t_2)=x_2,\dot{x}(t_2)=v_2\}}. \tag{2}$$

The total joint probability for the particle to be in $(x_1, v_1)$ at $t = t_1$ and in $(x_2, v_2)$ at $t = t_2$ is then given by the path (functional) integral [19, 20]

$$P\left(x_1, v_1, t_1 | x_2, v_2, t_2\right) = \int [\mathcal{D}x] P[x(t)]. \tag{3}$$

The functional integrand in Eq. (3), though complicated, is positive-definite, and thus, the numerical implementation of the path integral in Eq. (3) is well-defined. The physically interesting regime usually corresponds to the situation such that the thermal energy $k_BT$ is much less than the activation barriers. Under this condition, there exist some special paths whose contribution to the path integral in Eq. (3) dominates over all the other paths. To illustrate this more clearly, we introduce a length scale $d$, an energy scale $V_0$, and a time scale $\tau_0 \equiv \sqrt{md^2/V_0} \equiv 1/\omega_0$ for the system under consideration. Equation (3) can then be expressed in terms of dimensionless, scaled variables in the form,

$$P(x_1, v_1, t_1|x_2, v_2, t_2) = \int [\mathcal{D}x] \exp\left\{-\frac{1}{\lambda} I[x(t)]\right\}, \tag{4}$$

with the boundary conditions $\{x(t_1) = x_1, \dot{x}(t_1) = v_1\}$ and $\{x(t_2) = x_2, \dot{x}(t_2) = v_2\}$. The effective action

$$I[x(t)] = \int_{t_1}^{t_2} dt \zeta^2, \tag{5}$$

where

$$\zeta(t) = \ddot{x}(t) + \gamma\dot{x}(t) + \nabla V(x(t)) \tag{6}$$

and $x, t, V$ are all in their dimensionless form. The dimensionless friction parameter $\gamma$ is defined as $\gamma \equiv \eta/m\omega_0$ , and the dimensionless parameter $\lambda$ appearing in the exponential of the functional integrand is $\lambda = 4\gamma k_BT/V_0$.

Clearly, in the low temperature ($k_BT << V_0$) and/or underdamped regime ($\gamma << 1$), $\lambda \ll 1$, the minimal path of functional $I[x(t)]$ carries the dominant weight in the path integration of Eq. (4). Since the action functional $I$ is positive-definite, deviations from the minimal path carry negligible contributions to the path integral in Eq. (4). The minimal path is determined from the condition that the functional derivative $\delta I/\delta x(t)$ vanishes. This leads to the standard Euler-Lagrange equation for the minimal path

$$\ddot{\zeta}(t) - \gamma\dot{\zeta}(t) + \zeta \cdot \nabla\nabla V(x(t)) = 0. \tag{7}$$

Note again that all quantities are in a dimensionless form.

An obvious solution to Eq. (7) is $\zeta = 0$, *i.e.*,

$$\ddot{x}_d(t) + \gamma\dot{x}_d(t) + \nabla V(x_d(t)) = 0. \tag{8}$$

Integrating Eq. (8) forward with boundary conditions $x_d(t_M) = x_M, \dot{x}_d(t_M) = v_M$, for $t_f > t_M$, a minimal path $x_d(t)$ is obtained that characterizes deactivation from state $(x_M, v_M)$ at $t_M$ to state $(x_f = x_d(t_f), v_f = \dot{x}_d(t_f))$ at

$t_f$. Along this path $x_d(t)$ the energy of the system decreases monotonously and the action $I = \int_{t_M}^{t_f} dt\zeta(t)^2$ assumes its minimal value $I_d = 0$.

A less obvious solution to the minimal path Eq. (7) is that $\zeta = 2\gamma\dot{x}_a$, *i.e.*,

$$\ddot{x}_a(t) - \gamma\dot{x}_a(t) + \nabla V(x_a(t)) = 0. \tag{9}$$

Integrating Eq.(9) backward with boundary conditions $x_a(t_M) = x_M, \dot{x}_a(t_M) = v_M$, for $t_i < t_M$, a minimal path $x_a(t)$ is obtained that characterizes activation from $(x_i = x_a(t_i), v_i = \dot{x}_a(t_i))$ at $t_i$ to $(x_M, v_M)$ at $t_M$. Along this path $x_a(t)$, from $t_i$ to $t_f$, the energy of the system increases monotonously and the action $I = \int_{t_i}^{t_M} dt\zeta(t)^2$ assumes its minimal value $I_a = 4\gamma(E_M - E_i)$, where $E_M = v_M^2/2 + V(x_M)$ and $E_i = v_i^2/2 + V(x_i)$ are the energies in the dimensionless form at times $t_i$ and $t_M$, respectively.

It is worth noting that the deactivation path $x_d(t)$ is a minimum (not maximum) of the action functional $I[x(t)]$ and that so is the activation path $x_a(t)$. While it is straightforward to see $x_d(t)$ as a minimum, the fact that $x_a(t)$ is a minimum can be verified by expanding $I[x(t)]$ to the second order in an infinitesimal deviation $\delta x(t)$ from $x_a(t)$, *i.e.*,

$$I[x_a(t) + \delta x(t)] = I[x_a(t)] + \delta^{(1)}I + \delta^{(2)}I. \tag{10}$$

Here, the first order term $\delta^{(1)}I = 0$ by virtue of the Euler-Lagrange equation, (7), and the second order term can be manipulated into the following form

$$\delta^{(2)}I = \int_{t_M}^{t_f} dt\left[\delta\ddot{x}(t) - \gamma\delta\dot{x}(t) + \nabla\nabla V(x_a(t))\delta x(t)\right]^2, \tag{11}$$

which is postive definite.

Weaving the activation path $x_a(t)$ and deactivation path $x_d(t)$ together, we obtain the following three-configuration joint probability, within the minimal path approximation,

$$\begin{aligned} P\left(x_i, v_i, t_i | x_M, v_M, t_M | x_f, v_f, t_f\right) &= \delta(x_i - x_a(t_i|M))\delta(v_i - \dot{x}_a(t_i|M)) \\ &\exp\left[-(I_a + I_d)/\lambda\right]\delta(x_f - x_d(t_f|M))\delta(v_f - \dot{x}_d(t_f|M)). \end{aligned} \tag{12}$$

Here $(\cdot|M)$ indicates that the activation path $x_a(t)$ and the deactivation path $x_d(t)$ are joined at $t_M$, *i.e.*, $x_a(t_M) = x_d(t_M) = x_M$ and $\dot{x}_a(t_M) = \dot{x}_d(t_M) = v_M$. This three-configuration joint probability can be used, together with the initial (equilibrium or nonequilibrium) distribution $P(x_i, v_i)$, to conveniently evaluate the expectation values of physical observables.

It should be pointed out that Weaving the activation path and the deactivation path at $t_M$ brings in a discontinuity in $\zeta(t)$ of Eq. (6) and thus in the integrand of $I[x(t)] = \int_{t_i}^{t_f} dt\zeta^2$. However, both $x(t)$ and $\dot{x}(t)$ are continuous at $t_M$ and the $I[x(t)]$ integral is well-defined for such a

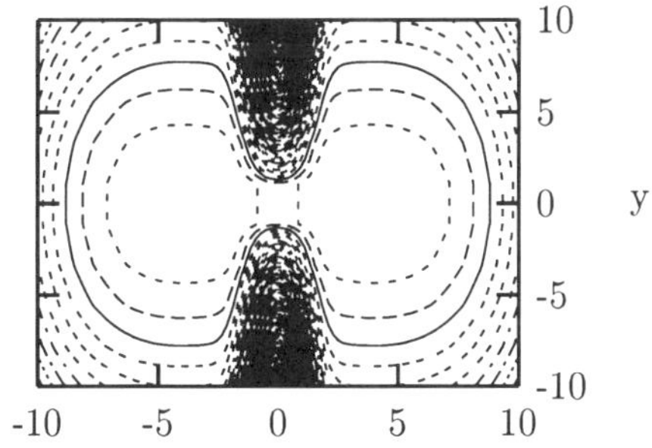


*Figure 1.* Contour plot of the external potential $V_{ex}(x, y)$.

discontinuity, where $\zeta(t)$ experiences a finite jump from $2\gamma\dot{x}_a(t_M)$ to 0. Therefore, it should not bring any artificial error into the evaluation of physical observables.

## 3. Three Lennard-Jones Particles in an External Potential

To illustrate the application of the minimal path approximation, we consider a cluster of three particles ($N = 3$) in two dimension ($D = 2$) interacting with one another in the form of Lennard-Jones potential, in its dimensionless form,

$$V_{LJ}(i, j) = 4\left(\frac{1}{r_{ij}^{12}} - \frac{1}{r_{ij}^{6}}\right), \tag{13}$$

where $r_{ij}$ is the distance between the $i$-th particle and the $j$-th particle. This cluster is subject to an external potential

$$V_{ex}(x, y) = \left(1 - (x/4)^2\right)^2 + \left(4 + 100\exp(-x^2)\right)(y/4)^2. \tag{14}$$

This external potential has two minima located at (−4,0) and (4,0), separated by a narrow passage located around the saddle point (0,0) as shown in the equipotential contour plot in Fig. 1. Relative to the two minima where $V_{ex}(\pm 4, 0) = 0$, the saddle point has a potential $V_{ex}(0, 0) = 1$. The configuration of lowest energy is when three particles form an equilateral triangle and reside in one of the two potential wells, *i.e.*, the left or the right configuration in Fig. 2.

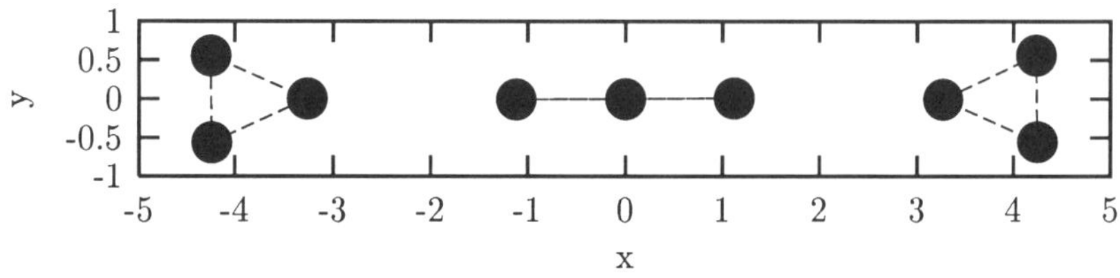


*Figure 2.* The initial (left), intermediate (center), and final (right) configurations of the cluster along the minimal path.

Considering activation of the cluster initially in the well located around (−4,0) into the other well located around (4,0), the minimal path of successful activation can be found by the following scheme. Choose an intermediate state $(x_M, v_M)$, then integrate the activation path equation, (9), backward from $t_M$ to $t_i$, arriving at the initial state $(x_i, v_i)$, and integrate the deactivation path equation, (8), forward from $t_M$ to $t_f$, arriving at the final state $(x_f, v_f)$. If the initial and final states thus obtained are distinctly different (*i.e.*, in different wells), a successful activation path is found. And the probability for this path is $\exp(-I_a/\lambda)$ (noting that $I_d = 0$). Thus, the dynamics information is obtained for each chosen intermediate state. The integration of Eqs. (8) and (9) need to be performed only for short durations in the presence of finite friction (damping) $\gamma$. The numerical effort required can be further reduced through random Monte-Carlo sampling of the intermediate points $(x_M, v_M)$ in the $(x, v)$ phase space.It is interesting to pay close attention to one minimal path of this three cluster activation whose action $I_a$ is the smallest among all paths of successful activation that corresponds to the lowest value of energy $E_M$ at the intermediate state $(x_M, v_M)$ (note that $I_a = 4\gamma(E_M - E_i)$). The energy and the x-coordinate of the center of mass of the cluster along this path are plotted in Figs. 3 and 4, respectively. In order to go along this minimal path, the cluster breaks one of the three "bonds" in the initial state and squeezes through the narrow passage located around $x = 0$ as shown in the center configuration in Fig. 2. This aspect is clearly seen in Fig. 5: The area of the triangle diminishes as the cluster comes into the narrow passage (in the intermediate state) and recovers afterward when it deactivates into the potential well in the right.

## 4. Summary

We have proposed an approximate solution to the Langevin equation for processes when a system initially in a stable/metastable state is being activated out of it and deactivating into another. A three-configuration joint

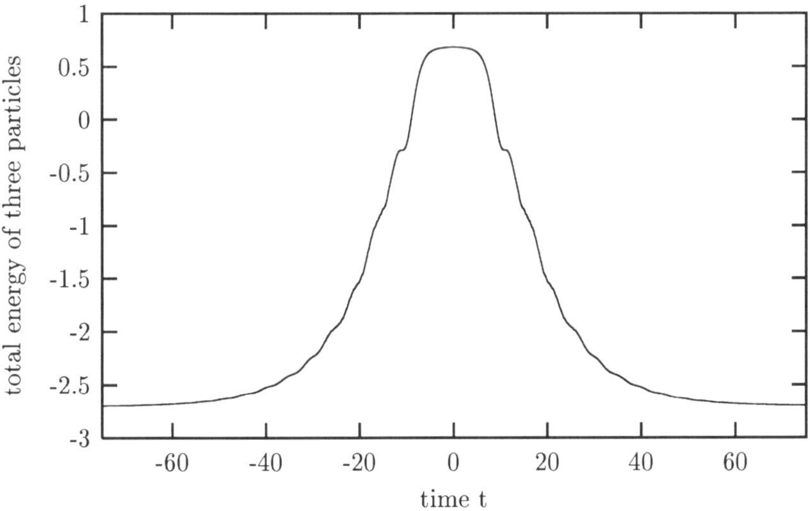


*Figure 3.* Energy of the cluster along the minimal path.

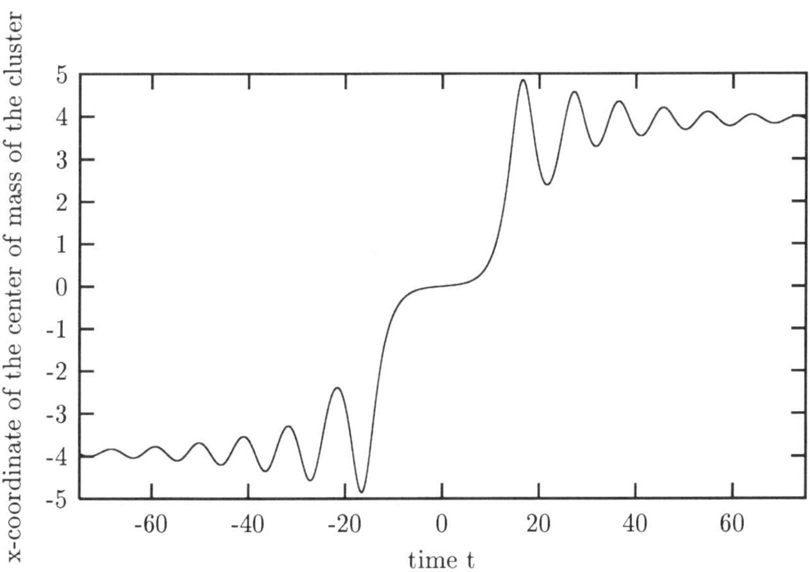


*Figure 4.* X-coordinate of the center of mass of the cluster along the minimal path of activation.

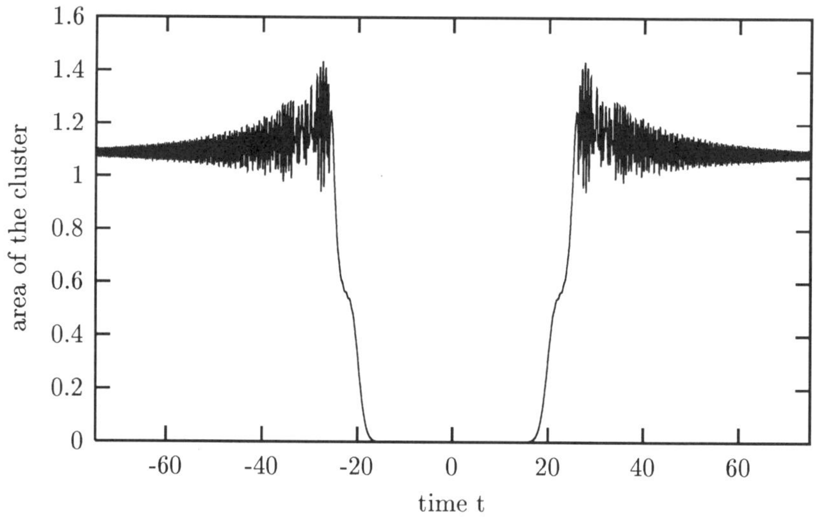


*Figure 5.* Area of the cluster along the minimal path of activation.

probability, for the system being in the initial state $(x_i, v_i)$ at time $t_i$, final state $(x_f, v_f)$ at time $t_f$, and, an intermediate state $(x_M, v_M)$ at time $t_M$, is obtained by connecting $(x_i, v_i)$ to $(x_M, v_M)$ with an activation path having the minimal "action" and connecting $(x_M, v_M)$ to $(x_f, v_f)$ with a deactivation path also having minimal "action". This minimal path approach is accurate at low temperatures and thus complements the existing methods that work better at higher temperatures. For a temperature not low enough, fluctuations around the minimal paths can be sampled numerically to improve the computation. As an example, we have studied the activation of a cluster of three Lennard-Jones particles over a narrow passage from one stable state to another in a two dimensional external potential. With moderate numerical efforts, we are able to identify the relevant activation mechanism and to locate new stable/metastable states.

## 5. Acknowledgments

Thanks are due to Professors Tapio Ala-Nissila and Karsten W. Jacobsen for helpful discussions.

## References

1. Gomer, R. (1990) *Rep. Prog. Phys.* **53**, 917.
2. Anderson, J. B. (1995) *Adv. Chem. Phys.* **71**, 381.

3. Keck, J. (1960) *J. Chem. Phys.* **32**, 1035.
4. Chandler, D. (1978) *J. Chem. Phys.* **68**, 2959.
5. Jonsson, H. Mills, G. and Jacobsen, K. W. (1998) in Berne, B. J., Ciccotti, G., and Coker, D. F., (eds.) *Classical and Quantum Dynamics in Condensed Phase Simulations* World Scientific, Singapore.
6. Trushin, O. S., Salo, P., and Ala-Nissila, T., (2000) *Phys. Rev.* B**62**, 1611.
7. *e.g.*,Voter, A. (1997) *Phys. Rev. Lett.* **78**, 3908;(1997) *J. Chem. Phys.* **106**, 4665.
8. Pratt, L. R. (1986) *J. Chem. Phys.* **85**, 5045.
9. Dellago, C.,Bolhuis, P. G.,Csajka, F. S. and Chandler, D. (1998) *J. Chem. Phys.* **108**, 1964.
10. Jacobsen, J., Jacobsen, K. W., and Sethna, J. P. (1997) *Phys. Rev. Lett.* **79**, 2843.
11. Chen, L. Y. and Ying, S. C. (1999) *Phys. Rev.* B**60**, 16965.
12. Mel'nikov, V. I. (1991) *Phys. Rep.* **209**, 1.
13. Hänggi, P., Talkner, P. and Borkovec, M. (1990) *Rev. Mod. Phys.* **62**, 251.
14. Risken, H. (1989) *The Fokker–Planck equation*, Springer-Verlag, Berlin.
15. Ferrando,R., Spadacini, R., and Tommei, G. E. (1993) *Phys. Rev.* E**48**, 2437.
16. Pollak, E. and Talkner, P. (1993) *Phys. Rev.* E**47**, 922.
17. Chen, L. Y. and Ying, S. C. (1993) *Phys. Rev. Lett.* **71**, 4361; Chen,L. Y., Baldan,M. R., and Ying,S. C. (1996) *Phys. Rev.* B**54**, 8856.
18. Schmid, A. (1982) *J. Low Temp. Phys.* **49**, 609.
19. Feynman, R. P. and Hibbs, A. R. (1965) *Quantum Mechanics and Path Integrals*, McGraw-Hill, New York.
20. Brandbyge, M., Hedegrd,P., Heinz, T. F., Misewich,J. A., and Newns,D. M. (1995) *Phys. Rev.* B**52**, 6042-6056.

# LONG JUMPS IN SURFACE DIFFUSION INVESTIGATED BY STOCHASTIC EQUATIONS

R. FERRANDO, R. SPADACINI, AND G.E. TOMMEI
*INFM, CFSBT/CNR, Dipartimento di Fisica dell'Università via Dodecaneso 33, 16146 Genova, Italy*

**Abstract.** The jump diffusion at crystal surfaces is studied by means of three stochastic equations, the Fokker-Planck, the Bhatnagar-Gross-Krook and the Fokker-Planck equation with memory. The equations are solved calculating the jump rate and the jump-length probability distribution. The results are then discussed and compared to each other.

## 1. Introduction

The diffusion of adatoms at crystal surfaces can be theoretically investigated mainly by different kinds of simulations [1-4] or by solving stochastic equations [5-13]. Although simulations are more apt to describe real systems, general physical conditions for the different mechanisms of diffusion, largely independent of the peculiarities of the systems, are more easily obtained from stochastic equations [5-7, 11-13].

Surface diffusion may proceed by jumps or quasi-continuously. Activated or jump diffusion is recovered at high potential barriers where the adatom spends the most part of its time in oscillating at the bottom of the wells, rarely escaping from them. In fact, in these conditions, the jump rate, $r_j$, can be properly defined, the inverse rate being the longest time scale in the physics of the problem. We demonstrated [10] that in rather general conditions, at least without memory effects, a very good definition of the rate, and then of the activated regime, can be given at barriers $E_a \geq 4k_BT$ and that $2k_BT$ is a limiting barrier for the jump regime.

We present here a survey of the single-particle diffusion in the activated regime, comparing results for the jump rate and for the jump-length probability distribution (JLPD) from three stochastic equations corresponding to different physical frameworks for the interaction between the adatom



*M.C. Tringides and Z. Chvoj (eds.),*
*Collective Diffusion on Surfaces: Correlation Effects and Adatom Interactions,* 295–304.

and the crystal surface: the Fokker-Planck (FPE) [5, 14], the Bhatnagar-Gross-Krook (BGK) [8, 14] and the Fokker-Planck with memory (MFPE) [9]. Results from the FPE and the BGK, here given in different ranges of the parameters with respect to the published ones [5, 6, 8], are reviewed together with their differences and analogies. New preliminary results about the MFPE are presented for the first time.

## 2. General Features and Theory

The most important quantity for single-particle diffusion is the self part of the dynamic structure factor $S_s(q,\omega)$,

$$S_s(q,\omega) = \frac{1}{2\pi}\int_{-\infty}^{+\infty} \langle \exp\{iq[x(t) - x(0)]\}\rangle \exp(-i\omega t)dt. \tag{1}$$

By Kubo formulas all the relevant statistical quantities, e.g., the diffusion coefficient, the spectrum of the velocity correlation, and the mean-square displacement are derived from $S_s$. A very important result [5] is that also the jump rate $r_j$ and the JLPD are obtained from $S_s$ in the periodic case; in fact defining the decay function $f(q)$ of $S_s(q,\omega)$ by the following limit,

$$f(q) = \lim_{\omega\to 0} \omega \left[\frac{S_s(q,0)}{S_s(q,\omega)} - 1\right]^{-1/2}, \tag{2}$$

the Fourier analysis of $f(q)$ furnishes [5] the probability $P(n)$ of a jump of length $n$ and the total jump rate $r_j$ as

$$P(n) = -\frac{2a}{\pi r_j}\int_0^{\frac{\pi}{a}} f(q)\cos(naq)dq, \tag{3}$$

$$r_j = \frac{a}{\pi}\int_0^{\frac{\pi}{a}} f(q)dq. \tag{4}$$

The dynamic structure factor $S_s$ of Eq. (1) is obtained by the solution of an appropriate stochastic equation for the probability density $f$.

For all the equations employed here, the interaction between the adparticle and the surface is divided into two parts, static and dynamic. The static part is pictured as a static periodic force $F(x)$ derived from an adiabatic periodic potential. Here we assume a 1D cosine for simplicity,

$$U(x) = -A\cos\left(\frac{2\pi x}{a}\right), \tag{5}$$

where $a$ is the lattice spacing. The dynamic part depends on the physics underlying the stochastic equation employed.

All the equations are solved by using the matrix continued fraction method (MCFM), the powerful numerical method due to Risken [5, 14], which directly furnishes the dynamic structure factor $S_s$. Normalized units are employed for the time scales and for the barrier by means of a characteristic time $\tau$ (the time to cross over a lattice cell at thermal velocity),

$$\tau = \frac{a}{2\pi}\sqrt{\frac{m}{k_BT}}, \tag{6}$$

and of $4k_BT$, respectively. Thus,

$$g = E_a/(4k_BT) \qquad \text{and} \qquad \omega_0 = \sqrt{2g} \tag{7}$$

are the normalized barrier and the small oscillation frequency at the well bottom, respectively. $g \geq 1$ fully defines the jump regime and $g \sim 0.5$ is the limiting barrier at least for the FPE and the BGK equation [10].

### 2.1. THE STOCHASTIC EQUATIONS

#### 2.1.1. *The Fokker-Planck equations: the MFPE and the FPE*

The framework for the MFPE is the generalized Langevin equation (GLE)

$$m\frac{dv}{dt} = F(x) - m\int_0^t dt'\; K(t-t')\; v(t') + \Gamma(t), \tag{8}$$

where the dynamic interaction adparticle-surface is pictured as a damping force with memory kernel $K$ and as a stochastic force $\Gamma(t)$. Due to the fluctuation-dissipation theorem,

$$\langle \Gamma(t)\Gamma(t')\rangle = mk_BTK(t-t'), \tag{9}$$

$\Gamma(t)$ is a colored noise and the process $(x, v)$ (position and velocity) is not Markovian. Assuming an exponentially decaying memory kernel [9],

$$K(t-t') = \eta\eta_1 \exp[-\eta_1(t-t')], \tag{10}$$

where $\eta$ is the static friction, and by adding a new variable $z$,

$$z = -\int_0^t dt'\; K(t-t')\; v(t') + \Gamma(t)/m, \tag{11}$$

a Markovian process $(x, v, z)$ is recovered. The GLE is then equivalent to the MFPE for the density probability $f(x, v, z, t)$:

$$\frac{\partial f}{\partial t} = -v\,\frac{\partial f}{\partial x} - \left(\frac{F(x)}{m} + z\right)\frac{\partial f}{\partial v} + \eta_1\frac{\partial}{\partial z}\left(\eta v f + z f + \frac{\eta\eta_1 k_BT}{m}\frac{\partial f}{\partial z}\right). \tag{12}$$

$S_s$ depends on three parameters $\gamma$, $\gamma_1$, and $g$ where $\gamma = \tau\eta$ is the normalized static friction, and $\gamma_1 = \tau\eta_1$ is the normalized memory decay parameter.

The Langevin equation of the Brownian model, where the friction force is $-m\eta v$ with a frequency-independent $\eta$, is recovered if $v$ and $x$ vary on a slower time scale with respect to the correlation time of the fluctuations, i.e., if $x$ and $v$ are "slow" variables. In this case, $K(t-t') \propto \eta\delta(t-t')$, the process $(x, v)$ is Markovian, $\Gamma(t)$ is a white noise and the FPE for $f(x, v, t)$ is obtained,

$$\frac{\partial f}{\partial t} = -v\frac{\partial f}{\partial x} - \frac{F(x)}{m}\frac{\partial f}{\partial v} + \eta\frac{\partial}{\partial v}\left(vf + \frac{k_B T}{m}\frac{\partial f}{\partial v}\right). \tag{13}$$

$S_s$ depends now on two parameters $\gamma$ and $g$.

As for the physical conditions for the validity of the usual Langevin equation, the adatom couples to the electronic excitations as well as to the lattice vibrations; due to the small electron mass, the assumptions underlying the FPE are valid for the electronic interaction, while the time scale separation is more difficult for phonons; in the continuum Debye model [15] the friction force is written as $-m\eta v$ if the frequencies associated to the motion of the adsobate are well below the highest phonon frequency. It seems, then, reasonable to assume that the MFPE reduces to the FPE when the mass of the adatom is large with respect to the mass of the substrate atoms.

From a slightly different point of view, $x$ and $v$ are "slow"variables if the energy exchanges adatom-surface satisfy the condition $\Delta E << k_B T$, that means "weak and frequent collisions." In other words, a large number of collisions occurs for the thermalization of the adatom in a well. This condition should be again valid for heavy adatoms, while the opposite condition, "strong and rare" collisions (see below the BGK equation), should be valid for light adatoms. The exponential memory kernel (10) leading to the MFPE (12) is an oversimplified kernel and moreover the memory depends on the system [16]; this heavily complicates the problem and the BGK equation may be considered as a short cut to overcome this difficulty. However, the physical conditions for the validity of one or the other of the BGK and MFPE models are not yet well understood.

Nevertheless, results from the FPE were in very good agreement with Bienfait et al. experiments and in good qualitative agreement with Frenken el al. experiments on premelting systems where diffusing atoms and substrate have the same mass [17].

If $v$ is a "fast" variable, i.e., if the friction $\gamma \to \infty$, the FPE tends to the Smoluchowski equation (SE) where $f$ is a function only of the position. In this case the jump regime is characterized by single jumps only [5, 17].

2.1.2. *The Bhatnagar-Gross-Krook kinetic equation*

The dynamic interaction is pictured in the framework of the Boltzmann equation. "Strong and rare" collisions are assumed; thermalization immediately occurs after each collision, that is, after each collision, the velocity is extracted from the Maxwell distribution $M(v)$ at the surface temperature $T$. In this case the Boltzmann collision integral can be linearized, giving the BGK-like equation for the density probability $f(x,v,t)$ [8],

$$\frac{\partial f}{\partial t} + v\frac{\partial f}{\partial x} + \frac{F(x)}{m}\frac{\partial f}{\partial v} = \eta\left[M(v)\int_{-\infty}^{+\infty} f(x,v,t)dv - f(x,v,t)\right], \tag{14}$$

where $\gamma = \tau\eta$ is the normalized collision frequency. $S_s$ depends on two parameters $\gamma$ and $g$.

As both the BGK equation and the FPE tend to the SE at $\eta \to \infty$, the parameter $\eta$, although having different meanings, plays the same role in both equations.

## 3. Results: The Total Jump Rate and the JLPD

The total jump rate $r_j$ and the JLPD are evaluated applying the MCFM [14] to calculate $S_s$ and then using Eqs. (2), (3), and (4).

As in many cases the rate is evaluated from the Transition State Theory (TST), where

$$r_{TST} = (\omega_{osc}/\pi)\exp(-E_a/k_BT), \tag{15}$$

the ratio $r_j/r_{TST}$, versus $\gamma$, is given in Fig. 1, for the FPE, at $g = 4$ (white circles) and $g = 1$ (white squares), and for the BGK equation at $g = 1$ (black circles). At these values for the barrier the diffusion is fully activated.

All curves show, at some $\gamma_t$, the characteristic Kramers turnover between two limiting behaviors at $\gamma \to 0$ (where $r_j \to 0$ as $\gamma$) and at $\gamma \to \infty$ (where $r_j \to 0$ as $1/\gamma$). From a comparison between the FPE values at $g = 1$ and $g = 4$, it can be seen that the value of $\gamma_t$ lowers as $g$ is increased; at $g = 1$, $\gamma_t^{FPE} = 0.7$, at $g = 4$, $\gamma_t^{FPE} = 0.5$. At the same time, the maximum value of the rate increases with $g$ and then the simple TST value, independent of $\gamma$, is approached, in a narrow intermediate range of friction, only at very large barriers. Calculations at high barriers, such as $g = 4$, or in the extremely underdamped regime, such as $\gamma \leq 0.1$, require too much computational time in the BGK model.

At the same barrier $g = 1$, both the FPE and the BGK equations give the same Smoluchowski limit at $\gamma \to \infty$, but the BGK model gives a jump rate lower and lower than the FPE value with decreasing $\gamma$. Furthermore, $\gamma_t^{BGK} = 0.9$ while $\gamma_t^{FPE} = 0.7$; the BGK turnover value is then shifted towards higher $\gamma$'s. At the same time, the maximum value of $r_j$ is lower and therefore the TST is even a worse approximation in the BGK model. In the FPE framework we have always found $\gamma_t < 1$.

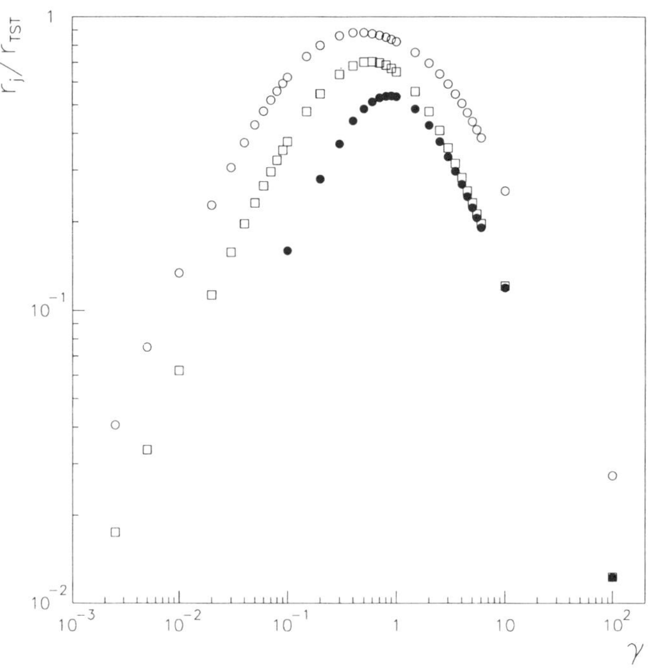


*Figure 1.* $r_j/r_TST$ versus $\gamma$. FPE results at $g = 4$ corresponding to an activation barrier $E_a = 16\ k_B\ T$ (white circles) and at $g = 1$ corresponding to $E_a = 4\ k_B\ T$ (white squares); BGK results at $g = 1$ (black circles).

In Fig. 2 the rate obtained from the MFPE, versus the memory parameter $\gamma_1$, at fixed barrier, $g = 1$, and static friction $\gamma = 1$ is plotted. These new and preliminary results present an interesting memory turnover at some $\gamma_{1,t}$ (here $\gamma_{1,t} \sim 1.5$) depending on the barrier and on the static friction. The exponentially decaying memory kernel induces oscillatory motions even without potential and without noise; in periodic potentials, at small $\gamma_1$, i.e., at sufficiently strong memory, the short-time dynamics of the adatom is characterized by two oscillatory motions: a low frequency memory-induced oscillation over several cells and an oscillatory motion at the well bottoms at higher frequency [9]. As a consequence, the spectrum of the velocity correlation function has two peaks at small $\gamma_1$. Due to the interplay between the peaks, the diffusion coefficient $D$ is never monotonic and has a maximum as a function of $\gamma_1$, at fixed $g$ and $\gamma$; the maximum is very pronounced in resonance conditions, that is when $\gamma > \sqrt{2}\,\omega_0$, and is located at $\gamma_1 \approx {\omega_0}^2/\gamma$ [9]. The resonance condition is not satisfied with the parameters of Fig. 2, where $\omega_0 = \sqrt{2}$; nevertheless the memory-induced oscillations cause the turnover in the rate. This will have interesting consequences on the JLPD as shown in the following.

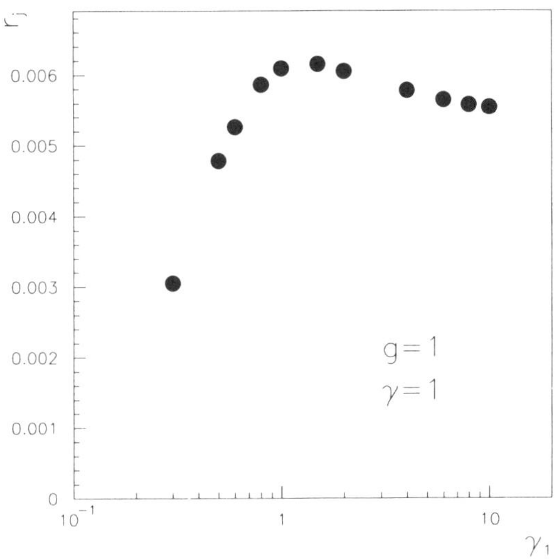


*Figure 2.* $r_j$ versus the memory parameter $\gamma_1$ at $g = 1$ and $\gamma = 1$.

The total probability of multiple jumps $P_m = 1 - P(1)$ and the double-jump probability $P(2)$ are plotted in Figs. 3 and 4, respectively, versus $\gamma$, for the FFE and the BGK models for the same conditions and symbols of Fig. 1.

The FPE and BGK results show the same qualitative behavior. $P_m$ is always decreasing with increasing $\gamma$, then $P(1)$ always increases and single jumps only are activated going towards the Smoluchowski limit. $P(2)$ (as well as all other $P(n)$) is small at very small $\gamma$; in fact, all jump lengths tend to become equally probable and small. At high $\gamma$, $P(2)$ (as well as all other $P(n)$ with $n \neq 1$) tends to zero.

Quantitatively the BGK results are different; at the same $\gamma$ and $g$, the BGK model gives a much larger proportion of long jumps. In fact, as shown from a comparison between Figs. 3 and 1, a significant percentage of long jumps appears only well below the turnover point in the rate, the latter being higher in the BGK model. From a physical point of view the Langevin dynamics is more efficient in thermalizing the velocity of the adatom and then long deterministic trajectories are less likely.

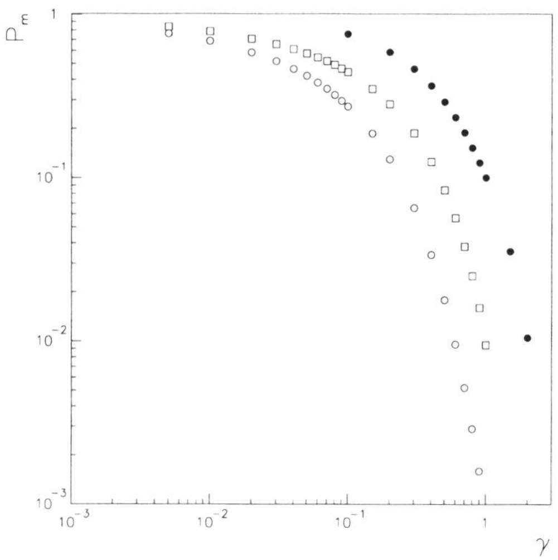


*Figure 3.* The total probability of multiple jumps $P_m = 1 - P(1)$ versus $\gamma$ for the FFE and the BGK model. Symbols as in Fig.1.

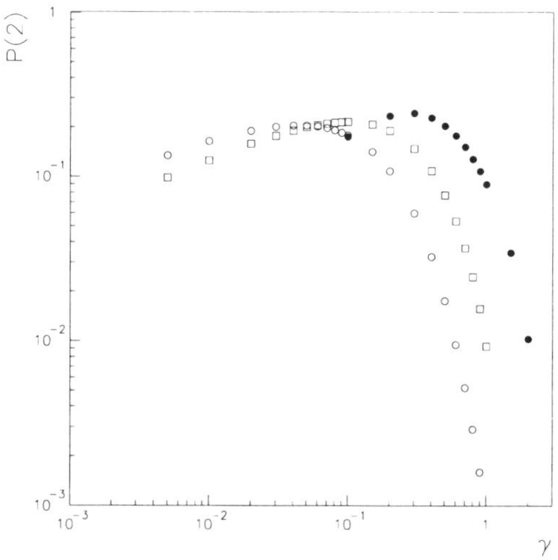


*Figure 4.* The double-jump probability $P(2)$, versus $\gamma$, for the FFE and the BGK models. Symbols as in Fig. 1.

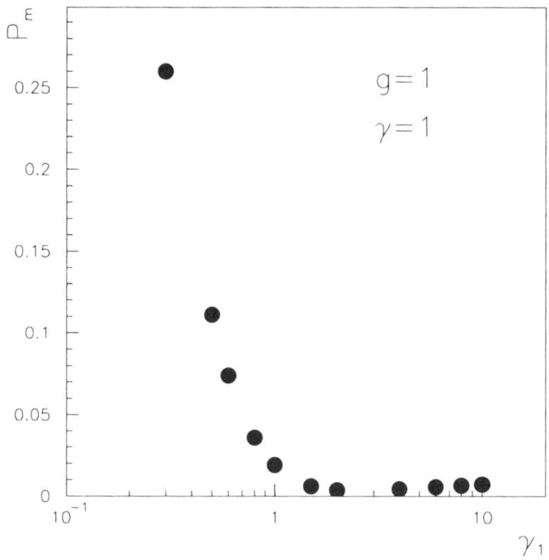


*Figure 5.* The total probability of multiple jumps $P_m = 1 - P(1)$ versus $\gamma_1$ for the MFFE model. Symbols as in Fig. 2.

The total probability of multiple jumps $P_m$ for the MFPE is given in Fig. 5 at $g = 1$ and $\gamma = 1$. Well below the memory turnover ($\gamma_{1.t} \sim 1.5$ in Fig. 2), a significant percentage of multiple jumps appears. At $\gamma_1 = 0.3$ the percentage of multiple jumps is more than 25%; the result is rather surprising because at static friction, $\gamma = 1$, the percentage of long jumps is zero in the absence of memory [5]. Once again, the probability of multiple jumps becomes significant below some turnover point in the rate, $r_j$; in this case, below the memory turnover.

## 4. Conclusions

We have shown that stochastic equations are apt to describe general properties about activated diffusion of single adatoms at crystal surfaces. The total jump rate and the jump-length probability distribution have been calculated from the FPE, the BGK equation and the FPE with memory. The FPE and the BGK dynamics give the same qualitative behavior of the calculated quantities and quantitative different results, long jumps being favored in the BGK case. The presence of memory in the FPE, inducing oscillations of the adatom on several cells, allows the adatom to perform long jumps in conditions where they are strictly forbidden in the absence of memory.

## References


1. Dobbs, K.D. and Doren, D.J. (1992) Dynamics of molecular surface diffusion: Origins and consequences of long jumps, *J. Chem. Phys.*, **97 5**, pp. 3722–3735.
2. Prévot, G., Cohen, C., Schmaus, D. and Pontikis, V. (2000) Non-isotropic surface diffusion of lead on Cu(110): a molecular dynamics study, *Surf. Sci.*, **459**, pp. 57–68.
3. Papageorgiou, D.G. and Evangelakis, G.A. (2000) Multiple excitations and self-diffusion processes on and near the Cu(110) surface by molecular dynamics simulations, *Surf. Sci.*, **461**, pp. L543–L549.
4. Ferrando, R. (1996) Correlated jump-exchange processes in the diffusion of Ag on Ag(110), *Phys. Rev. Lett.*, **76 22**, pp. 4195–4198; Montalenti, F. and Ferrando, R. (1999) Jumps and concerted moves in Cu, Ag, and Au(110) adatom self-diffusion, *Phys. Rev. B*, **59 8**, pp. 5881–5891.
5. Ferrando, R., Spadacini, R. and Tommei, G.E. (1993) Kramers problem in periodic potentials: Jump rate and jump lengths, *Phys. Rev. E*, **48 4**, pp. 2437–2451.
6. Ferrando, R., Spadacini, R., Tommei, G.E. and Caratti, G. (1994) Correlation functions in surface diffusion: the multiple-jump regime, *Surf. Sci.*, **311**, pp. 411–421.
7. Caratti, G., Ferrando, R., Spadacini, R. and Tommei, G.E. (1998) The Kramers problem in 2D-coupled periodic potentials, *Chem. Phys.*, **235**, pp. 157–170; (1996) Noise-activated diffusion in the egg-carton potential, *Phys. Rev. E*, **54 5**, pp. 4708–4721.
8. Ferrando, R., Spadacini, R. and Tommei, G.E. (1993) Diffusion in a periodic potential in the strong collision limit, *Chem. Phys. Lett.*, **202 3,4**, pp. 248–252; Ferrando, R., Montalenti, F., Spadacini, R. and Tommei, G.E. (1999) Long-jump probabilities in a BGK model for surface diffusion, *Chem. Phys. Lett.*, **315**, pp. 153–157; (2000) Long jumps in the strong-collision model, *Phys. Rev. E*, **61 6**, pp. 6344–6350.
9. Caratti, G., Ferrando, R., Spadacini, R., Tommei, G.E. and Zelenskaya, I. (1998) Resonant diffusion in periodic systems with memory, *Chem. Phys. Lett.*, **290**, pp. 509–513 and references therein.
10. Ferrando, R., Spadacini, R. and Tommei, G.E. (1994) Separation of time scales in rate problems at very low barriers, *Chem. Phys. Lett.*, **224**, pp. 308–312.
11. Chen, L.Y. and Ying, S.C. (1999) Solution of the Langevin equation for rare events rates using a path-integral formalism, *Phys. Rev. B*, **60 24**, pp. 16965–16971.
12. Beenakker, J.J.M. and Krylov, S.Yu. (1998) Jump length distribution in molecule-on-substrate diffusion, *Surf. Sci. Lett.*, **411**, pp. L816–L821.
13. Ala-Nissila, T. and Ying, S.C. (1992) Theory of classical surface diffusion, *Prog. Surf. Sci.*, **39 3**, pp. 227–323.
14. Risken, H. (1989) *The Fokker-Planck Equation*. Springer, Berlin.
15. Persson, B.N.J. (1998) *Sliding Friction-Physical Principles and Applications*. Springer, Berlin.
16. Cucchetti, A. and Ying, S.C. (1996) Memory effects in the frictional damping of diffusive and vibrational motion of adatoms, *Phys. Rev. B*, **54 5**, pp. 3300–3310.
17. Ferrando, R., Spadacini, R. and Tommei, G.E. (1993) Theory of diffusion in premelting systems, *Il Nuovo Cimento*, **15 D 2-3**, pp. 557–563 and references therein.


We acknowledge financial support from the Italian MURST under the project N. 9902112831

# ANALYTICAL KINETIC THEORY OF SINGLE-PARTICLE AND COLLECTIVE SURFACE DIFFUSION

S. YU. KRYLOV
*Institute of Physical Chemistry, RAS, Moscow, Russia*

**Abstract.** This paper generalizes the results of a recently developed kinetic theory of surface diffusion conceptually different both from the popular transition state hopping model and from the Fokker-Plank kinetic model (or its equivalent Generalized Langevin approach). Following the first principles of statistical mechanics, the theory is based on the evolution of the particle's probability density, and takes into account the possibility of a finite change in adparticle energy due to its interaction with the substrate excitations. We show how, in this general way under some simplifying assumptions, one can reach a relatively simple, analytical description of surface diffusion that goes far beyond the abilities of the TST model. Examples are given by the occurrence of long jumps and "anomalous" pre-exponential factors of the diffusion coefficient in the low-density limit. Moreover, the theory allows, for the first time, kinetic treatment of surface diffusion at finite occupancy, resulting in a new sight on mechanisms determining the density dependence of collective diffusivity. Some puzzling aspects of surface cluster diffusion (gliding) can also be elucidated.

## 1. Introduction

In the theoretical description of surface diffusion, great progress has so far been achieved [1, 2] in quantifying the "potential" part of the problem, e.g., *ab initio* calculations of intersite potential barriers, potential energy surface, and detailed spectrum of atomic vibrations in the system. Being undoubtedly of principal—and sometimes decisive—importance, information of this type is, however, not yet sufficient to complete the diffusion picture: one still has to account, in some way or another, for the statistical part of the problem. In this respect, the most popular way [2-4] is to use a lattice gas



*M.C. Tringides and Z. Chvoj(eds.),*
*Collective Diffusion on Surfaces: Correlation Effects and Adatom Interactions*, 305–316.

model, with diffusion jump rates derived from the transition state theory (TST). As far as temperature ($k_BT$) is small enough with respect to the potential surface corrugation ($V_m = V_{\max} - V_{\min}$), the lattice gas approximation is, of course, reasonable. However, the TST approximation is a very strong oversimplification (a fact which is often underestimated in the literature), because it is based on quasi-equilibrium treatment of the system and ignores the role of the particle's coupling to the substrate excitations in the dynamics of jumps. It comes as no surprise that such theories fail to explain some important details like, say, the "anomalous" prefactors of the diffusion coefficient often observed in practice [5].

The first principles of nonequilibrium statistical mechanics [6] prescribe how to describe surface diffusion in terms of a kinetic theory. This approach follows evolution of the probability density of particle(s) residing, in our case, in the surface adsorption potential and interacting with the substrate bath. If $k_BT << V_m$, such an approach can be simplified to a discrete, lattice-gas-like description, but with jump rates microscopically justified.

Kinetic treatments of this type, in application to surface diffusion, have so far been less popular, and nearly all, with very few exceptions [7], have been restricted to the Fokker-Plank (FP) model of particle–substrate-bath coupling [4, 8]. On the one hand, FP approach is rather complicated mathematically, so that it generally requires numerical calculations, and its use is usually restricted to the low density limit only. On the other hand, the approach is as yet not general, in particular, it implies diffusive motion in phase space, and hence it (tacitly) assumes the characteristic change of the particle energy ($\Delta$) and momentum caused by the substrate excitations to be small. In fact, for particle-phonon interactions, one expects $\Delta$ to be of the order of the Debye temperature ($T_D$) of the substrate, $\Delta \sim k_BT_D$. This quantity is not much smaller, but often even larger than the temperature of the experiment, $\Delta \geq k_BT$. Earlier the role of finite $\Delta$ was realized in the theory of adsorption, desorption, and accommodation [9, 10].

In this paper we generalize and extend the results of a series of recent works [11-17] in which a new kinetic approach to the theory of surface diffusion has been developed. The possibility of finite-value energy exchange ($\Delta$) between the particle and substrate excitations is taken into account from the start. Assuming $k_BT \leq \Delta << V_m$, the kinetic equation for the one-particle distribution function can be reasonably simplified, in order to reach relatively simple analytical solutions both for the diffusion coefficient and for jump length distribution. This allows one to clearly visualize (in contrast to speculative interpretations of computed results) some inherent nonequilibrium features of surface diffusivity that lie far beyond the predictions of TST-based models. This is, in particular, the possibility of a strong deviation of the diffusion prefactor from the "normal" value (as given by

TST). More importantly, the developed kinetic theory allows for the first time a generalization to the case of finite occupancy, resulting in a new sight on mechanisms determining the density dependence of collective diffusion. Moreover, the theory can be extended to highlight recent intriguing observations [18] of gliding diffusivity of large surface clusters. The discussion below is mainly restricted to a one-dimensional case for which a more or less rigorous analytical treatment is possible.

## 2. Outline of the Theory

### 2.1. KINETIC EQUATION

From the point of view of nonequilibrium statistical mechanics (physical kinetics [6]), a natural way to describe surface diffusion is to consider evolution of the one-particle distribution function (probability density) $f(v, x, t)$ of adatoms residing in a periodic potential field $V(x)$ of the surface and interacting with thermal excitations of the substrate. Neglecting memory effects, the kinetic equation can generally be written in the form

$$\frac{\partial f}{\partial t} + v\frac{\partial f}{\partial x} - \frac{1}{m}\frac{dV}{dx}\frac{\partial f}{\partial v} = I_S\{f\} + I_M\{f\} , \tag{1}$$

with $v$ and $m$ the velocity and mass of the particle, $I_S\{f\}$ a dissipative term taking into account the interaction with the substrate bath, and $I_M\{f\}$ a term accounting for the effect of interparticle interactions. Assuming for simplicity the particle–substrate-bath coupling to be independent of $x$, $I_S\{f\}$ can generally be written as

$$I_S\{f\} = \int f^{\pm}(E', x)W(E' \to E)dE' - f^{\pm}(E, x)\int W(E \to E')dE' , \tag{2}$$

with $E = mv^2/2+V(x)$ the total energy of the particle, and $W(E' \to E)$ the probability of transition per unit time between levels $E'$ and $E$ accompanied by creation or annihilation of phonons (or other thermal excitations) in the solid. In view of the detailed balance principle, the transition probability obeys the reciprocity relation

$$\exp(-E/k_BT)W(E \to E') = \exp(-E'/k_BT)W(E' \to E) . \tag{3}$$

Microscopic derivation of (3) is concerned with an averaging over all degrees of freedom of the solid which is considered as a thermal bath [19].

If the characteristic value ($\Delta$) of $|E' - E|$ is sufficiently small, the general expression (2) can be reduced to the diffusive Fokker-Plank form used by many authors [4, 8]. As a matter of tradition which dates back to Kramers [20] (rather than a matter of microscopical foundation), the FP model, together with its analog the generalized Langevin approach (GL), are sometimes believed to be one of the most sophisticated ways to analyze activated surface phenomena [20]. The discussion of applicability of FP/GL theory, e.g., in connection with memory effects, is usually restricted to the relation between characteristic frequencies of the system [21]. However, the question of the role of finite $\Delta$ in diffusion kinetics remains open.

An alternative way of treating $I_S\{f\}$, as proposed in [7], is to use a BGK-like model, a rather simple model which implies a very large value of $\Delta$, and which cannot be derived from the master equation ( 2). Both in FP and BGK variants, the kinetic Eq. (1) remains complicated and generally (except in limiting cases of strong and weak friction) requires numerical solution [7, 8].

Instead of exactly solving essentially approximate equations, one may try to develop an approximate treatment of the problem on the basis of "exact" Eqs. (1) and (2). Our aim is to achieve a relatively simple analytical treatment of surface diffusion that would keep—in contrast to TST—more or less correct dynamics of adparticle–substrate-bath interaction.

### 2.2. PHYSICAL PICTURE

As long as $k_BT << V_m$, adparticles spend most of the time in bound states ($E < V_m$) in the potential wells, so that one observes a lattice gas. Jumps between the adsorption sites result from excitation of adparticles to unbound states ($E > V_m$) where they have enough energy to overcome the potential barrier. Motion in unbound states is thus the key requirement for diffusion to take place, though population of these states is exponentially small. Instead of introducing in *ad hoc* manner (TST) hypothetical transition state from which the particle will necessarily fall down to the neighboring adsorption well, kinetic Eq. (1) relates particle's preferences with its coupling to the substrate excitations (friction). Depending on friction, an excited (unbound) particle can prefer to fall back into the well of its origin (i.e., no jumps occur), or it can fly over several lattice distances until it will be trapped (a long jump). Furthermore, once being trapped, the particle can fall down the well, but it can also be excited again (re-trapping). Basically, this is the physics behind any kinetic approach, including the FP model.

Remarkably, this picture, though very complicated in its mathematical description [4, 8] when $\Delta$ is small, can be substantially simplified in case of finite $\Delta$, namely when $k_BT < \Delta << V_m$. Expecting $\Delta \sim k_BT_D$ (in view of multiphonon character [22] of interaction $\Delta$ can be even larger than $k_BT_D$), one may believe this condition to cover nearly all temperature intervals of interest in surface diffusion experiments. If $k_BT < \Delta$, transitions downward in energy are much more probable than upward transitions, as follows from (3). Consequently, re-trapping is not very important and can be neglected. Furthermore, one can neglect unbound-unbound transitions ($E, E' > V_m$) in a narrow zone of width about $k_B T$ above $V_m$ (a "conduction zone" responsible for intersite jumps). Consequently, the effective friction experienced by an unbound particle will be determined by the unbound-bound transitions only. With these simplifications, Eq. (1) with (2) turns out to be analytically solvable, in order to reach both jump length probabilities and (with one additional assumption) the diffusion coefficient.

### 2.3. JUMP LENGTH DISTRIBUTION

Using the above simplifications, one can easily obtain the jump length distribution from a formal solution of Eq. (1) with (2) for the evolution of an unbound particle [13, 14]:

$$p_k = (1-\alpha)^{k-1}\alpha \ , \ \alpha = 1 - \exp(-t_{fl}/\tau) \ . \tag{4}$$

Here $p_k$ is the probability of a jump of length $k$ in units of the lattice spacing $l$, and $\alpha$ can be interpreted as the probability of being trapped in a well. $t_{fl}$ is the mean time of the (unbound) particle flight between two saddle points, while $\tau$ is determined by the summed frequency of downward transitions from a certain energy state $E$,

$$\tau^{-1} = \int_{E'<E} W(E \to E')dE' \ , \tag{5}$$

which is independent of $E$ since $\Delta << V_m$.

The jump length distribution (4) is completely determined by a dimensionless parameter $t_{fl}/\tau$ which characterizes the effective friction experienced by an unbound particle (one should notice a certain analogy and an essential difference with the usual definition of friction as introduced in FP/GL theory). The time-of-flight $t_{fl}$ is explicitly expressed [12, 14] in terms of $V(x)$, and it can be easily calculated for any given potential. This quantity is typically of the order of $10^{-12}\,s$ (in the 1D case), it depends on the shape of the potential, and it only weakly depends on temperature (weaker than $T^{-1/2}$). As far as the particle-surface relaxation time $\tau$ is concerned, it is not a simple problem to calculate this quantity under actual conditions of multiphonon coupling to the substrate [10]. However, information about analogous quantities of the same physical nature is available from extensive studies of various molecule-surface interaction phenomena (see [10] and references therein). Accordingly, one expects $\tau$ to be an increasing function of temperature, with typical values around $10^{-13}$–$10^{-12}\,s$.

With these characteristic values of the parameters involved, the result (4) predicts the occurrence of extended jumps to vary from system–to–system between situations where it is negligible and where it is of the order of several tens percent. In the case of two-dimensional surface diffusion, the situation is complicated by the fact that the corresponding time-of-flight parameter $t_{fl}^{(2D)}$ is typically 1 to 2 orders of magnitude larger than $t_{fl}^{(1D)}$ [15]. This increase is caused by the multipassage character of adparticle motion in the unbound states due to the coupling between the two translational degrees of freedom in the surface potential field. In contrast to the 1D case, an excited (unbound) particle still has to find a narrow "window" (at saddle point) to be able to escape from a two-dimensional potential cell. The physics manifested here is analogous to that discussed qualitatively in [23] . Since $t_{fl}$ enters the theory as a quantity determining (in its competition with $\tau$) the effectiveness of particle-substrate relaxation, the above topological effect can be treated as an increase in the effective friction. As a result, the occurrence of extended jumps in the 2D case should be small.

The result (4) readily suggests simple explanations [13, 14] for the occurrence of double and triple jumps recently observed in direct experiments for quasi-one-dimensional surface diffusion in $Pd/W(211)$ [24] and $Pt/Pt(110)$ [25] systems. As an important feature, we note here a very sharp exponential dependence on the effective friction parameter in (4). Not only does it lead to a simple explanation [13] for the apparent additional activation barrier for double hops [25], but it also suggests a key [13] to understanding an extremely strong decay of long jump with decreasing temperature as observed for $Pd/W(211)$ [24]. This last result was difficult to explain in the framework of FP/GL theory.

### 2.4. DIFFUSION COEFFICIENT

Calculation of the diffusion coefficient is a more complicated problem and it needs an additional assumption. Let us now assume that population of bound states ($E < V_m$) is as in equilibrium. The statement seems obvious for "deep" states near the bottom of the wells, where the particles spent a time sufficiently large for equilibration. Since re-trapping is neglected (see above), the assumption seems reasonable also for all bound states. (Note, this is a partial reminiscence of TST, where populations of all energy levels, including unbound states, are taken as in equilibrium.) With this and the above assumptions, and using the reciprocity relation (3 ), the general expression (2) can be easily transformed for unbound particles to a simple form $I_S|_{E \geq V_m} = -(f^{\pm} - f_0)\tau^{-1}$, with $f_0$ the equilibrium distribution, and with $\tau$ determined by (5). Note, this is not a BGK-like model, but a direct consequence of the master Eq. (2) for unbound particles. Now the kinetic

Eq. (1), with $I_M\{f\} = 0$, can be easily solved analytically with respect to particle transport on a macroscopic scale [12, 14]. The diffusion coefficient in the low density limit (when the collective and tracer diffusion coefficients coincide) is written as

$$D = D_0 \exp(-\frac{V_m}{k_B T}) \ , \ D_0 = 2 \left(\frac{t_{fl}}{\tau}\right)^{-1} l^2 \nu \ , \tag{6}$$

with $l$ the period of the potential $V(x)$ and $\nu$ the frequency of adparticle vibration at the bottom of the potential wells.

In addition to the trivial dependence on $l^2$, $\nu$ and $\exp(V_m/k_B T)$ (as predicted by TST), there appears in a natural way an effective friction parameter $t_{fl}/\tau$, the same as in (4). The physics behind the appearance of this parameter in (6) is clear. If the particle–bath coupling (as described by $t_{fl}/\tau$) is strong, an excited (unbound) particle prefers to fall back into the well of its origin, so that the jump rate and hence $D$ are small. In the opposite case of a weak coupling, $D$ is large as a result of contribution of long jumps, see (4), although the jump rate in this case is small again [14].

The classical result of the transition state approximation is formally retrieved by (6) in the particular case when $t_{fl}/\tau = 2$ (which of course is not expected to have any special physical meaning in view of microscopically unjustified assumptions used in TST).

Strictly speaking, the above assumption of equilibrium in the wells is restrictive, and it bounds the result (6) to the case of not too strong effective friction, presumably $t_{fl}/\tau < 10$ [14]. Otherwise, expression (6) is expected to strongly underestimate diffusivity. In the case of a very strong coupling to the substrate excitations, one cannot assume equilibrium in the wells, but at least local equilibrium, and then one has to account for particle flux inside the wells (in analogy with Smoluchowski regime in FP/GL theory [20]).

## 3. On the Problem of Anomalous Prefactors

One old puzzle in the field of surface diffusion is concerned with a strong discrepancy between the theoretical expectation and experimental results for the preexponential factor ($D_0$) of the diffusion coefficient (as observed in the low coverage regime). Experimental data for $D_0$ obtained for various systems [5] are scattered in an interval that spans more than 8 orders of magnitude around a "normal" value $D_0 \sim l^2\nu \sim 10^{-4} - 10^{-3}\, cm^2/s$, resulting from all TST-based theories. While in the framework of TST, neither accounting for realistic potential energy surface nor addressing detailed spectra of atomic vibrations in the system [2, 26] could produce a significant change in the theoretical prediction.

Kinetic theory, in addition to TST result for $D_0$, brings out an essential dependence of diffusivity (6) on the effective friction parameter $t_{fl}/\tau$, which—in contrast to $l$ and $\nu$—can vary significantly from one system to another. From this point of view, the possibility of a significant deviation of the prefactor from the "normal" value is not really a puzzle [15]. Our analytical result (6 ) allows us to recognize several physical reasons for that. *First*, this is an expected (although hardly quantifiable microscopically) dependence of the relaxation time $\tau$ on the spectrum of surface phonons (and hence on morphology of the surface) and, possibly, on other types of the substrate excitations. If only phonons are involved into the interaction, typical values [10] of $\tau$ are around $10^{-13} - 10^{-12}\,s$, but one expects a decrease of $\tau$ (up to two orders of magnitude) in cases when electron-hole pairs also contribute to the relaxation. *Second*, we point out the expected variation of $\tau$ with temperature. Thus, as suggested by experiments on molecule-surface energy accommodation, there can be an order of magnitude change [10] in the relaxation time in a temperature interval $100 - 300\,K$. *Third*, preexponential factor produced by the kinetic theory essentially depends, via $t_{fl}$, on the shape of the potential energy surface and its dimensionality. In the one-dimensional case $t_{fl}^{(1D)} \sim 10^{-12}\,s$, but in the case of a two-dimensional surface diffusion, the corresponding time-of-flight parameter $t_{fl}^{(2D)}$ is typically 1 to 2 orders of magnitude larger, as discussed above.

Note, this last effect can cause a systematic difference between the prefactors of diffusion on surface terraces and over surface steps, since motion over the activation barrier is in these cases effectively two- and one-dimensional, respectively [15]. Should only $t_{fl}$ be concerned, one expects $D_0^{(\text{down steps})}$ to be up to two orders of magnitude larger than $D_0^{(\text{terrace})}$. Remarkably, this prediction directly corresponds to recent observations for the $Ag/Ag(111)$ system obtained using four different methods [27].

## 4. Collective Diffusion at Finite Occupancy

The principal achievement of the approach proposed is the possibility to develop, for the first time, a kinetic theory of surface diffusion at finite occupancy. Averaging the distribution function over one potential unit of the length $l$, the kinetic Eq. (1) can be reduced to a form characteristic for (low dimensional) gases. This gives the opportunity to use a collision integral $I_M\{f\}$ of the type introduced by Enskog to take into account the finite size ($\sigma$) of the molecules [11]. Assuming the adparticles to interact as hard particles, the expression for the collective diffusion coefficient can be written [12, 14] as

$$D_c(n) = DY(n)Z(n)\ , \tag{7}$$

with $D$ given by (6). The two factors given by

$$Y(n) = (1 - n\sigma)^{-2} \ , \ Z(n) = \left(1 + 2\xi \frac{\tau}{t_{fl}} \frac{nl}{1 - n\sigma}\right)^{-1} \tag{8}$$

determine the dependence of the diffusivity on density $n$. The factor $\xi$ can vary between 0 and 1 depending on $\sigma/l$ (for truly hard particles); $\xi = 1/2$ is expected to be a good approximation.

The physics behind the factor $Y$ in (7) and (8) can be treated in two ways. In the course of the kinetic derivation, $Y$ originates from the nondissipative part of the collision integral $I_M\{f\}$ and reflects an effect of the finite size of adparticles on diffusivity in the presence of a density gradient. However, $Y$ can also be derived [14] directly from the equation of state of the system (without any additional assumptions at all) as a thermodynamic factor [3, 4] that relates the particle fluxes under the action of density and pressure gradients. One observes in this way the microscopic essence of the thermodynamic factor, the quantity usually introduced in $D(n)$ for formal reasons only. One observes also an essential difference between the thermodynamic factors for the system of hard particles considered here and for the lattice gas model with site exclusion [3, 4], where $Y = 1/(1 - \theta)$, with $\theta$ the occupancy ($\theta = nl$).

The factor $Z$ in (7) and (8) has an essentially kinetic nature, because it originates from the dissipative part of the collision integral $I_M\{f\}$ in (1). Physically, it describes an additional contribution to particle-substrate friction concerned with interaction (collisions) between the adparticles.

The developed theory allows us [14] to describe collective diffusion also in traditional (but, in fact, more complicated) terms of jump rates. In this way, the factor $Z$ is seen as an effect of the reduction of the effective jump rates due to adparticle interactions (in some analogy to the site exclusion effect [3]), while $Y$ reflects an anisotropy of the jump rates in the presence of the density gradient.

While $Y$ gives rise to an increase of the diffusivity with increasing density, $Z$ has the opposite effect, but they do not compensate each other. For instance, for values of the parameters realistic for physisorbed particles, the net effect is an increase of $D$ with increasing density, in accordance with experiments [12]. Furthermore, the kinetic approach based on the hard particle approximation leads to a variety of trends in the behavior of diffusivity as a function of occupancy. Depending on the value of the effective friction parameter $t_{fl}/\tau$, expressions (7) and (8) predict either a monotonous increase of $D(\theta)$ or nonmonotonous dependence with a minimum [14].

Further development of kinetic theory is needed in order to incorporate the finite-distance interaction between adparticles into the kinetic scheme. This should bring up a dependence of the effective activation barrier on occupancy (the physics that is introduced by lattice gas models in an *ad hoc* fashion [3]). Note that the density dependent activation barrier will be an additional mechanism with respect to the two discussed above, but not the only one determining the density dependence of diffusivity. In view of their generality, these two mechanisms—the collisional relaxation (as described by $Z$) and the gradient-induced anisotropy of mobility ($Y$)—will manifest themselves in any physical system.

## 5. Gliding Diffusion of Surface Clusters

Surface mobility of adparticle clusters (islands) can be determined by several mechanisms [28] concerned with the individual motion of atoms (or groups of atoms) of the cluster. One of the most exciting experimental observations in the field is concerned with the gliding-like motion of a two-dimensional cluster as a whole. Such a behavior was recently reported for activated diffusion of close-packed hexagonal clusters $Ir_7$ and $Ir_{19}$ on an $Ir(111)$ surface [18]. Not only is this collective effect curious in itself, there are at least three puzzles in the observations: (i) the pre-exponential factor of the cluster diffusion coefficient is unusually large (about 4 orders of magnitude larger than that for one-particle diffusivity); (ii) there is a significant contribution of long jumps for large ($Ir_{19}$) clusters; and (iii) there is no expected correlation between the large prefactor and the occurrence of long hops. The developed kinetic approach allows us to explain all these observations attributing them to the manifestation of two interesting effects [16, 17].

Since the cluster moves as a whole, it can be considered as a particle with internal degrees of freedom. In this case the above results (6) and (4) can be directly applied, with the parameters involved still to be specified. Calculating the total cluster-surface interaction $V^{(c)} = \sum_n V^{(1)}(x_n)$, with $V^{(1)}(x_n)$ the interaction of its $n$th atom with the surface, one observes [16] the center of mass of the cluster to move in a periodic "potential" (with period $l$) whose corrugation depends on an instant internal configuration and, hence, to oscillate in time. Since in our case intra-cluster interaction is obviously strong with respect to adatom-surface interaction, it is indicated to average $V^{(c)}$ over fast internal vibrations. The diffusion barrier (for the cluster as a whole) is found to be $V_m^{(c)} = NV_m^{(1)} - \delta V_{\text{stat}} - p(N)k_BT$, with $N$ the number of atoms in the cluster, and $\delta V_{\text{stat}}$ a contribution concerned with (static) misfit between the cluster and substrate lattices. The third term in $V_m^{(c)}$ reflects the effect of "dynamical" misfit caused by the internal vibrations, a kind of entropic effect. Since this contribution is proportional to temperature, it will manifest itself not in the activation energy of the cluster diffusion, but in the preexponential factor, $D_0^{(c)} \propto \exp[p(N)]$. The dynamical misfit factor $p(N)$ can be calculated exactly for a long linear array of particles with nearest-neighbor interactions. The result obtained predicts a very strong increase of the prefactor with increasing $N$ [16], and it readily explains the high experimental values observed [18] for $Ir_7$ and $Ir_{19}$ clusters.

According to the general formulae (4), the cluster jump length distribution (for the gliding motion) is determined by corresponding friction parameter $t_{fl}^{(c)} / \tau^{(c)}$. The time-of-flight $t_{fl}^{(c)}$ remains close to the value characteristic for single adatoms, because both the potential corrugation and mass increase with $N$. For the relaxation time $\tau^{(c)}$ this is not the case. For a single adatom, generally speaking all phonon modes contribute to the relaxation. For a large cluster the situation is essentially different by the fact that the most energetic short-wave phonon modes, with wave length smaller than the linear size of the cluster, can effectively interact with intra-cluster vibrations, but not with the cluster's motion as a whole. The latter is significantly coupled to long-wave phonon modes only, and such a discrimination increases with $N$. As a result of this size effect, $\tau^{(c)}$ should be an increasing function of $N$. Note that in view of a strong exponential dependence of the jump length distribution (4) on the relaxation time, even a small increase of $\tau^{(c)}$ can lead to quite a significant increase in the occurrence of extended hops. One arrives at a conclusion that might seem paradoxical: *larger clusters are inclined to exhibit longer jumps.* This readily explains the observations [18]. For $Ir_{19}$ the occurrence of double hops observed, $p_2^{(19)}/p_1^{(19)} \approx 0.2$, corresponds to $t_{fl}^{(19)}/\tau^{(19)} \approx 1.6$ in (4). For $Ir_7$ clusters long hops are rare, $p_2^{(7)}/p_1^{(7)} \approx 0.04$, but the corresponding value of the friction parameter is only twice as large, $t_{fl}^{(7)}/\tau^{(7)} \approx 3.2$. This suggests a difference in the relaxation times by a factor of 2, quite a reasonable value to be attributed to the difference in the linear sizes of $Ir_7$ and $Ir_{19}$.

This paper concludes a series of works started with professor J. J. M. Beenakker (The Netherlands), who passed away in July, 1998, and whose scientific spirit permanently stimulates the author. Thanks are also due to M. C. Tringides for attracting the author's interest to some of the problems considered, and to L. J. F. Hermans for helpful discussions.

## References

1. Feibelman, P.J. (1989) Theory of adsorbate interactions, *Annu. Rev. Phys. Chem.* **40**, 261–290.
2. Ratsch, C. and Scheffler, M. (1998) Density-functional theory calculations of hopping rates of surface diffusion, *Phys. Rev.* B**58**, 13163–13166.
3. Gomer, R. (1990) Diffusion of adsorbates on metal surfaces, *Rep. Prog. Phys.* **53**, 917–1002.
4. Barth, J.V. (2000) Transport of adsorbates at metal surfaces: From thermal migra-

tion to hot precursors, *Surf. Sci. Rep.* **40**, 75–149.
5. Seebauer, E.G. and Allen, C.E. (1995) Estimating surface–diffusion coefficients, *Prog. Surf. Sci.* **49**, 265–330.
6. Lifshitz, E.M. and Pitaevskii, L.P. (1981) *Physical Kinetics*, Pergamon, Oxford.
7. Ferrando, R., Spadacini, R. and Tommei, G.E. (1993) Diffusion in a periodic potential in the strong collision limit, *Chem. Phys. Lett.* **202**, 248–252.
8. Ferrando, R., Spadacini, R. and Tommei, G.E. (1993) Kramers problem in periodic potentials – jump rate and jump length, *Phys. Rev.* E**48**, 2437–2451.
9. Kreuser, H.J. and Gortel, Z.W. (1986) *Physisorption Kinetics,* Springer Ser. Phys. Chem., Springer, Berlin.
10. Borman, V.D., Krylov, S.Yu. and Prosyanov, A.V. (1988) Theory of nonequilibrium phenomena at gas–surface interface, *Sov. Phys. JETP* **67**, 2110–2121.
11. Beenakker, J.J.M. and Krylov, S.Yu. (1997) One-dimensional surface diffusion: Density dependence in a smooth potential, *J. Chem. Phys.* **107**, 4015–4023.
12. Krylov, S.Yu., Prosyanov, A.V. and Beenakker, J.J.M. (1997) One dimensional surface diffusion: Density dependence in a corrugated potential, *J. Chem. Phys.* **107**, 6970–6979.
13. Beenakker, J.J.M. and Krylov, S.Yu. (1998) Jump length distribution in molecule-on-substrate diffusion, *Surf. Sci.* **411**, L816–L821.
14. Krylov, S.Yu., Beenakker, J.J.M. and Tringides, M.C. (1999) On the theory of surface diffusion: kinetic versus lattice gas approach, *Surf. Sci.* **420**, 233–249.
15. Krylov, S.Yu. and Tringides, M.C. (1999) Anomalous preexponential factors in surface diffusion: A kinetic look into the problem, ECOSS-18, *Europhys. Conf. Abst.* **23**G, abstract Tu1440.
16. Krylov, S.Yu. (1999) Surface gliding of large low-dimensional clusters, *Phys. Rev. Lett.* **83**, 4602–4605.
17. Krylov, S.Yu. (2000) Failure of 1D models for Ir island diffusion on Ir(111) – Krylov replies, *Phys. Rev. Lett.* **85**, 1581–1584.
18. Wang, S.C., Kurpick, U. and Ehrlich, G. (1998) Surface diffusion of compact and other clusters: Ir-x on Ir(111), *Phys. Rev. Lett.* **81**, 4923–4926.
19. Maksimov, L.A. and Ilyin, A.V. (1974) *Physical Kinetics*, MPhTI, Moscow.
20. Hänggi, P., Talkner, P. and Borkovec, M. (1990) Reaction-rate theory – 50 years after Kramers, *Rev. Mod. Phys.* **62**, 251–341.
21. Cucchetti, A. and Ying, S.C. (1996) Memory effects in the frictional damping of diffusive and vibrational motion of adatoms, *Phys. Rev.* B**54**, 3300–3310.
22. Hofmann, F., Toennies, J.P. and Manson, J.R. (1997) The transition from single phonon to multiphonon energy transfer in atom-surface collisions, *J. Chem. Phys.* **106**, 1234-1247.
23. Chen, L.Y., Baldan, M.R. and Ying, S.C. (1996) Surface diffusion in the low-friction limit: Occurrence of long jumps, *Phys. Rev.* B**54**, 8856–8861.
24. Senft, D.C. and Ehrlich, G. (1995) Long jumps in surface diffusion – one-dimensional migration of isolated adatoms, *Phys. Rev. Lett.* **74**, 294–297.
25. Linderoth, T.R., *et al.* (1997) Surface diffusion of Pt on Pt(110): Arrhenius behavior of long jumps, *Phys. Rev. Lett.* **78**, 4978–4981.
26. Kürpick, U., Kara, A. and Rahman, T.S. (1997) Role of lattice vibrations in adatom diffusion, *Phys. Rev. Lett.* **78**, 1086–1089.
27. Roos, K.R. and Tringides, M.C. (2000) Determination of interlayer diffusion parameters for Ag/Ag(111), *Phys. Rev. Lett.* **85**, 1480–1483.
28. Jensen, P. (1999) Growth of nanostructures by cluster deposition: Experiments and simple models, *Rev. Mod. Phys.* **71**, 1695–1735.

# ADATOM ISLAND DIFFUSION ON METAL FCC(100) SURFACES

O. S. TRUSHIN * , J. HIRVONEN , J. HEINONEN , P. SALO , M. ALATALO AND T. ALA-NISSILA
*Helsinki Institute of Physics and Laboratory of Physics, Helsinki University of Technology, P.O. Box 1100, FIN-02015 HUT, Espoo, Finland*

I. KOPONEN
*Department of Physics, University of Helsinki, P.O. Box 9, FIN-00014 University of Helsinki, Finland*

AND

J. MERIKOSKI
*Department of Physics, University of Jyväskylä, P.O. Box 35, FIN-40351 Jyväskylä, Finland*

**Abstract.**

We study the energetics and atomic mechanisms of diffusion of adatom islands on fcc(100) metal surfaces. For small islands, we perform detailed microscopic calculations using semi-empirical embedded-atom model and glue potentials in the case of Cu and Al, respectively. Combining systematic saddle-point search methods and molecular statics simulations allows us to find all the relevant transition paths for island motion. In particular, we demonstrate that there are novel many-body mechanisms such as internal row shearing which can, in some cases, control the island dynamics. Next, we show how using the master equation formalism, diffusion coefficients for small islands up to about five atoms, can be calculated analytically from the knowledge of the transition rates. In the case of larger islands, we perform large-scale kinetic Monte Carlo simulations based on microscopic single particle energetics. These results show that, as a function of the island size, there is a crossover from periphery dominated mass transport to a regime controlled by the motion of vacancies inside the island.



*M.C. Tringides and Z. Chvoj (eds.),*
*Collective Diffusion on Surfaces: Correlation Effects and Adatom Interactions*, 317–325.

## 1. Introduction

Mobility of 2D clusters (islands) on solid surfaces plays an important role in crystal growth processes. Therefore, understanding how islands move on surfaces is essential for the construction of realistic models for thin film growth [1]. Experimental methods, such as field ion microscopy [2] and scanning tunneling microscopy [3], can provide direct microscopic information about atomic configurations of the islands. However, usually these methods do not give any conclusive information on the microscopic mechanisms of diffusion processes and thus theoretical studies are needed.

Single adatom diffusion mechanisms on metal surfaces are mostly well understood. However, island diffusion mechanisms have turned out to be nontrivial. For the case of fcc(100) surfaces, experiments suggest that the diffusion of small Rh adatom clusters on surfaces takes place by the sequential displacements of edge atoms [4]. These findings have been confirmed in recent embedded atom model (EAM) calculations by Shi *et al.* [5] for Ag, Cu, and Ni, up to the size of $s = 9$ atoms. They found a new mechanism for cluster diffusion through concerted dimer shearing which gives better agreement with experiments for the oscillatory behavior of diffusion as a function of cluster size. The idea of dimer shearing is based on bond-counting arguments which work in many cases for fcc(100) metal surfaces [5, 6, 7].

Regarding large islands, recent experiments have shown that in metal-on-metal systems, even large clusters can have significant mobility at moderate temperatures [8]. This is due to such basic transport mechanisms as evaporation-condensation (EC), terrace (or vacancy) diffusion (TD), and single-atom diffusion along the periphery (PD) [9]. However, the analysis of experimental results for the size-dependence of island diffusion is nontrivial because of crossover effects between the different mechanisms, as well as because of the presence of other microscopic mechanisms, such as vacancy-mediated ripening [10]. Computer simulations of island diffusion are also difficult because of the inherently slow dynamics of larger islands.

In this review, we consider the diffusion of adatom islands on metal fcc(001) surfaces at the two extremes. We first show how the energetics and dynamics of small islands (up to less than about 20 atoms) can be understood from microscopic calculations. Second, we show how the microscopic information can be used to unravel the dominant mass transport mechanisms of very large islands through kinetic Monte Carlo computer simulations.

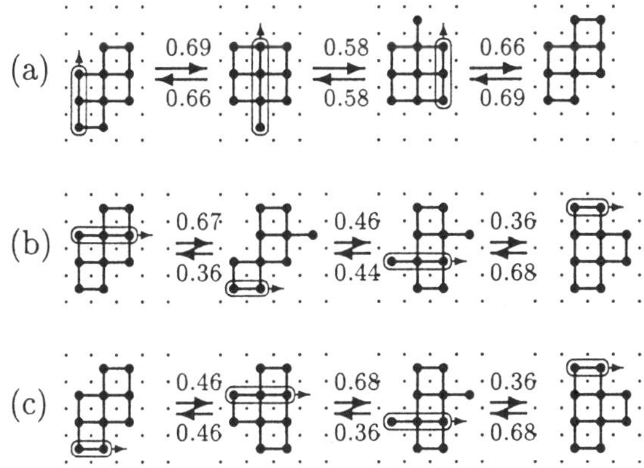


*Figure 1.* Series of processes for the CM motion of a ten-atom Cu cluster. The RLS is a trimer shearing inside the cluster with $E_a = 0.67$ eV in (b). The last step is not shown in (b) and (c).

## 2. Small Islands

### 2.1. ENERGETICS OF COPPER AND ALUMINUM ISLANDS

The dimer-shear mechanism suggests that other concerted many-particle diffusion mechanisms might exist and play a role on fcc(100) surfaces. We have recently shown that this is indeed the case [11]. Our calculations have been made possible by the evolution of novel systematic methods for finding transition paths (saddle points) for many-particle systems [12, 13]. We have applied such methods and studied the energetics and stability of small Cu islands up to $s = 10$ atoms on the Cu(100) surface using the EAM potential [14] of Ref. [5][1]. We have shown that there exist competing many-particle mechanisms which previously have been overlooked. In particular, the dimer-shear mechanism is just a special case of a more general row-shear mechanism that controls the energetics of some smaller islands. This row shearing may occur inside the island with an activation barrier $E_a$ lower than at the periphery. Summary of the main results for islands up to $s = 10$ are listed in Table I, and a series of transitions for a ten-atom island are shown in Fig. 1.

For the case of Cu(100), we have also studied possible transition paths for the interlayer transport of Cu adatoms at low temperatures. To this

[1]Within the EAM (and effective medium theory [7]) potential the energetics of Ag and Ni islands should be qualitatively similar to Cu.

TABLE 1. Equilibrium shapes of small Cu islands up to ten atoms, with the corresponding rate limiting steps (RLS) with activation energies $E_a$ for the center-of-mass (CM) motion (boldfaced) and dissociation listed. In every case the RLS involves only one saddle point. We also show the corresponding values from Ref. [5] when available.

| processes | $E_a$ (eV) | $E_a$ (eV) in Ref. [5] |
|---|---|---|
|  | **0.488** | 0.503 |
|  | 0.484, **0.820** | 0.494, – |
| 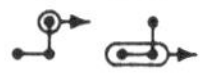 | 0.548, **0.621** | 0.552, – |
| 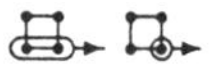 | **0.686**, 0.818 | 0.688, 0.815 |
| 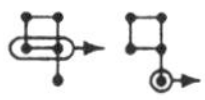 | **0.528**, 0.547 | – , 0.551 |
| 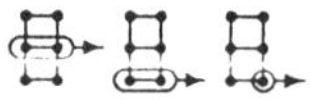 | **0.701**, 0.714, 0.834 | – , 0.713, 0.835 |
| 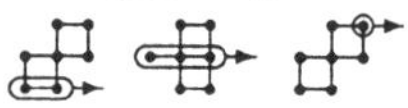 | 0.454, **0.499**, 0.836 | 0.45, – , – |
| 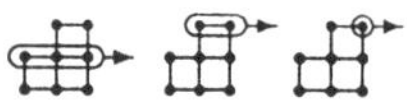 | **0.634**, 0.757, 0.838 | – , 0.758, 0.838 |
|  | **0.810**, 0.821, 0.947 | – , 0.857, 0.975 |
| 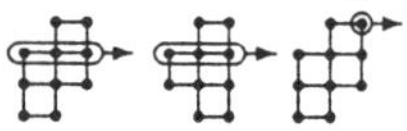 | **0.667**, 0.682, 0.846 | – , – , – |

end, we placed a Cu adatom on top of a small island of 17 atoms on the Cu(100) surface and searched for possible low energy paths. As a result we found a new mechanism for embedding an adatom inside the island simultaneously through a concerted internal dimer shearing near a kink site with an activation barrier $E_a = 0.43$ eV. Remarkably, within the EAM potential this energy barrier was lower than that of a single adatom jump on a flat surface (0.49 eV) which is needed before the adatom can cross the step edge via the exchange process near a kink site (0.36 eV) [15]. The embedding mechanism thus provided a possible important channel for an adatom to descend from atop of an island that could contribute to layer-by-layer growth at low temperatures.

Single adatom diffusion on Al(100) has also been intensively studied during the recent years [16, 17]. This system represents a rare example of a surface where the exchange diffusion mechanism dominates over simple jump events, as predicted by first-principles calculations [18]. To study Al(100), we have used the Ercolessi-Adams glue potential [19]. It has been fitted on the basis of a force-matching method within first-principles calcu-

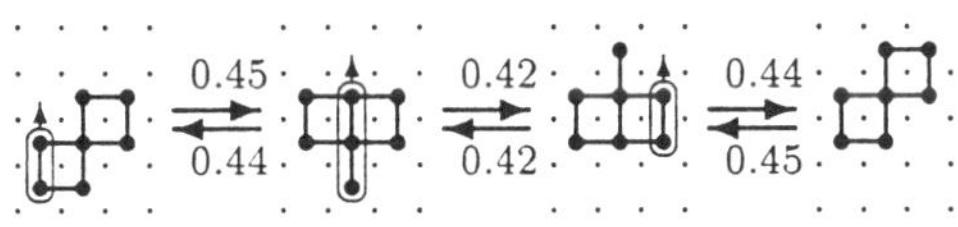


*Figure 2.* A series of processes for the CM motion of a seven-atom Al island. Activation energies are in eV. The RLS has $E_a = 0.45$ eV.

lations [19]. For Al adatom diffusion on the Al(100) surface, this potential gives energy barriers of 0.53 eV for jump and 0.56 eV for exchange. The first barrier is very close to recent first-principles calculations using the generalized gradient approximation which give 0.51 eV [17]. Although the barrier for exchange does not compare quantitatively with the first-principles results, we use this potential here to illustrate a case where exchange and jump for single adatoms have almost the same barriers, in contrast to, *e.g.*, Ag or Cu, where exchange is not believed to play a role at lower temperatures [7].

The case of a seven-atom island on Al(100) is shown in Fig. 2. Perhaps surprisingly, we find processes very similar to those occurring on Cu(100): dimer shearing at the periphery (0.44–0.45 eV), and internal three-atom row shearing (0.42 eV). The main conclusion here is that with the glue potential, we do not find low-energy exchange processes for the seven-atom cluster. All such processes have activation energies about or higher than 0.8 eV. They include a three-atom exchange (0.85 eV), and other more complicated multiparticle exchange processes whose barriers are in excess of 1 eV. There are also many competing mechanisms for vacancy formation with barriers around 1 eV. Similar processes have been described by Henkelman and Jónsson using another semi-empirical potential [20].

## 2.2. ANALYTIC EXPRESSIONS FOR THE DIFFUSION COEFFICIENTS

For small islands, the master equation method [21, 22] can be used to calculate the diffusion coefficient $D$ analytically. This method is based on the Markovian approximation of independent transitions between different configurations. With all the relevant transition rates known, $D$ can be thus obtained at all temperatures where the Markovian approximation holds. This method has been used for calculating diffusion coefficients of small islands up to $s = 6$ using transition processes in which single atom jumps occur, and also considering dimer shearing in some cases [22]. We have recently extended these calculations to the case of a five-atom island with dimer shearing included [23]. In Fig. 3 we show a comparison between kinetic

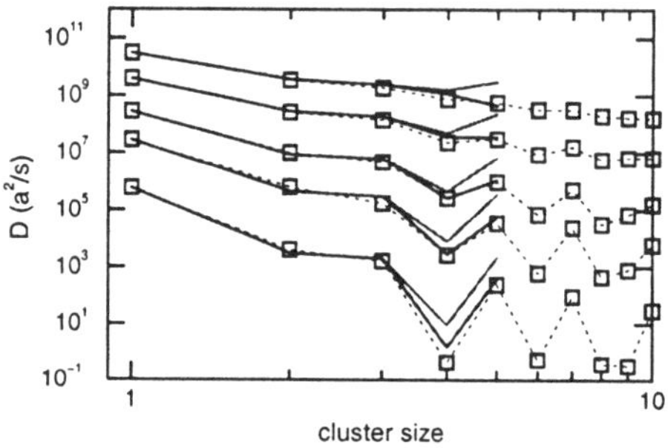


*Figure 3.* Cu adatom island diffusion coefficients $D$ *vs.* $s$ up to $s = 10$ at $T = 300$, 400, 500, 700, and 1000 K (from bottom to top). Squares are Monte Carlo data from Ref. [6], and lower and upper solid lines indicate master equation results without and with dimer shearing, respectively.

Monte Carlo simulations [6] (with single-particle processes only), master equation results with single-particle processes only, and master equation results with dimer shearing included.

## 3. Diffusion of Large Islands

The central issue in the diffusion of large adatom islands on fcc(001) surfaces concerns the role of different mass transport mechanisms, and the corresponding behavior of $D$. For $s \gg 1$, it is expected to follow

$$D \propto e^{-\beta E_L} s^{-\alpha},$$

where $\beta = 1/(k_B T)$ and $E_L$ is an effective energy barrier for island diffusion. In the presence of only one of the three mass transport mechanisms proposed by Khare and Einstein [9], the scaling exponent $\alpha$ is expected to be 1/2 (EC), 1 (TD), or 3/2 (PD) (see Fig 3. in Ref. [9]). When both the TD and the PD mechanisms are present, clear dependence of $\alpha$ on $s$ should be observed, and finally one should always find $\alpha = 1$ for $s \to \infty$.

We have recently carried out a large-scale kinetic Monte Carlo (MC) simulation study of Cu adatom islands on the Cu(001) surface [6]. The parametrization of the model is based on our previous EMT calculations of Cu adatom energetics on flat and vicinal Cu surfaces [7]. In the model, only single-particle transition events have been considered, since we do not expect the many-particle mechanisms to play a role for the large clusters considered here [11]. For the smallest islands, the MC results compare well with the analytic calculations of $D$ using the master equation formalism, as shown in Fig. 3.

To be able to simulate large islands at large time scales, we employed the Bortz-Kalos-Lebowitz (BKL) algorithm [24] using a binary tree structure.

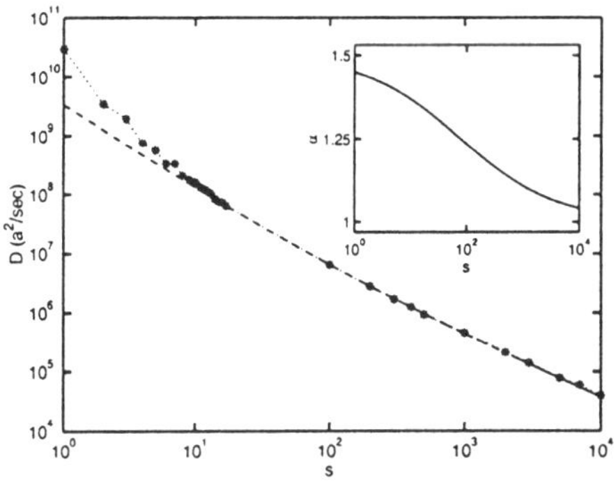


*Figure 4.* Adatom island diffusion coefficient $D$ *vs.* $s$ for $1 \leq s \leq 10^4$ at $T = 1000$ K ($a \approx 3.5$ Å is the lattice constant of copper). Stars denote the results of our simulations, and the dashed line is a fit to Eq. (36) from Ref. [9] ($R_{st} = 5.0 \times 10^{-2}$ and $R_{sv} = 5.0 \times 10^{-4}$). Error bars are of the size of the symbols or smaller. The inset shows the effective exponent $\alpha$ using this fit.

In Fig. 4 we show the results of the simulations for $D(s)$ at $T = 1000$ K, up to $s = 10000$ atoms. This high-temperature case reveals that the values of $\alpha$ reported in the relevant experiments [8, 25] are most likely due to crossover effects. Indeed, our low temperature simulations corresponding to the experiments are in excellent agreement with the data for Cu(001) [6], and show that there exists a very long crossover towards $\alpha = 1$ for this system. Larger *effective* values of $\alpha$ may thus be obtained if only relatively small island sizes are considered as in the experiments [8].

The second main result we have been able to obtain is that this crossover is actually due to PD dominated diffusion changing over to TD dominated case, where the microscopic mechanism for the TD process comes from *vacancy diffusion* within large islands. There is actually a very simple physical explanation for this. The Arrhenius barriers for the TD and PD mechanisms are almost the same. However, with increasing island size the kink density at the periphery remains constant which implies that it becomes more and more difficult for PD to transport mass around the cluster. Due to the fact that vacancies are easily created at kink sites, the TD mechanism is not affected and it takes over as $s$ increases.

## 4. Summary and Conclusions

In summary, by combining a variety of methods ranging from microscopic semi-empirical calculations to large-scale kinetic Monte Carlo simulations, we have been able to obtain a rather complete picture of adatom island diffusion on some fcc(001) metals. For both Cu/Cu(100) and Al/Al(100) there exist several previously overlooked many-particle mechanisms that play an important role. This means that the CM motion of islands up to size ten and even larger can be activated more easily than previously assumed. In the case of much larger islands, there is a crossover from periphery-dominated mass transport to vacancy diffusion within the islands as their size increases. This leads to a well-defined scaling exponent $\alpha = 1$ asymptotically as $s \to \infty$. It would be of great interest to carry out careful first-principles calculations to test the validity of the semi-empirical calculations on which the parametrizations are based here (see *e.g.* [26]).

## 5. Acknowledgments

This work has been in part supported by the Academy of Finland through its Center of Excellence program. O.S.T. acknowledges partial support from the Russian Foundation for Basic Research (grant no. 98–02–16581). We also wish to thank Jarmo Pirhonen (CSC, Espoo, Finland) for the visualization of video animations of small islands that can be found at `http://www.fyslab.hut.fi/~pts/cu_ene.html`.

*On leave from The Institute of Microelectronics, Russian Academy of Sciences, Universitetskaya 21, Yaroslavl 150007, Russia.

## References

1. Barabási A.-L. and Stanley H. E. (1995), *Fractal Concepts in Surface Growth*, Cambridge University Press, Cambridge.
2. Erlich G. (1974) *CRC Crit. Rev. Solid State Mater. Sci.* 4 205; Gomer R. (1990) *Rep. Prog. Phys.* **53** 917.
3. Binning G., Fuchs H. and Stoll E. (1986) *Surf. Sci.* **169** L295.
4. Kellogg G. L. (1994) Oscillatory behavior in the size dependence of cluster mobility on metal surfaces: Rh on Rh(100), *Phys. Rev. Lett.* **73** 1833-1836.
5. Shi Z.-P., Zhang Z., Swan A. K., and Wendelken J. F. (1996) Dimer shearing as a novel mechanism for cluster diffusion and dislocation on metal(100) surfaces, *Phys. Rev. Lett.* **76** 4927-4930.
6. Heinonen J., Koponen I., Merikoski J., and Ala-Nissila T. (1999) Island diffusion on metal fcc(100) surfaces, *Phys. Rev. Lett.* **82** 2733-2736.
7. Merikoski J., Vattulainen I., Heinonen J., and Ala-Nissila T. (1997) Effect of kinks and concerted diffusion mechanisms on mass transport and growth on stepped metal surfaces, *Surf. Sci.* **387** 167-182.
8. Pai W. W., Swan A. K., Zhang Z., and Wendelken J. F. (1997) Island Diffusion and Coarsening on Metal (100) Surfaces, *Phys. Rev. Lett.* **79** 3210-3213.
9. Khare S. V. and Einstein T. L. (1998) Unified view of step-edge kinetics and fluctuations, *Phys. Rev. B* **57** 4782-4797.
10. Hannon J. B., Klünker C., Giesen M., Ibach H., Bartelt N. C., and Hamilton J. C. (1997) Surface Self-Diffusion by Vacancy Motion: Island Ripening on Cu(001),

*Phys. Rev. Lett.* **79** 2506-2509.
11. Trushin O. S., Salo P., and Ala-Nissila T. (2000) Energetics and many-particle mechanisms of two-dimensional cluster diffusion on Cu(100) surfaces, *Phys. Rev. B* **62** 1611-1614.
12. Jónsson H., Mills G. and Jacobsen K. W. (1998) in: *Classical and Quantum Dynamics in Condensed Phase Simulations*, Eds. B. J. Berne, G. Ciccotti, and D. F. Coker, World Scientific, Singapore.
13. Munro L. J. and Wales D. J. (1999) Defect migration in crystalline silicon, *Phys. Rev. B* **59** 3969-3980, and references therein.
14. Foiles S. M., Baskes M. I., and Daw M. S. (1986) Embedded-atom-method functions for the fcc metals Cu,Ag,Au,Ni,Pd,Pt, and their alloys, *Phys. Rev. B* **33** 7983-7990.
15. Trushin O. S., Kokko K., Salo P. T. , Hergert W., and Kotrla M. (1997) Step roughening effect on adatom diffusion, *Phys. Rev. B* **56** 12135-12138.
16. R. Stumpf and M. Scheffler (1996) *Ab initio* calculations of energies and self-diffusion on flat and stepped surfaces of Al and their implications on crystal growth, *Phys. Rev. B* **53** 4958-4973.
17. Feibelman P. J. (1999) Scaling of hopping self-diffusion barriers on fcc(100) surfaces with bulk bond energies, *Surf. Sci.* **423** 169-174.
18. Feibelman P. J. (1990) Diffusion path for an Al adatom on Al(001), *Phys. Rev. Lett.* **65** 729-732.
19. Ercolessi F. and Adams J. B. (1994) Interatomic potentials from first-principles calculations: the force-matching method, *Europhys. Lett.* **26** 583-588.
20. Henkelman G. and Jónsson H. (1999) A dimer method for finding saddle points on high dimensional potential surfaces using only first derivatives, *J. Chem. Phys.* **111** 7010-7022.
21. Titulaer U. M. and Deutsch J. M. (1982) *J. Chem. Phys.* **77** 472.
22. Sanchez J. R. and Evans J. W. (1999) Diffusion of small clusters on metal (100) surfaces: Exact master-equation analysis for lattice-gas models, *Phys. Rev. B* **59** 3224-3233.
23. Hirvonen J. *et al.* (2000), unpublished .
24. Bortz A. B., Kalos M. H., and Lebowitz J. L. (1975) J. Comp. Phys. **17**, 10 ; Kotrla M. (1996) *Comp. Phys. Commun.* **97**, 82 .
25. Bardotti L., Bartelt M. C., Jenks C. J., Stoldt C. R., Wen J.-M., Zhang C.-M., Thiel P. A., and Evans J. W. (1998) *Langmuir* **14**, 1487 ; Stoldt C. R., Jenks C. J., Thiel P. A., Cadilhe A. M., and Evans J. W. (1999) *J. Chem. Phys.* **111**, 5157.
26. Hansen U., Vogl P., and Fiorentini V. (1999) Quasiharmonic versus exact surface free energies of Al: A systematic study employing a classical interatomic potential, *Phys. Rev. B* **60** 5055-5064.

# PATHS, BARRIERS, AND PREFACTORS FOR ADATOM DESCENT FROM AG CLUSTERS ON AG(111)

TALAT S. RAHMAN[1,2], ABDELKADER KARA[1], ALTAF KARIM,[1] AND AHLAM AL-RAWI[1]

[1]*Department of Physics, Cardwell Hall,*
*Kansas State University, Manhattan, Kansas 66506, USA*

AND

[2]*Laboratory of Physics, Helsinki University of Technology,*
*P.O. Box 1100, FIN-02015 HUT, Espoo, Finland*

**Abstract.** We have calculated the energetics and the dynamics for the diffusion of Ag adatoms which land on top of small two-dimensional Ag clusters on Ag(111), using realistic many-body interaction potentials. Purely energetic considerations of the descent of the adatoms from the islands show exchange mechanism to dominate over hopping and the process to favor the formation of 100-microfacetted steps (A-type) over the 111-microfacetted ones (B-type). Accompanying molecular dynamics simulations validate these findings at low temperatures, but show a reversal in the trend above room temperature making the formation of B-type step more probable. Calculations of the diffusion coefficient confirm that the pre-exponential factor for the processes leading to the formation of the A and B type steps are significantly different.



*M.C. Tringides and Z. Chvoj (eds.),*
*Collective Diffusion on Surfaces: Correlation Effects and Adatom Interactions,* 327–338.

## 1. Introduction

The era of scanning tunneling microscopy (STM) has brought about a tremendous surge in the investigation of growth processes on metal surfaces. A variety of experimental studies of homo- and heteroepitaxy on the (100), (110) and (111) surfaces of several transition metals has led to interesting observations. Among these the case of homoepitaxial growth on fcc(111) surfaces has been particularly intriguing, since the innate geometry of the surface forces the steps formed by nucleating two-dimensional islands, to take two types of configurations. Monoatomic closed-packed steps formed on this surface may have either a (100)-microfacet (A-type) or a (111)-microfacet (B-type). The relative abundance of one type of step over the other may thus lead to very distinct features in the growth patterns. Since the local structural, dynamical and electronic properties of these steps may be different from each other, it is not surprising that experimental observations of homoepitaxial growth on several fcc(111) surfaces display very distinct characteristics. In the case of Pt(111), for example, the relative abundance of the A-type or the B-type steps have been found to depend on the surface temperature [1]. The case of Ir(111) is somewhat different and it may be argued that both types of steps are equally probable [2], although the issue is still debatable. Several theoretical papers have also addressed the issue of the relative energetics of the A and B type steps [3]. Theoretical calculations have also examined the energetics of various diffusion processes on fcc(111) surfaces [4, 5, 6]. The intricacies of the processes revealed by these studies have established the need for studying the details of the potential energy surface available to the diffusing atoms as accurately and as extensively as feasible. The object of our work here is to carry out an extensive study of the energetics of an initial stage of formation of steps on Ag(111) surface. For this purpose we have carried out an examination of the potential energy surface experienced by an adatom as it attempts to descend from a small two dimensional cluster of Ag atoms already nucleated on Ag(111). As we shall see, the investigation of the activation energy barriers and diffusion paths for these atoms in low coordinated sites allow us to make some general observations of homoepitaxial growth on fcc(111) surfaces. Since our aim is to predict the minimum energy path for the diffusion of the adatom, we have constructed a very fine grid for the calculation of the total energy of the system. As an *ab initio* calculation of this type would be formidable with the computational resources available to us, we have used reliable manybody interaction potentials from the embedded atom method for the purpose.

Another major undertaking in the present paper is the examination of the characteristics of diffusion process as a function of the surface temperature. While most theoretical calculations of adatom diffusion focus their attention on the activation energy barriers and assume the diffusion prefactors to be 'normal' and the same for all relevant processes, we find no reasons to do so. In fact, we expect the potential energy surface to undergo modulations with temperature and the characteristics of surface vibrations to play a role in the diffusion process. We have thus carried out molecular dynamics (MD) simulations of the system of interest here for the temperature range 300 K to 700 K. Interestingly, our results show that predictions of diffusion coefficients made from considerations of surface energetics alone are not enough to give us reliable results at finite temperatures. In particular, we show that diffusion preexponential factors for the formation of A and B type steps may differ significantly from each other.

In the next section we provide some details of the theoretical technique that we have used. This is followed by a presentation of the results and their discussion.

## 2. Details of the Calculations

For adsorbed two-dimensional islands on metal surfaces, the relaxation patterns near and around the corners may be quite complex [7], as a result of the low local atomic coordination and symmetry as compared to that on flat terraces. These complexities in atomic relaxations can also be expected to affect the potential energy surface in the vicinity of the islands and be reflected in the structural and dynamical phenomena at these sites. Of interest to the work here is the descent of adatoms which land on top of small two-dimensional islands on fcc(111) surface (see Fig. 1). To determine the diffusion path for the adatom accurately, we need to evaluate the potential energy surface near the corners and step edges accurately and reliably. With this in mind we have used a two dimensional drag/grid method for the purpose. According to this method, we minimize the total energy of a N-atom system by allowing (3N-2) degrees of freedom to freely relax while holding 2 degrees of freedom of the atom involved in the diffusion process fixed. In the cases studied here, these two degrees of freedom correspond to the x and y components contained in the surface plane. The grid used in these calculations consisting of 100x100 points allows a smooth representation of the full potential energy surface available to the adatom and an *a priori* determination of the diffusion paths. The usage of a sophisticated graphic applications package (XFARBE) further facilitates the extraction of the energy barriers and the minimum energy path for diffusion.

While we expect the calculated energy barriers to give us reliable predictions for the adatom diffusion paths at very low temperatures (at which the assumption of a static lattice is still valid), there is still the question about the relative values of the diffusion preexponential for the different competing processes. To check the validity of the method for determination of diffusion paths at finite temperatures, we supplement the grid calculations with molecular dynamics simulations carried out for the temperature range 300 K - 700 K, for the case of a Ag adatom descending a small island on Ag(111). These simulations are done in a microcanonical ensemble in which the number of particles, the volume, and the total energy are conserved (NVE). These NVE runs were preceded by simulations in the canonical ensemble for which the number of particles, the volume, and the temperature are kept constant (NVT). This first step is essential to produce a thermalized system at the desired temperature. The molecular dynamics cell consists of 9 layers of fcc(111) with 144 atoms per layer, on top of which we place the 4-atom island and the adatom above it. The two bottom layers of the slab are kept rigid. Using a time step of 1.5 femtosecond, the (NVE) simulations were run until the adatom has diffused down.

To calculate the structure and the dynamics of the systems at hand, we use interaction potentials based on the embedded atom method (EAM) [8] which, despite being empirical, have had success in predicting several surface properties of the fcc metals Ag, Cu and Ni [8]. For example, surface relaxations [9], the dispersion of the surface phonons [10], the vibrational dynamics and thermodynamics of vicinal surfaces [11, 12], and also self-diffusion [13] on these transition metal surfaces appear to be in reasonable agreement with experiments where available. We have used here the parameterization by Voter and Chen [14] in which experimental values of the bond length and the binding energy of the diatomic molecule are also included, in addition to several bulk properties, in extracting the interaction potentials. Energy minimization of the system were performed via the conjugated gradient method.

## 3. Results of Calculations of the Surface Energetics

We now turn to the descent of a Ag adatom from a small island on Ag(111). In Fig.1 we depict the two possible scenarios when an adatom initially deposited on a 4-atom island descends via an exchange process involving one of the island atoms. Depending on which of the two available fcc sites the exchanged island atom will occupy, the long step edge (3 atoms) of the final 5-atoms island will be exhibit a (111) microfacet (B-type) or a (100) microfacet (A-type), as indicated in the figure.

The determination of the diffusion paths and energy activation barriers was done by calculating the total energy of the constrained system with the exchanged atom placed at all point on the grid and allowing all other atoms to relax, as described above. The grids used in these calculations, for

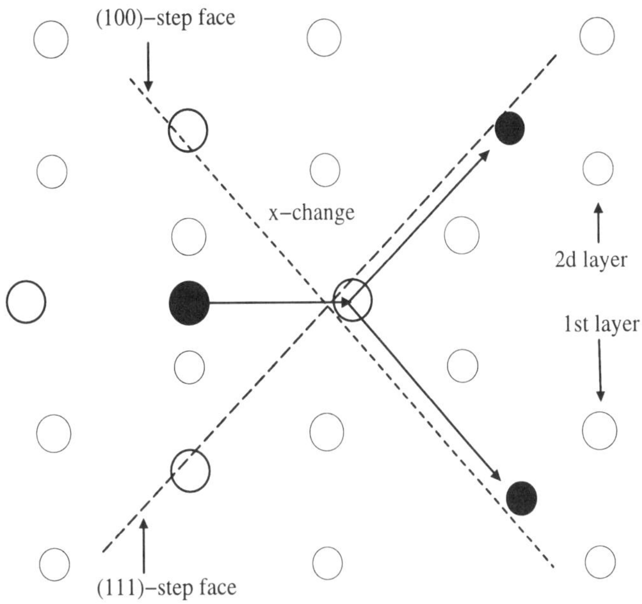


*Figure 1.* Formation of A or B type step face via exchange mechanism

the two scenarios by which the adatom could undergo a simple exchange process with one of the cluster atoms, are shown in Fig. 2. The collection of the total energy of the system evaluated with the exchanged atom at points of the grid are then used to create a contour plot from which one can extract the minimum energy paths, saddle points and the energy minima.

The relevant information obtained from the contour plots for the two possible exchange events is shown in Fig. 3. The top half of the figure concentrates on the minimum energy paths that would be taken by the adatom and the exchanged atom to eventually lead to the nucleation of a B-type step, while the bottom half illustrates the same for the formation of an A-type step. Quite clearly both sets of paths show concerted motions of the two atoms. Also, the energy at the saddle points (Eb) for the top set of minimum energy paths is higher than that for the bottom ones. This is very interesting because the main difference in the two sets of paths arises from the differences in the local environments because of the locations of the atoms in the layer below. In order to land at the fcc site in the top half of the figure, the minimum energy path takes the exchanged atom to

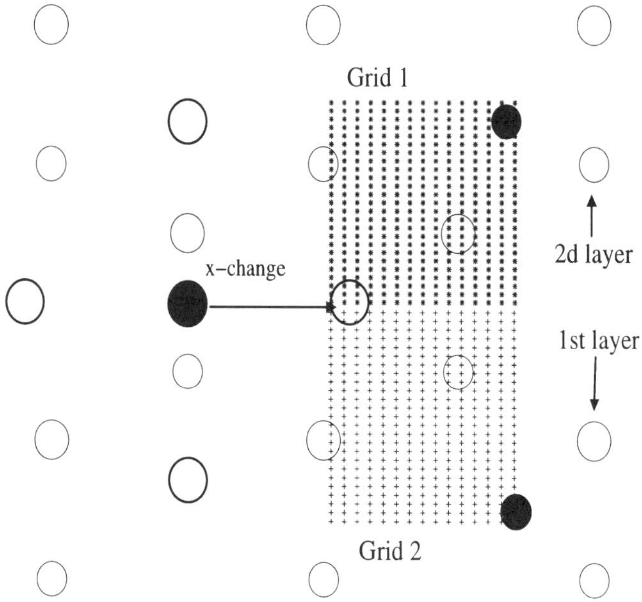


*Figure 2.* The grid

the vicinity of a hcp site. This hcp site falls also in the minimum energy path of the adatom itself as it proceeds to descend and exchange with the atom below. It is worth remembering that the potentials used here (EAM) are based on coordination and as a consequence every atom would tend to maximize it's coordination at all times. Thus, the exchanged atom, en route to the fcc site, and the adatom both want to go to their final positions via the hcp site. This competition for the hcp site affects the total energy of the system and makes the process energetically unfavorable, as compared to the other possibility.

We turn now to the event in the bottom half of Fig. 3, which would lead to the formation of an A-type step. In this case we see that the exchanged atom, en route to the fcc site, finds an hcp site on the way. Since the adatom is always far away from this hcp site, there is no competition. Here the first saddle point is at an energy of 52 meV, beyond which the system actually settles into an equilibrium state that is about 40 meV lower than the initial one. Eventually, the exchanged atoms sees a second barrier of about 86meV which it has to overcome to make the A-step for this 5-atom flat island.

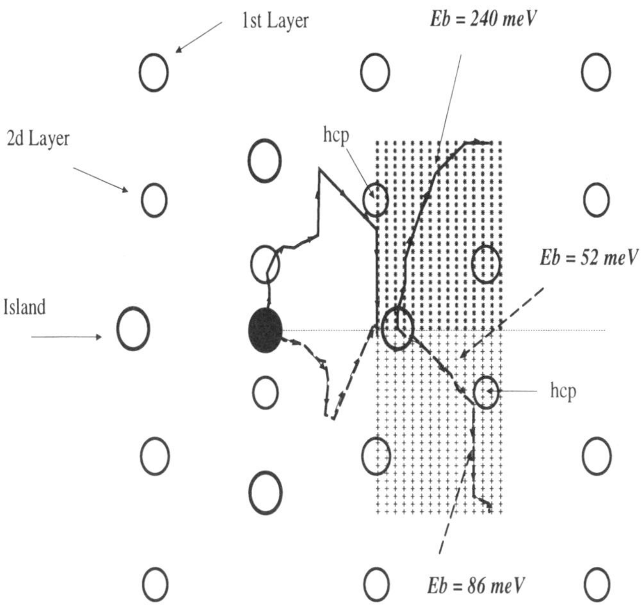


*Figure 3.* Diffusion paths and barriers

Fig. 4 is a schematic representation of the energy levels for the two events, taking the total energy of the system to be zero when the adatom is on top of the 4-atom island (start 4+1). To follow the path to form a B-step, we encounter a saddle point with an energy of 240 meV above the zero, while to form the A-step, one encounters first a saddle point of energy 52 meV, then a minimum energy level that is lower than the starting one, and finally a saddle point 86 meV above this minimum.

In summary, we have shown above that when an adatom descends from a small island via an exchange process on Ag(111), the atoms involved prefer to arrive at a final fcc site via an hcp site. If the two atoms compete for the same hcp site, the barrier to diffusion is higher than if there is no such competition. This argument is quite general and should be applicable to other fcc(111) surfaces for which local coordination dictates the local structure and dynamics. If it is further assumed that preexponential factors are irrelevant, and that activation barriers calculated for a static lattice control the diffusion process, the above arguments would predict the formation of A-type step in this adatom descent. Further, from the simple calculation

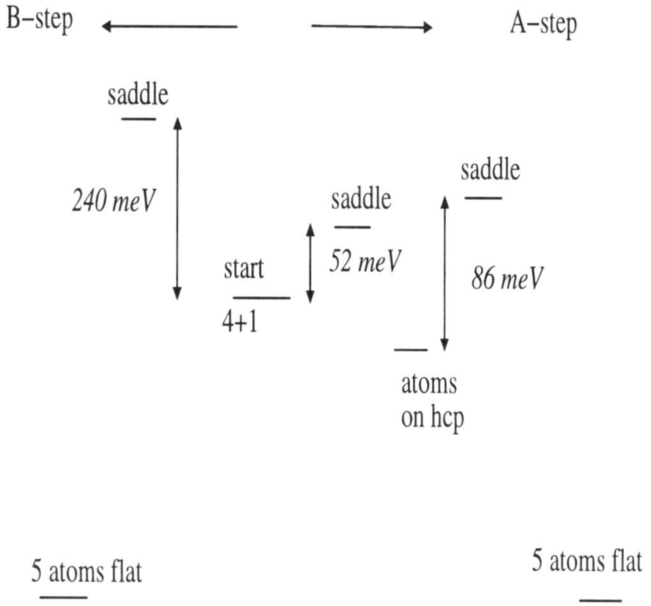


*Figure 4.* Schematics of the barrier energies

presented below, relative rates of formation of the A and B type steps, as a function of the surface temperature, could also be deciphered.

Assuming that the diffusion processes considered here follow an Arrhenius behavior, the diffusion coefficients, $D_A$, and $D_B$, to form the A and B type steps are, respectively, given by:

$$\mathrm{D_A} = D_{0A} exp(-E_{bA}/k_B T)$$
$$\mathrm{D_B} = D_{0B} exp(-E_{bB}/k_B T)$$

with $E_{bA}$ and $E_{bB}$ the activation energies to form A and B type step-edges, respectively, 86 and 240 meV.

If we now make the usual assumption that both processes have the same preexponential factor (i.e., $D_{0A} = D_{0B}$), then, at 300 K, the ratio of $D_A/D_B$ is 473:1, in favor of the probability of formation of an A-type step. Even at 500 K, the formation of A-type is favored by a factor of 40.

## 4. Results of Molecular Dynamics Simulations

From the discussion in the previous section, one would expect the formation of A-type edges to be dominant at all temperatures when an ad-atom descends from a 4-atom island. To test this prediction, we have performed MD simulations of the system at 300, 500, and 700 K, collecting the statistics for each temperature for a large number of initial conditions.

Table 1 shows the number of simulations that ended up with an A-type edge and those that ended up with B-type. At 300 K, the total number of simulations is 50 and only 3 B-type edges were formed. While this result is in qualitative agreement with predictions above from considerations of energetics for a static lattice, the probability of finding B-type steps is non-negligible. The situation changes dramatically when the temperature is increased to 500 K, at which 43 out of 81 simulations ended with a B-type edge. At 700 K, 58 out 100 events resulted in A-type step edges. The probabilities (P) associated with these processes are listed in Table 2. The differences in the calculated ratios of the formation probability of the A and B type steps in Table 1, from the predictions in section 3, indicate that for Arrhenius behavior, the prefactors associated with the two processes cannot be the same.

In order to determine these prefactors, we have calculated the diffusion coefficient for the two processes leading to the formation of either A or B type step, from the descent time (or residence time of the adatom prior to the descent). Table 2 shows the average descent time for the 2 processes at 300, 500, and 700 K. This descent time (DT) is then used to get the frequency of the event and, hence, the diffusion coefficient using:

$$D = \nu \Gamma l^2 / 2\alpha$$

with l the distance traveled during the exchange process, $\Gamma$ the frequency of the event, $\alpha$ the dimensionality (2 in this case), and $\nu$ the number of possibilities to form an A-type or B-type (2 in this case).

Table 3 shows the diffusion coefficient for the formation of an A-type and B-type edges at different temperatures. It shows that A-type is one order of magnitude more frequent than the B-type at 300 K and a dominance by B-type at 700 K with a crossing around 500 K.

In Fig. 5, we plot the log of these diffusion coefficients as function of $1/k_B$T. Both processes present an Arrhenius behavior. This plot also allows

TABLE 1. MD statistics for 4+1 island

| Temperature | Number of A-type | Number of B-type | Total |
|---|---|---|---|
| 300 K | 47 | 3 | 50 |
| 500 K | 38 | 43 | 81 |
| 700 K | 42 | 58 | 100 |

TABLE 2. Descending times (DT) and formation probabilities (FP) for A and B types

| Temperature | DT for A-type | DT for B-type | P for A-type | P of B-type |
|---|---|---|---|---|
| 300 K | 250x$10^{-12}$ sec | 3000x$10^{-12}$ sec | 0.94 | 0.06 |
| 500 K | 15x$10^{-12}$ sec | 13x$10^{-12}$ sec | 0.47 | 0.53 |
| 700 K | 3x$10^{-12}$ sec | 2x$10^{-12}$ sec | 0.42 | 0.58 |

us to extract the energy barriers and prefactors. For the formation of an A-type edge by adatom descent, we find an energy barrier of 197 meV and a prefactor of 3.28x$10^{-3} cm^2/sec$, while for the B-type step-edge the energy barrier is 335 meV and the prefactor 6.35x$10^{-2} cm^2/sec$. Note that the barrier energies extracted from the Arrhenius plot representing the MD results at finite temperatures is about 100 meV *higher* than those calculated with a static system using the grid method. This energy is higher than that associated with the phonons and may be related to frictional forces associated with the island rubbing the substrate. Further studies are needed before we can establish the exact reasons for the differences in the energy barriers in the static and the dynamic calculations. We also find that the prefactor associated with the formation of a B-type edge to be about 20 times larger than that associated with the formation of an A-type edge. This difference in the prefactor is responsible for the inversion in the preference

TABLE 3. Diffusion coefficients (D) for A and B types

| Temperature | D for A-type $cm^2$/sec | D for B-type $cm^2$/sec |
|---|---|---|
| 300 K | 1.69x$10^{-6}$ | 1.40x$10^{-7}$ |
| 500 K | 2.82x$10^{-5}$ | 3.25x$10^{-5}$ |
| 700 K | 1.42x$10^{-4}$ | 2.13x$10^{-4}$ |

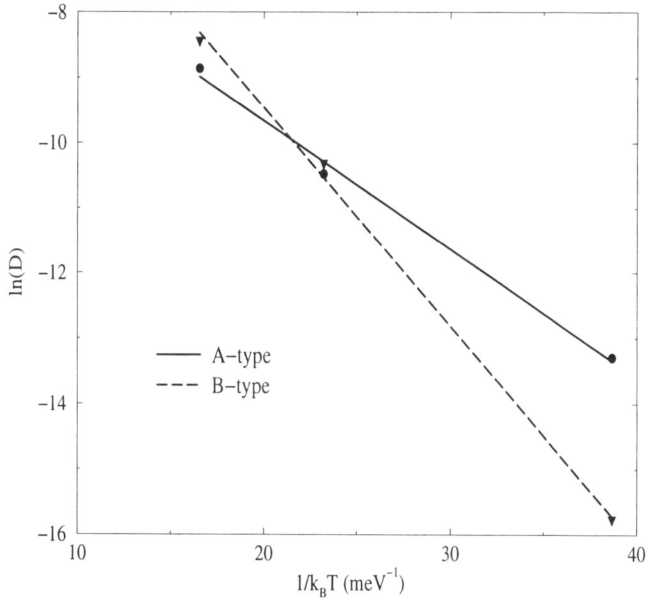


*Figure 5.* Arrhenius plot of the diffusion coefficient of processes leading to A and B type steps

from A-type step formation to B-type at temperatures above 300 K.

## 5. Summary

In this paper we have examined the energetics and the dynamics of adatom descent from small two dimensional Ag islands on Ag(111). From considerations of the total energy of the system, we find exchange to be the dominant mechanism for the diffusion of the adatom. Through the usage of a fine computational grid which allows us to map out the potential energy surface in great detail, we find that the minimum energy path for the adatom exchange process is complex. In this coupled motion of the two atoms, both try to maximize their coordination by involving an hcp site in the minimum energy path to diffusion. As a result the process that would lead to the formation of B-type step becomes energetically unfavorable and the A-type step is expected to dominate the growth on this surface. On the other hand, our accompanying molecular dynamics simulations show that the probability of the formation of B-type steps is non-negligible at room temperature and significant at higher temperatures. In addition, these dynamical studies show that the preexponential factors for the two diffusion processes differ by a factor of 20. We also find the barrier to diffusion to also depend on temperature. These results call into question the application of solely energetic considerations in drawing conclusions about processes at finite temperatures. Further work is in progress to place these conclusions on stronger grounds.

## 6. Acknowledgments

This work was supported in part by NSF Grant No. EEC-0085604. We would like to thank Tarun Souradeep for helpful discussions and Chandana Ghosh for help with implementation of the graphics application package.

## References

1. T. Michely, M. Hohage, M. Bott, and G. Comsa, *Phys. Rev. Lett.* **70**, 3943 (1993).
2. T. Fu and T. T. Tsong, *Phys. Rev. B* **61**, 4511 (2000).
3. P. Feibelman, *Phys. Rev. B* **62**, 17020 (2000), and references therein.
4. A. Bogicevic et al., *Phys. Rev. Lett.* **81**, 637 (1998).
5. A. Bogicevic et al., *Phys. Rev. Lett.* **83**, 2608 (1999).
6. F. Maca, M. Kotrla, and O.S. Trushin, *Surf. Sci.* **454-456**, 579 (2000).
7. A. Kara, T.S. Rahman, to be published.
8. For a review about the embedded atom method see: M. S. Daw, S. M. Foiles and M. I. Baskes, *Mat. Sci. Rep.* **9**, 251 (1993).
9. K.P. Bohnen and K.M. Ho, *Surface Science Reports* **19**, 99 (1993).
10. T. S. Rahman, in *Condensed Matter Theories*, eds. J. Clark, K. Shoaib, and A. Sadiq, Vol. **9**, Nova, NY (1994), p299.
11. A. Kara, S. Durukanoglu, and T.S. Rahman, *J. Chem. Phys.* **106**, 2031 (1997).
12. S. Durukanoglu, A. Kara, and T.S. Rahman, *Phys. Rev. B* **55**, 13894 (1997).
13. U. Kürpick, A. Kara, and T.S. Rahman, *Phys. Rev. Lett.* **78**, 1086 (1997); U. Kürpick and T.S. Rahman, *Surf. Sci.* **383**, 137 (1997).
14. A. F. Voter and S. P. Chen, *Proc. Mat. Res. Soc.* **82**, 175 (1986).

# CORRELATED DIFFUSION OF ADATOMS AND ISLANDS ON THE LOW INDEXED SURFACES OF NOBLE METALS AND ALLOYS

G.A. EVANGELAKIS
*University of Ioannina, Department of Physics, Solid State Division, Ioannina, P.O.Box 1186, Ioannina, 45110 GREECE*

The diffusion processes of adatoms and islands on the low indexed noble metal and alloy surfaces were studied by Molecular Dynamics Simulations, based on semi-empirical potential models in analogy to the Tight Binding Theory in the Second Moment Approximation. Several systems were considered, like Cu adatoms and Cu or Au islands/adlayers on the Cu(110) and Cu(111) surface, respectively, and Cu or Au adatoms on the $Cu_3Au(001)$ face. Various diffusion processes, that were monitored and analyzed, were found for these systems. It comes out that collective excitations may play an important role not only in the diffusivity of the adsorbates, but also in the structural and dynamic behavior of the substrate. In addition, it was found that the diffusive behavior of islands is characterized by the high mobility of the peripheral atoms and their different relaxed positions from the other island atoms. The surface and half monolayer island (0.5ML) relaxation were also found to play an important role in the stability and diffusivity of adlayers deposited on the 0.5ML island. Moreover, it was found that the presence of Cu or Au adatoms on the $Cu_3Au(001)$ surface stimulates diffusion processes that can be related to wetting and disordering phenomena occurring at higher temperatures.

## 1. Introduction

The diffusion of adatoms and small islands on metallic surfaces has been the object of many studies in the last decade since it plays a key role in many technologically important applications like deposition, catalysis, thin film growth, surface coating, etc. Indeed, phenomena such as surface coarsening, adatom and island coalition, adsorbate segregation or surface wetting and alloying, are some of the processes primarily dependent on the adatom's and island's diffusive behavior. Focusing on the low index surfaces of noble metals, single adatom self-diffusion on the low-index copper surfaces has been extensively studied both experimentally [1-4] and theoretically [5-17] during



*M.C. Tringides and Z. Chvoj (eds.),*
*Collective Diffusion on Surfaces: Correlation Effects and Adatom Interactions*, 339–349.

the last few years. From these studies, it has been found that adatom diffusion takes place mainly via hopping mechanisms and exchange-type processes.

The hetero-diffusion of adatoms on metallic substrates is a decisive process for the initial stages of thin film growth and morphology. Focusing on noble metals systems, it was found that surface alloying takes place during hetero-atom deposition, depending on coverage and deposition conditions. Examples of such systems refer to Au on Cu low-index surfaces [18-25], Ag on Cu(001) [18,26], Au on Ag(110) [25,27], Pt on Cu(111) [28], and Cu on Pt(111) [29]. It appears that alloy formation arises by adsorbate replacement of the substrate atoms. In the case of Au atoms adsorption on pure Cu surfaces, it was found that $Cu_3Au$ alloy is formed. More specifically, it is found that $Cu_3Au$ type alloy surface is formed by the adsorption of Au atoms [21,24] on pure Cu faces. Most of the processes involved in these phenomena are characterized by correlated diffusion mechanisms.

## 2. Models and Computational Details

We performed standard MD simulations in the isothermal canonical ensemble using the Nose thermostat and slab geometry. For the atomic interactions we adopted the semi-empirical potential models in analogy to the Tight-Binding theory in the Second Moment Approximation in conjunction with a Born-Mayer functional for the repulsive and attractive contributions, respectively. The adatoms or 2D islands were deposited on the surfaces without initial kinetic energy and the systems were let to equilibrate for some picoseconds before taking nanosecond statistical averages. The rates of the various diffusion processes were calculated by counting the number of events that occurred within the simulation length, while direct computations of the diffusion coefficients were done with the use of the Kubo formula of the velocity auto-correlation function.

## 3. Results

### *3.1 Adatom dynamics and correlated diffusion*

Adatom self-diffusion processes on the Cu(110) face have been studied extensively the last decade. The interest is motivated from the fact that this surface exhibits both roughening and pre-melting transitions well below the melting point [30-33]. From these studies the different reaction paths for in-channel diffusion have been found among which a metastable intermediate adatom position permitting the migration of the particles along the potential hills towards the [001] direction. Very good work has been done on this subject by Ferrando et al. [34,35]. Concerted motion of dimmers has also been observed and analyzed enlightening the richness with which diffusion takes place on this surface [36].

At higher temperatures (above Ts~0.6Tm) diffusion becomes very important. This was related with the massive creation of adatoms and with the initiation of multiple diffusion events and it was interpreted as the mechanism

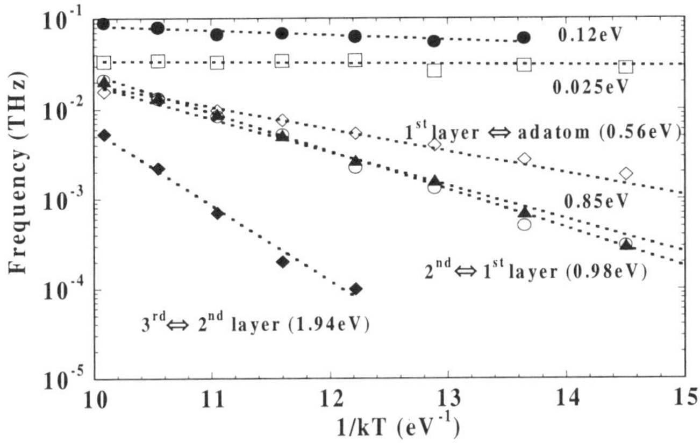


Fig. 1 Arrhenius diagram for the various adatom self-diffusion processes happening on the Cu(110) surface.

for explaining the roughening and pre-melting phenomena happening at higher temperatures. However, it has been also found that while the number of adatoms was exponentially increasing with temperature, the frequency of the various diffusion processes on the surface saturated above Ts, indicating that the driving force for activation was related to collective excitations. Indeed, we have shown recently that above Ts all diffusion mechanisms, including interlayer mass transport that is also initiated at the high temperature region, follow Arrhenius type behavior and obey the Meyer-Neldel (MN) rule [37,16].

According to this rule the pre-exponential factor of an Arrhenius process compensates for the decrease due to the energetic requirements of a given process. It can be shown that the slope of this diagram is related to phonon excitations [38]. This can be seen in Fig. 1 where we show the Arrhenius plot of the various diffusion processes and in Fig. 2 in which we present the corresponding MN diagram. It comes out that the corresponding MN energy is about 3-4 times the basic excitation, which is of the order of 25meV for the case of copper.

We conclude therefore, that above Ts, all diffusion processes studied are multi-phonon activated, demonstrating the role of collective dynamics not only in the concerted movement of many atoms, but even for the single adatom dynamics.

Another interesting case of single adatom diffusion is the existence of exchange type diffusion on the closed packed fcc surface, the (111) face [39-

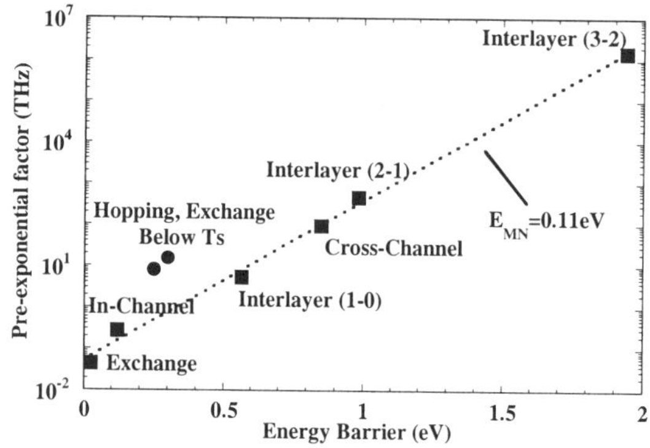


Fig. 2 Meyer-Neldel diagram for the various surface self-diffusion processes on the Cu(110) surface.

41]. On this surface, the exchange mechanism is found very demanding in energetic requirements and therefore is not expected as one of the main

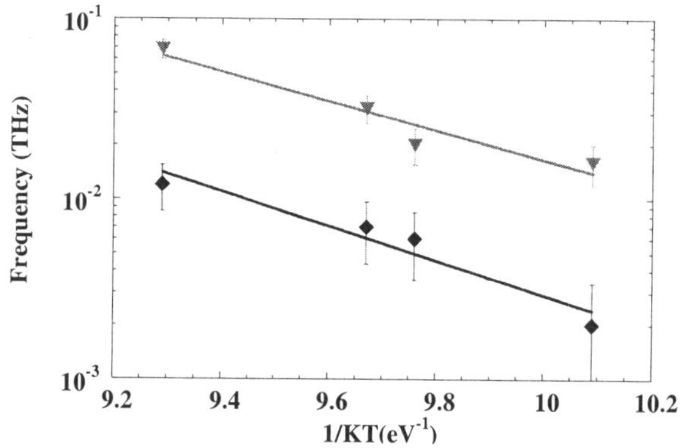


Fig. 3 Arrhenius diagram of the simple and double exchange processes happening on the Ag(111) surface.

diffusion processes for this surface. The existence of this diffusion mechanism is very important for phenomena like segregation, wetting, and surface alloying. Indeed, taking into account only energetic considerations (zero K calculations) the energy required is several eV. However, the mechanism involves more than one particle and therefore the motion is correlated. The preexponential factor of the process may compensate for the high migration energy of the mechanism. Indeed, we found very high pre-exponential factors (of the order of $10^{18}$ THz) for the two types of exchange process we studied, the double exchange (triangles) and simple exchange mechanism (diamonds), the Arrhenius diagrams are shown in Fig.3. In Fig. 4 we present a sequence of snapshots showing the simple exchange mechanism in the case of Cu adatom on the Cu(111) surface. Similar phenomena have been observed in systems like Au/Cu(111) and Ag/Ag(111).

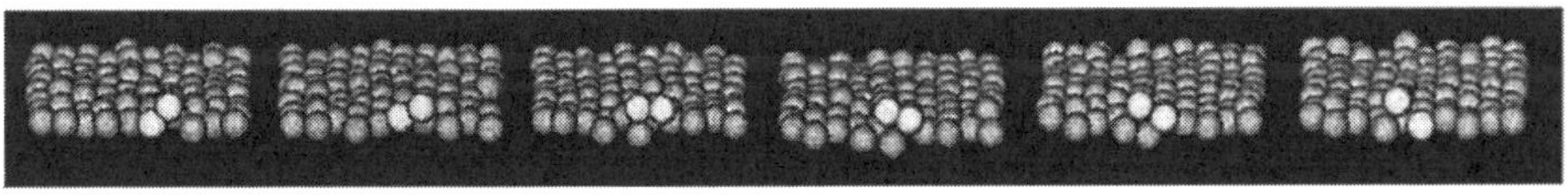

Fig. 4 Sequence of snapshots showing an exchange processes taking place on the Ag(111) surface.

We conclude therefore that the exact knowledge of the energy barrier of a specific process is not sufficient for a correct description of the diffusive behavior, even in the simple case of simple adatom diffusion. It must be pointed out that the results shown in this communication are all based on a semi-empirical potential model and therefore one must be very careful when interpreting finite temperature results. However, the existence of the exchange mechanism in this example is at least indicative of the necessity of the semi-empirical potentials and points out that exact static calculations, although very useful, are by far not enough to describe the diffusive behavior of a real material.

### *3.2 Diffusion of small 2D islands*

The very early stage of epitaxy is governed by the dynamics and the diffusion of small 2D islands that are usually formed by the coalescence of adatoms or bunches. Starting from a simple case, we studied the homo-epitaxy of small islands on the Cu(111) surface. On equilibrated surfaces, we put different numbers of adatoms, randomly distributed, and we let the system evolve for several picoseconds. For all temperatures studied, (we covered a range from room temperature up to 900K) we found that the adatoms combine to form small fast diffusing bunches that coalesce in large two-dimensional islands. In addition, it came out that the diffusion of small 2D islands, that takes place with a variety of diffusion mechanisms, exhibits Arrhenius behavior from where we

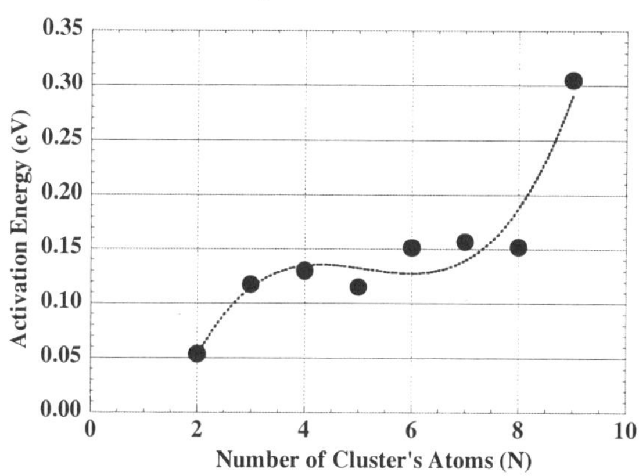


Fig. 5 Variation of the migration energy the migration energy of small 2D clusters on the Cu(111) surface

deduced the corresponding migration energies. The variation of this quantity with the island's size is depicted in Fig. 5.

As it can be seen from this picture, the energy barrier for diffusion is practically the same (0.15eV) for clusters between three and eight atoms, to reach the value of 0.3eV for a nine-atom cluster. Above this size and up to coverage of 0.5ML,

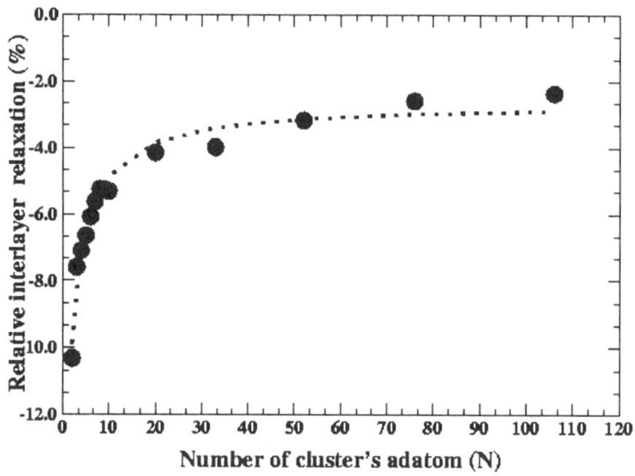


Fig.6. Relative relaxation with respect to the bulk interlayer spacing as a function of cluster atoms.

it has been found that island's diffusion, dominated by the circulation of its peripheral atoms, has energetic requirements that remain unchanged and close to 0.35eV [17]. Of course this is an effective energy barrier of the various diffusion processes happening and does not correspond to a specific diffusion mechanism. Among the diffusion processes, the variety of which depends on the size of the island, we distinguish those happening via concerted movement of the island's atoms. Indeed, it is worth noting that while it is rather expected to have translation of small clusters (dimmers or trimmers) by correlated movements, the migration of larger islands usually is accompanied by cluster reshaping. Interestingly, within the time scale of our simulations, we observed several correlated and concerted movements even for a nine-particle island.

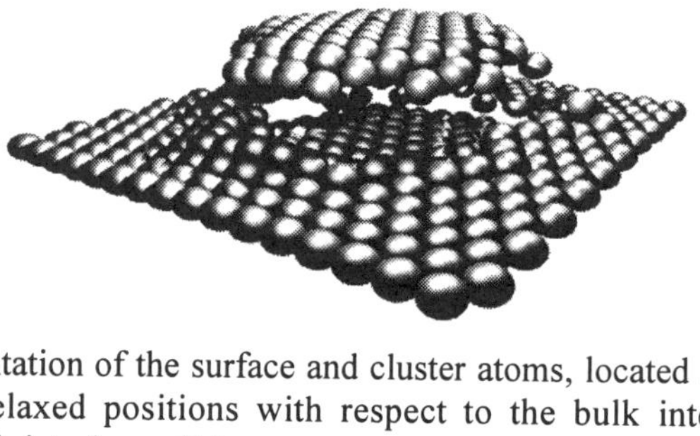

Fig. 7. Representation of the surface and cluster atoms, located at sites according to their relative relaxed positions with respect to the bulk intelayer distance and separated by their interlayer distance.

The presence of an island on this surface seems to be related also with surface and island relaxation, while we found that small clusters are responsible for new phonon modes and the alteration of some of the existing surface vibrational modes. In Fig. 6, we present the relative relaxed positions of the cluster atoms as a function of the island's size. As can be seen, the relative relaxed positions of the island atoms exhibit strong contraction (about –10% for the dimmer's case) that quickly recover to attain –2% for islands greater than 10 atoms. In addition, the cluster leaves its fingerprint on the surface by exalting the surface atoms lying below the peripheral cluster atoms and contracting all surface atoms underneath the island. This effect can be viewed in Fig. 7 in which we show the surface and cluster atoms located at sites according to their relative relaxed positions with respect to the bulk interplanar distance and separated by their interlayer distance.

The presence of the small islands on the surface is also manifested by the changes in the vibrational behavior of the later. To demonstrate this we integrated the phonon spectral densities (SD) for the cases of several islands deposited on the surface, calculated in the [110] direction, over two separate frequency ranges: a) from 0 to 3.5THz, region in which the new phonon modes attributed to the presence of the islands are located and b) from 3.5 to 7.5THz, a region dominated by the surface and bulk vibrational character. As we can see in Fig. 8, the small cluster's vibrational contribution is dominating in the low energy region (circles) to become, after an oscillation for an eight-atom island,

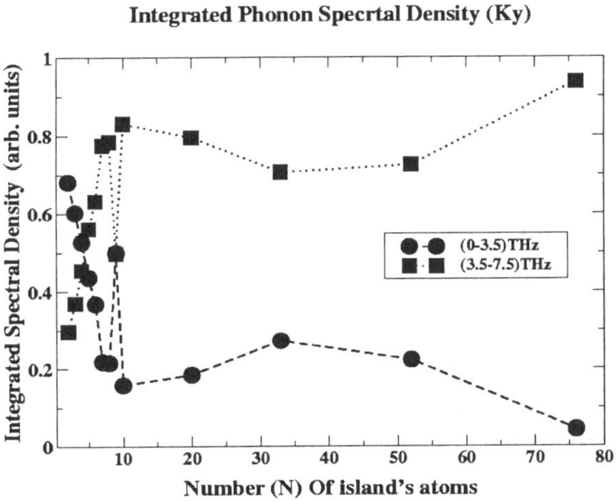


Fig. 8. Integrated phonon spectral densities of the modes due to the islands (circles) and those due to the surface, as a function of the number of cluster

insignificant for islands greater than ten atoms, while surface and bulk contributions (squares) exhibit an anti-symmetric behavior.

### *3.3 Correlated diffusion of Cu or Au adatoms on the $Cu_3Au(001)$ surface*

Another very interesting case of correlated diffusion processes that are found to play a very important role in the surface behavior concerns the case of adatoms on noble metal alloy faces. We studied the Cu and Au adatom diffusion processes on the $Cu_3Au(001)$ surface, often used as a model system for the understanding of the behavior of $A_3B$ type alloys. Focusing our attention only on the correlated diffusion processes, we describe a case that we found to play an important role not only in the diffusive behavior of this surface but also in the surface order and the wetting and segregation phenomena. The process concerns an exchange type mechanism (SLE) that takes place with the participation of a second layer Cu atom. The mechanism is a four-atom process and is presented in Fig. 9 in a sequence of snapshots. As can be seen from this figure, the Cu adatom residing in the hollow position, moves on top of a surface Cu atom entering in the pure Cu second layer by popping a Cu atom into an adatom position. The process is found to take place for both Cu and Au adatoms, the second one appearing as being unstable, segregating via this mechanism and wetting the surface layer. Besides the originality of the process found to happen quite often, what is also important is the change in the ordering of the surface layer.

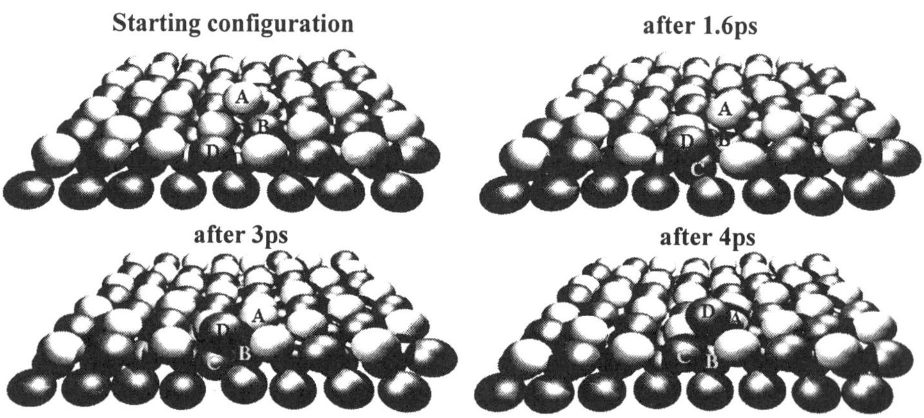


Fig. 9. SLE mechanism of Cu adatom on $Cu_3Au(001)$ at 550K

Other diffusion processes, even more complex, have been found to occur on this surface, most of which are taking place via correlated diffusion mechanisms [41]. An interesting case is the creation of diffusive vacancies that is stimulated by the presence of a Cu adatom on the same surface. The formation of the vacancy is taking place by successive displacements of several atoms and the simultaneous incorporation of another surface atom in the second pure Cu layer. This highly correlated process has an important contribution in the rearrangement of the surface atoms and can be related to the order-disorder transition occurring at higher temperatures.

## 4. Concluding Remarks

Summarizing, in this communication we tried to point out the importance of correlated processes in the diffusion of adsorbates. We found that the diffusion of 2D islands is characterized by the high mobility of the peripheral atoms, while small islands (up to nine atoms) are found to migrate via concerted movements. The migration energy of island's atoms increases with cluster size to attain a value of 0.15eV for islands between three and eight atoms and 0.34eV for clusters bigger than 12 atoms.

In addition, the presence of an island on the surface is manifested by changes in the surface relaxation and phonon modes. Interestingly, we found that the relaxed positions and phonon modes due to islands follow the same trend with the migration energy. Moreover, it was found that correlated diffusion mechanisms of adatoms on the $Cu_3Au(001)$ surface stimulate processes that can be related to wetting and local disordering phenomena.

## Acknowledgments

The help of E. Vamvakopoulos, Ch. Lekka, D. Pantelios and V. Papathanakos is gratefully acknowledged. Some of the figures showing diffusion mechanisms have been done using the 'VMD-Visual Molecular Dynamics' package, Humphrey W., Dalke A., Schulten K., *J. Molecular Graphics*, **14** (1996) 33.

## 5. References

1. J.J. deMiguel, A. Sanchez, A. Cebollada, J.M. Gallego, J. Ferron, S. Ferrer, *Surf. Sci.* **189/190** (1987) 1062.
2. M. Breeman, D.O. Boerma, Surf. Sci. 269/270 (1992) 224; *Surf. Sci.* **278** (1992) L110.
3. H.-J. Ernst, F. Fabre, J. Lapujoulade, *Phys. Rev.* **B46** (1992) 1929.
4. H. Durr, J.F. Wendelken, J.-K. Zuo, *Surf. Sci.* **328** (1995) 1929.
5. C.L. Liu, J.M. Cohen, J.B. Adams, A.F. Voter, *Surf. Sci.* **253** (1991) 334.

6. G.W. Jones, J.M. Marcano, J.K. Nørskov, J.A. Venables,. *Phys. Rev. Lett.* **65** (1990) 3317.
7. L. Hansen, P. Stoltze, K.W. Jacobsen, J.K. Nørskov, *Phys. Rev.* **B44** (1991) 6523.
8. L.S. Perkins, A.E. DePristo, *Surf Sci.* 294 (1993) 67; *Surf. Sci.* **325** (1995) 169.
9. J.E. Black, Z.J. Tian, *Phys. Rev. Lett.* **71** (1993) 444.
10. K.-D. Shiang, *Phys. Lett.* **A180** (1993) 444.
11. Ch. Lee, G.T. Barkema, M. Breeman, A. Pasquarello, R.Car, *Surf .Sci.* **L306** (1994) 575.
12. M. Karimi, T. Tomkowski, G. Vidali, O. Biham, *Phys. Rev.* **B52** (1995) 5364.
13. G.A. Evangelakis, N.I. Papanicolaou, *Surf. Sci.* **347** (1996) 376.
14. G.C. Kallinteris, G.A. Evangelakis, N.I. Papanicolaou, *Surf. Sci.* **369** (1996) 185.
15. G.A. Evangelakis, G.C. Kallinteris, N.I. Papanicolaou, *Surf .Sci.* **394** (1997) 185.
16. D.G. Papageorgiou and G. A. Evangelakis, *Surf. Sci.* **461** (2000) L543
17. G.A. Evangelakis, E. Vamvakopoulos, D. Pantelios, N.I. Papanicolaou, *Surf. Sci.* **425**, (1999), L393
18. P.W. Palmberg, T.N. Rhodin, *J. Chem. Phys.* **49** (1968) 134.
19. G.W. Graham, *Surf. Sci.* **184** (1987) 137.
20. J.C. Hansen, J.A. Benson, W.D. Clendening, M.T. Mc Ellistrem, J.G. Tobin, *Phys. Rev.* **B36** (1987) 6186.
21. Z.Q. Wang, Y.S. Li, C.K.C. Lok, J. Quinn, F. Jona, *Solid State Comm.* **62** (1987) 181.
22. J.C. Hansen, J.G. Tobin, *J. Vac. Sci. Tech.* **7** (1989) 2475.
23. D.D. Chambliss, S. Chiang, *Surf. Sci..* **264** (1992) L187.
24. S.M. Foiles, *Surf. Sci.* **191** (1987) 329.
25. T.J. Raeker, A.E. DePristo, *Surf. Sci.* **248** (1991) 134.
26. P.T. Sprunger, E. Laegsgaard, F. Bessenbacher, *Phys. Rev.* **B54** (1996) 8136.
27. E.S. Hirschorn, D.S. Lin, E.D. Hansen, T.-C. Chiang, *Surf. Sci.* **323** (1995) L299.
28. U. Schoder, R. Linke, J-H. Boo, K. Wandelt, *Surf. Sci.* **352** (1996) 211.
29. N.T. Barrett, R. Belchou, J. Thiele, *Surf. Sci.* **351** (1995) 776.
30. D. Gorse, J. Lapujoulade, *Surf. Sci.*, **162** (1985) 847
31. V. Rosato, V. Pontikis, G. Ciccotti, *Phys. Rev.* **B33**, (1986) 1860
32. A.P. Baddorf, E.W. Plummer, *Phys. Rev. Lett.*, **66** (19991) 2770
33. T.R. Linderoth, S. Horch, E. Laegsgaard, I. Stensgaard, F. Besenbacher, *Phys. Rev. Lett.*, **78** (1997) 4978
34. F. Montalenti, R. Ferrando, *Surf. Sci.*, **433-435**, (1999), 445
35. F. Montalenti, R. Ferrando, *Phys. Rev. Lett.*, **82**, (1999), 1498

36. T.R. Linderoth, S. Horch, L. Petersen, S. Helveg, E. Laegsgaard, I. Stensgaard, F. Besenbacher, *Phys. Rev. Lett.*, **82** (1999) 1494
37. G. Boisvert, L. Lewis, A. Yelon, *Phys. Rev. Lett.*, **75** (1995) 469
38. A.Yelon, B. Movaghar, H.M. Branz, *Phys. Rev.* **B46**, (1992) 12244
39. P.J. Feibelman, J.S. Nelson, G.L. Kellogg, *Phys, Rev*, **B49**, (1994), 10548
40. P.J. Feibelman, *Surf. Sci.*, **313**, (1994), L801
41. Ch. Lekka, G.A. Evangelakis, *Surf. Sci.*, **473**, (2001) 39

# AUTHOR INDEX


| | Page |
|---|---|
| Al-Rawi, Ahlam | 327 |
| Ala-Nissila, T. | 47, 273 |
| Altalo, M. | 317 |
| Asscher, M. | 129 |
| Besenbacher, F. | 201 |
| Bokun, G. S. | 157 |
| Chab, Vladimir | 11 |
| Chen, L. Y. | 285 |
| Chovoj, Z. | 143 |
| Cohen, C. | 107 |
| Conrad, E. H. | 23 |
| Danziger, I. N. | 129 |
| Dudek, M. R. | 179 |
| A. Evangelakis | 339 |
| Feibelman, Peter J. | 213 |
| Ferrando, R. | 237, 295 |
| Fichthorn, Kristen A. | 225 |
| Frenken, J. W. M. | 191 |
| Graham, Z. P. | 1 |
| Granato, E. | 273 |
| Groda, Ya. G. | 157 |
| Heinonen, J. | 317 |
| Hirvonen, J. | 317 |
| Hjelt, T. | 47 |
| Horn Von Högen, M. | 23 |
| Jastrabik, L. | 71 |
| Kara, Abdelkader | 327 |
| Karim, Altaf | 327 |
| Kerner, G. | 129 |
| Koponen, I. | 317 |
| Kotrla, Miroslav | 247 |
| Krug, Joachim | 247 |
| Krukowski, Stanislaw | 59 |
| Krylov, S. Yu. | 305 |
| Kammlar, K. | 23 |
| Kuusela, E. | 47 |
| Lægsgaard, E. | 201 |
| Lahtinen, J. M. | 47 |
| Loburets, A. T. | 97 |
| Markov, Ian | 259 |
| Menzel, A. | 23 |
| Merikoski, J. | 317 |
| Métois, J. J. | 33 |
| Montalenti, F. | 237 |
| Morgenstern, K. | 201 |
| Mutombo, Pingo | 11 |
| Nash, P. L. | 285 |
| Naumovets, A. G. | 97 |
| Nieto, F. | 71, 83 |
| Prévot, G. | 107 |
| Rahman, S. Talat | 327 |
| Salo, P. | 317 |
| Savchenkov, Alexei | 11 |
| Scheffler, Matthias | 225 |
| Schmaus, D. | 107 |
| Schukrinov, Pavel | 11 |
| Seneko, N. B. | 97 |
| Slezak, Jiri | 11 |
| Smilauer, Pavael | 247 |
| Somfai, E. | 191 |
| Spadacini, R. | 295 |
| Stoyanov, S. S. | 33 |
| Tarasenko, A. A. | 71, 83 |
| Toennies, J. P. | 1 |
| Tommei, G. E. | 295 |
| Tonchev, V. | 33 |
| Tringides, M. | 23, 179 |
| Trushin, O. S. | 317 |
| Turski, Lukasz A. | 59 |
| Uebing, C. | 71, 83 |
| Van Albada, S. B. | 191 |
| Van Gastel, R. | 191 |
| Van Saarloos, W. | 191 |
| Vattulainen, I. | 47, 169 |
| Vedula, Yu. S. | 97 |
| Vikhrenko, V. S. | 157 |
| Voss, N. | 23 |
| Xiao, Xudong | 117 |
| Ying, S. C. | 47, 273, 285 |
| Zaluska-Kotur, M. A. | 59 |
| Zhao, W. | 129 |

# SUBJECT INDEX

| | *Page* |
|---|---|
| 1-dimension decay | 208 |
| 3-dimensional islands | 113, 135 |
| Ab-initio calculations | 216 |
| Activation dynamics | 287, 339 |
| Activation energies | 1, 329, 343 |
| Adatom diffusion | 239, 319 |
| Adatom interaction | 165 |
| Adsorbate-adsorbate interaction | 157, 227 |
| Adsorbed monolayer | 72 |
| Adsorption isotherms | 78 |
| Alkali diffusion | 98, 129 |
| Alkali metals | 129 |
| Aluminum islands | 321, 323 |
| Autocorrelation function | 24, 47, 85 |
| Block transformation | 75 |
| Boltzmann-Matano method | 67, 98, 175, 179, 184 |
| Boundary lubrication | 277 |
| Bunching (of steps) | 34 |
| Carbon monoxide diffusion | 117 |
| Chain diffusion | 243 |
| Chemical diffusion coefficient | 157 |
| Chemical potential | 157, 262 |
| Cluster diffusion | 293, 316, 319 |
| Coadsorbed systems | 102, 134 |
| Collective diffusion coefficient | 174, 314 |
| Computational physics | 193 |
| Computer simulations | 107 |
| Concentration gradient | 145 |
| Contact potential difference | 98 |
| Copper islands | 322 |
| Copper on Si(111) | 18 |
| Copper | 107, 193, 215, 321, 341 |
| Correlation function (time) | 16, 276 |
| Coverage profile | 98 |
| Coverage-dependent diffusion | 100, 139, 185 |
| Coverage-grating | 118, 130 |
| Critical slowing down | 87 |
| Crystal-vapor equilibrium | 38 |
| Density functional Theory (DFT) | 230 |
| Dimer diffusion | 242 |
| Driven diffusion | 64 |
| Effective Medium Theory (EFT) | 210 |

Electronic interactions 228
Embedded Atom Method (EAM) 193, 200
Energy barriers 246
Energy resolution 3
Entropic repulsion 42
Evaporation Condensation (EC) 26
Exact enumeration method 179
Exchange process 266, 334
First-order phase transitions 90
Fokker-Planck equations 299
Frenkel-Kontorova model 281
Friction coefficient 7
Germanium on silicon 264
Gliding diffusion 316
Gold Au(110) 239
Gold on Si(111) 17
Growth morphology 215
Gruber-Mullins model 25
Hard circles on surfaces 52
Homoepitaxy 251
Hydrodynamic limit 151, 172
Impurities 257
Incoherent scattering 24
Indium on Cu(001) 194
Interactions 66, 71
Island decay mode 205
Island density 254, 267
Island morphology 252, 346
Jump diffusion coefficient 166
Jump Length Probability Distribution (JLPD) 301
Jump rate 90
Jump vector distributions 197, 311
Kawasaki hopping 157
Kelvin probe 131
Kinetic theory 307
Kinetics 265
Kink diffusion 30
Kink-formation energies 217
Kubo-Green formula 48, 85, 151, 298
Langevin equation 158, 160, 278, 287, 300
Langmuir gas 161
Large island diffusion 324
Laser Induced Thermal Desorption (LITD) 129, 135
Lateral interactions 133
Lattice gas model 59, 72, 83, 146, 158, 164, 170, 180, 231
Lead on Cu(100) 108

Lead on Si(111) 14
Lead Pb(111) 216
Leapfrog diffusion 242
LEED 23
Lennard-Jones potential 291
Long jumps 297, 311
Long-range interactions 230
Macroscopic diffusion 13, 130
Memory effects 49, 153, 157, 305
Memory expansion method 49
Metal surfaces 319
Metal/semiconductor 12
Metallic films 99, 215
Microscopic diffusion 17
Minimal path approximation 288
Missing-row-reconstruction 239
Molecular dynamics 110, 337, 341
Monolayer 72
Monte Carlo Simulations 63, 71, 79, 83, 84, 121, 179, 211, 249, 319
Mori's projection operator 147
Navier-Stokes equation 54
Nickel on Ni(110) 3
Non-equilibrium diffusion 146, 157, 161, 169, 179
Non-equilibrium statistical physics 141
Nonlinear diffusion 275, 276
Nucleation 252
Order parameter 63, 187
Ordered phases 60, 74, 111
Oswald ripening 206
Oxide surfaces 137
Oxygen on W(110) 50, 59, 170
Parallel computer 86
Particle lifetime 193
PEEM 13
Phase diagram 60
Phase transitions 86, 90, 111, 283
Platinum Pt(111) 216
Polymer diffusion 51
Potassium 129, 139
Power law decay 50, 208, 338
Prefactor of diffusion 118, 213
Profile evolution 65, 98, 108, 184
Quasi helium atom scattering 1
Rare-earth adsorbates 102
Rate equations 255
Real-space renormalization group (RSRG) 73

Reflection Electron Microscopy (REM) 34
Rutherford Back Scattering (RBS) 108
Scanning Tunneling Microscope 12, 193, 194, 203
Schwoebel barrier 117, 138
Second harmonic (SH) diffraction 136
Semiconductor heteroepitaxy 261
Seven by seven (7x7-Si(111)) 19
Silicon Si(111) 34, 268
Silver Ag(110) 203
Silver Ag(111) 228, 329, 343
Single particle diffusion 307, 329, 341
Sliding friction 275
Sodium on Cu(001) 6
Sodium on Pt(111) 7
Sodium/copper 6
Sodium/platinum 7
Soliton diffusion 99
Step edge barrier 265, 328
Step edges 25
Step formation 216
Step type A-B on fcc (111) 121, 217, 329
Step-edge diffusion (SD) 26
Stepped Si(100) 25
Stepped surfaces 23, 122
Step-step interaction 33, 44
Stochastic equations 297
Strained Ag(111) 230
Stranski-Krastanov growth 263
Structure factor 146, 172, 298
Submonolayer growth 249
Sulfur 215
Sulphur-decorated Cu-clusters 221
Superconductor Type II 282
Surface dynamics 144
Surface free energy 33
Surface phase transitions 99
Surfactants 251, 261
Susceptibility 81
Temperature Programmed Desorption (TPD) 132
Temporal LEED spectroscopy 23
Terrace Diffusion (TD) 26
Thermodynamic factor 11, 157, 277
Time-of-flight spectra 4
Time-of-flight 311
Total jump rate 301
Tracer diffusion coefficient 50, 173, 189, 329

Transition State theory (TST) 246
Transparent steps 43
Transport coefficients 158, 161
Trapping of step edges 137
Trio interactions 231
Two-dimensional islands 203
Type II superconductors 282
Vacancy diffusion 198
Velocity correlation 54
Wetting layer 113, 263
Work function 98, 134